Introduction to Mixed Modelling

Introduction to Mixed Modelling

Beyond Regression and Analysis of Variance

N. W. Galwey
Genetic Analysis, GlaxoSmithKline, UK

John Wiley & Sons, Ltd

Other Wiley Editorial Offices

John Wiley & Sons Inc., 111 River Street, Hoboken, NJ 07030, USA

Jossey-Bass, 989 Market Street, San Francisco, CA 94103-1741, USA

Wiley-VCH Verlag GmbH, Boschstr. 12, D-69469 Weinheim, Germany

John Wiley & Sons Australia Ltd, 42 McDougall Street, Milton, Queensland 4064, Australia

John Wiley & Sons (Asia) Pte Ltd, 2 Clementi Loop #02-01, Jin Xing Distripark, Singapore 129809

John Wiley & Sons Canada Ltd, 6045 Freemont Blvd, Mississauga, ONT, L5R 4J3, Canada

Wiley also publishes its books in a variety of electronic formats. Some content that appears in print may not be available in electronic books.

Library of Congress Cataloging-in-Publication Data:

Galwey, Nick.
 Introduction to mixed modelling : beyond regression and analysis of
variance / Nicholas W. Galwey.
   p. cm.
 Includes bibliographical references and index.
 ISBN-13: 978-0-470-01496-7 (acid-free paper)
 ISBN-10: 0-470-01496-2 (acid-free paper)
 1. Multilevel models (Statistics) 2. Experimental design. 3. Regression
analysis. 4. Analysis of variance. I. Title.
 QA276.G33 2006
 591.5–dc22

                                                            2006023991

1005664328

British Library Cataloguing in Publication Data

A catalogue record for this book is available from the British Library

ISBN-13: 978-0-470-01496-7 (HB)

Typeset in 10/12pt Times by Laserwords Private Limited, Chennai, India
Printed and bound in Great Britain by Antony Rowe Ltd, Chippenham, Wiltshire
This book is printed on acid-free paper responsibly manufactured from sustainable forestry
in which at least two trees are planted for each one used for paper production.

Contents

Preface

This book is intended for research workers and students who have made some use of the statistical techniques of regression analysis and analysis of variance (anova), but who are unfamiliar with *mixed models* and the criterion for fitting them called *residual maximum likelihood* (REML, also known as *restricted maximum likelihood*). Such readers will know that, broadly speaking, regression analysis seeks to account for the variation in a response variable by relating it to one or more explanatory variables, whereas anova seeks to detect variation among the mean values of groups of observations. In regression analysis, the statistical significance of each explanatory variable is tested using the same estimate of residual variance, namely the residual mean square, and this estimate is also used to calculate the standard error of the effect of each explanatory variable. However, this choice is not always appropriate. Sometimes, one or more of the terms in the regression model (in addition to the residual term) represents *random variation*, and such a term will contribute to the observed variation in other terms. It should therefore contribute to the significance tests and standard errors of these terms, but in an ordinary regression analysis, it does not do so. Anova, on the other hand, does allow the construction of models with additional random-effect terms, known as block terms. However, it does so only in the limited context of balanced experimental designs.

The capabilities of regression analysis can be combined with those of anova by fitting to the data a mixed model, so called because it contains both fixed-effect and random-effect terms. A mixed model allows the presence of additional random-effect terms to be recognised in the full range of regression models, not just in balanced designs. Any statistical analysis that can be specified by a general linear model (the broadest form of linear regression model) or by anova can also be specified by a mixed model. However, the specification of a mixed model requires an additional step. The researcher must decide, for each term in the model, whether effects of that term (e.g. the deviations of group means from the grand mean) can be regarded as a random sample from some much larger population, or whether they are a fixed set. In some cases this decision is straightforward; in others, the distinction is subtle and the decision difficult. However, provided that an appropriate decision is made (see Chapter 6, Section 6.3), the mixed model specifies a statistical analysis which is of broader validity than regression analysis or anova, and which is nearly equivalent to those methods (though slightly less powerful) in the special cases where they are applicable.

It is fairly straightforward to specify the calculations required for regression analysis and anova, and this is done in many standard textbooks. For example, Draper and

Smith (1998) give a clear, thorough and extensive account of the methods of regression analysis, and Mead (1988) does the same for anova. To solve the equations that specify a mixed model is much less straightforward. The model is fitted – that is, the best estimates of its parameters are obtained – using the REML criterion, but the fitting process requires recursive numerical methods. It is largely because of this burden of calculation that mixed models are less familiar than regression analysis and anova: it is only in about the last two decades that the development of computer power and user-friendly statistical software has allowed them to be routinely used in research. This book aims to provide a guide to the use of mixed models that is accessible to the broad community of research scientists. It focuses not on the details of calculation, but on the specification of mixed models and the interpretation of the results.

The numerical examples in this book are presented and analysed using two statistical software systems, namely:

- GenStat, distributed by VSN International Ltd, Hemel Hempstead, via the web site http://www.vsni.co.uk/products/genstat/. Anyone who has bought this book can obtain free use of GenStat for a period of 12 months. Details, together with Genstat programs and data files for many of the examples in this book, can be found at www.wiley.com/go/mixed-modelling (as can the solutions to the end of chapter exercises).

- R, from The R Project for Statistical Computing. This software can be downloaded free of charge from the web site http://www.r-project.org/.

GenStat is a natural choice of software to illustrate the concepts and methods employed in mixed modelling because its facilities for this purpose are straightforward to use, extensive and well integrated with the rest of the system, and because their output is clearly laid out and easy to interpret. Above all, the recognition of random terms in statistical models lies at the heart of GenStat. GenStat's method of specifying anova models requires the distinction of random-effect (block) and fixed-effect (treatment) terms, which makes the interpretation of designed experiments uniquely reliable and straightforward. This approach extends naturally to REML models, and provides a firm framework within which the researcher can think and plan. Despite these merits, GenStat is not among the most widely used statistical software systems, and most of the numerical examples are therefore also analysed using the increasingly popular software R. Development of this software is taking place rapidly, and it is likely that its already-substantial power for mixed modelling will increase steadily in the future.

I am grateful to Mr Peter Lane, Dr Aruna Bansal and Dr Caroline Galwey for valuable comments and suggestions on parts of the manuscript of this book, and to the participants in the GenStat Discussion List for helpful responses to many enquiries. (Access to this lively forum can be obtained via the GenStat User Area at web site http://www.vsni.co.uk/products/genstat/user/) I am particularly grateful to Dr Roger Payne for an expert and sharp-eyed reading of the entire manuscript. Any errors or omissions of fact or interpretation that remain are the sole responsibility of the author. I would also like to express my gratitude to the many individuals and organisations who have given permission for the reproduction of data in the numerical examples presented. They are acknowledged individually in their respective places, but the high level of support that they have given me should be recognised here.

1

The need for more than one random-effect term when fitting a regression line

1.1 A data set with several observations of variable *Y* at each value of variable *X*

One of the commonest, and simplest, uses of statistical analysis is the fitting of a straight line, known for historical reasons as a *regression line*, to describe the relationship between an *explanatory variable*, *X*, and a *response variable*, *Y*. The departure of the values of *Y* from this line is the *residual variation*, which is regarded as random, and it is natural to ask whether the part of the variation in *Y* that is explained by the relationship with *X* is significantly greater than this residual variation. This is a simple *regression analysis*, and for many data sets it is all that is required. However, in some cases, several observations of *Y* are taken at each value of *X*. The data then form natural groups, and it may no longer be appropriate to analyse them as though every observation were independent: observations of *Y* at the same value of *X* may lie at a similar distance from the line. We may then be able to recognise two sources of random variation, namely:

- variation among groups

- variation among observations within each group.

This is one of the simplest situations in which it is necessary to consider the possibility that there may be more than a single *stratum* of random variation – or, in the language of mixed modelling, that a model with more than one *random-effect term* may be required. In this chapter we will examine a data set of this type, and explore how the usual regression analysis is modified by the fact that the data form natural groups.

Introduction to Mixed Modelling: Beyond Regression and Analysis of Variance   N. W. Galwey
© 2006 John Wiley & Sons, Ltd

We will explore this question in a data set that relates the prices of houses in England to their latitude. There is no doubt that houses cost more in the south of England than in the north: these data will not lead to any new conclusions, but they will illustrate this trend, and the methods used to explore it. The data are displayed in a spreadsheet in Table 1.1. The first cell in each column contains the name of the variable held in that column. The variables 'latitude' and 'price_pounds' are *variates* – lists of observations that can take any numerical value, the commonest kind of data for statistical analysis. However, the observations of the variable 'town' can only take certain permitted values – in this case, the names of the 11 towns under consideration. A variable of this type is called a *factor*, and the exclamation mark (!) after its name indicates that 'town' is of this type. The towns are the groups of observations: within each town, all the houses are at nearly the same latitude, and the latitude of the town is represented by a single nominal value in this data set. In contrast, the price of each house is potentially unique.

Before commencing a formal analysis of this data set, we should note its limitations. A thorough investigation of the relationship between latitude and house price would take into account many factors besides those recorded here – the number of bedrooms in each house, its state of repair and improvement, other observable indicators of the desirability of its location, and so on. To some extent such sources of variation have been eliminated from the present sample by the choice of houses that are broadly similar: they are all 'ordinary' houses (no flats, maisonettes, etc.), and all have 3, 4 or 5 bedrooms. The remaining sources of variation in price will contribute to the residual variation among houses in each town, and will be treated accordingly. We should also consider in what sense we can think of the latitudes of houses as 'explaining' the variation in their prices. The easily measurable variable 'latitude' is associated with many other variables, such as climate and distance from London, and it is probably some of these, rather than latitude per se, that have a causal connection with price. However, an explanatory variable does not have to be causally connected to the response in order to serve as a useful predictor. Finally, we should consider the value of studying the relationship between latitude and price in such a small sample. The data on this topic are extensive, and widely interpreted. However, this small data set, illustrating a simple, familiar example, is highly suitable for a study of the *methods* by which we judge the significance of a trend, estimate its magnitude, and place confidence limits on the estimate. Above all, this example will show that in order to do these things reliably, we must recognise that our observations – the houses – are not mutually independent, but form natural groups – the towns.

1.2 Simple regression analysis. Use of the software GenStat to perform the analysis

We will begin by performing a simple regression analysis on these data, before considering how this should be modified to take account of the grouping into towns. The standard linear *regression model* (Model 1.1) is

$$y_{ij} = \beta_0 + \beta_1 x_i + \varepsilon_{ij} \tag{1.1}$$

Table 1.1 Prices of houses in a sample of English towns, and their latitudes
Data obtained from an estate agents' web site in October 2004.

	A	B	C
1	town!	latitude	price_pounds
2	Bradford	53.7947	39950
3	Bradford	53.7947	59950
4	Bradford	53.7947	79950
5	Bradford	53.7947	79995
6	Bradford	53.7947	82500
7	Bradford	53.7947	105000
8	Bradford	53.7947	125000
9	Bradford	53.7947	139950
10	Bradford	53.7947	145000
11	Buxton	53.2591	120000
12	Buxton	53.2591	139950
13	Buxton	53.2591	149950
14	Buxton	53.2591	154950
15	Buxton	53.2591	159950
16	Buxton	53.2591	159950
17	Buxton	53.2591	175950
18	Buxton	53.2591	399950
19	Carlisle	54.8923	85000
20	Carlisle	54.8923	89950
21	Carlisle	54.8923	90000
22	Carlisle	54.8923	103000
23	Carlisle	54.8923	124950
24	Carlisle	54.8923	128500
25	Carlisle	54.8923	132500
26	Carlisle	54.8923	135000
27	Carlisle	54.8923	155000
28	Carlisle	54.8923	158000
29	Carlisle	54.8923	175000
30	Chichester	50.8377	199950
31	Chichester	50.8377	299250
32	Chichester	50.8377	350000
33	Crewe	53.0998	77500
34	Crewe	53.0998	84950
35	Crewe	53.0998	112500

(*Continued overleaf*)

Table 1.1 (*continued*)

	A	B	C
36	Crewe	53.0998	140000
37	Durham	54.7762	127950
38	Durham	54.7762	157000
39	Durham	54.7762	169950
40	Newbury	51.4037	172950
41	Newbury	51.4037	185000
42	Newbury	51.4037	189995
43	Newbury	51.4037	195000
44	Newbury	51.4037	295000
45	Newbury	51.4037	375000
46	Newbury	51.4037	400000
47	Newbury	51.4037	475000
48	Ripon	54.1356	140000
49	Ripon	54.1356	152000
50	Ripon	54.1356	187950
51	Ripon	54.1356	210000
52	Royal Leamington Spa	52.2876	147000
53	Royal Leamington Spa	52.2876	159950
54	Royal Leamington Spa	52.2876	182500
55	Royal Leamington Spa	52.2876	199950
56	Stoke-on-Trent	53.0041	69950
57	Stoke-on-Trent	53.0041	69950
58	Stoke-on-Trent	53.0041	75950
59	Stoke-on-Trent	53.0041	77500
60	Stoke-on-Trent	53.0041	87950
61	Stoke-on-Trent	53.0041	92000
62	Stoke-on-Trent	53.0041	94950
63	Witney	51.7871	179950
64	Witney	51.7871	189950
65	Witney	51.7871	220000

where

x_i = value of X (latitude) for the ith town,

y_{ij} = observed value of Y ($\log_{10}$(house price in pounds)) for the jth house in the ith town,

β_0, β_1 = constants to be estimated, defining the relationship between X and Y,

ε_{ij} = the residual effect, i.e. the deviation of y_{ij} from the value predicted on the basis of x_i, β_0 and β_1.

Note that in this model the house prices are transformed to logarithms, because preliminary exploration has shown that this gives a more linear relationship between latitude and price, and more uniform residual variation. The model is illustrated graphically in Figure 1.1.

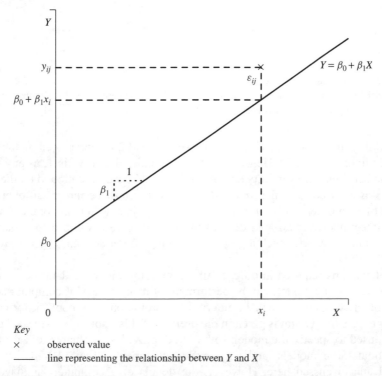

Figure 1.1 Linear relationship between an explanatory variable X and a response variable Y, with residual variation in the response variable.

The model specifies that a sloping straight line is to be used to describe the relationship between latitude and log(house price). The *parameters* β_0 and β_1 specify, respectively, the intercept and slope of this line. *Estimates* of these parameters, $\hat{\beta}_0$ and $\hat{\beta}_1$ respectively, are to be obtained from the data, and these estimates will define the *line of best fit* through the data. An estimate of each of the ε_{ij}, designated $\hat{\varepsilon}_{ij}$, will be given by the deviation of the ijth data point, in a vertical direction, from the line of best fit. The parameter estimates chosen are those that minimise the sum of squares of the $\hat{\varepsilon}_{ij}$. It is assumed that the *true* values ε_{ij} are independent values of a variable E which is *normally distributed with mean zero and variance* σ^2. The meaning of this statement, which can be written in symbolic shorthand as

$$E \sim N(0, \sigma^2),$$

is illustrated in Figure 1.2. The area under this curve between any two values of E gives the probability that a value of E will lie between these two values. For example,

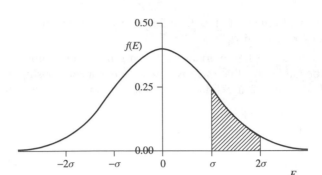

Figure 1.2 A normal distribution with mean zero and variance σ^2.
$f(E) = $ probability density of E. Total area under curve $= 1$. Hatched area $= $ probability that a value of E is greater than or equal to σ and less than 2σ.

the probability that a value of E will lie between σ and 2σ, represented by the hatched area in the figure, is 0.1359. Hence the total area under the curve is 1, as any value of E must be between minus infinity and plus infinity. The variable plotted on the vertical axis, $f(E)$, is referred to as the *probability density*. It must be integrated over a range of values of E in order to give a value of probability, just as human population density must be integrated over an area of land in order to give a value of population. For the reader unfamiliar with such regression models, a fuller account is given by Draper and Smith (1998).

The calculations required in order to fit a regression model to data (i.e. to estimate the parameters of the model) can be performed by many pieces of computer software, and one of these, GenStat, will be referred to throughout this book. Information on obtaining access to GenStat is given in the preface of this book. The GenStat command language, used to specify the models to be fitted, provides a valuable tool for thinking clearly about these models, and the GenStat statements required will therefore be presented and discussed here. However, the details of a computer language should not be allowed to distract from the statistical concepts that are our central topic. We will therefore note only a few key points about these statements: a full introduction to the GenStat command language is given in Section 1.3 of GenStat's *Introduction guide* (Payne *et al.*, 2003). This is available *via* GenStat's graphical user interface (GUI), which also gives access to the full specification of the language.

The following statements, in the GenStat command language, import the data into the GenStat environment and fit Model 1.1:

```
IMPORT \
    'Intro to Mixed Modelling\\Chapter 1\\house price, latitude.xls'; \
    SHEET = 'Sheet1'
CALCULATE logprice = log10(price_pounds)
MODEL logprice
FIT [FPROB = yes; TPROB = yes] latitude
```

The IMPORT statement specifies the file that contains the data, and makes the data available to GenStat. The CALCULATE statement performs the transformation to logarithms, and stores the results in the variate 'logprice'. The MODEL statement specifies the response variable in the regression model (Y, logprice), and the FIT statement specifies the explanatory variable (X, latitude). The *option setting* 'FPROB = yes' indicates that when an F statistic is presented, the associated probability is also to be given (see below). The option setting 'TPROB = yes' indicates that this will also be done in the case of a t statistic. The same operations could be specified – perhaps more easily – using the menus and windows of GenStat's GUI: the use of these facilities is briefly illustrated in Section 1.12, and fully explained by Payne *et al.* (2003).

A researcher almost always receives the results of statistical analysis in the form of computer output, and the interpretation of this, the extraction of key pieces of information and their synthesis in a report are important statistical skills. The output produced by GenStat is therefore presented and interpreted here. That from the FIT statement is shown in the box below.

Regression analysis

Response variate: logprice
Fitted terms: Constant, latitude

Summary of analysis

Source	d.f.	s.s.	m.s.	v.r.	F pr.
Regression	1	0.710	0.70955	20.55	<.001
Residual	62	2.141	0.03453		
Total	63	2.850	0.04524		

Percentage variance accounted for 23.7
Standard error of observations is estimated to be 0.186.

Message: the following units have large standardized residuals.

Unit	Response	Residual
1	4.602	−2.72
17	5.602	2.46

Message: the residuals do not appear to be random; for example, fitted values in the range 5.009 to 5.074 are consistently smaller than observed values and fitted values in the range 5.162 to 5.231 are consistently larger than observed values.

Estimates of parameters

Parameter	estimate	s.e.	t(62)	t pr.
Constant	9.68	1.00	9.68	<.001
latitude	−0.0852	0.0188	−4.53	<.001

This output begins with a specification of the model fitted. Note that the fitted terms include not only the explanatory variable, latitude, but also a constant, although none was specified: any regression model includes a constant by default (β_0 in this case).

Next comes an analysis of variance (anova) table, which partitions the variation in log(house price) between the *terms* in Model 1.1, namely:

- the effect of latitude (represented in the row labelled 'Regression' in the anova table), and

- the residual effects (represented in the row labelled 'Residual').

After the names of the terms, the next two columns of the anova table hold the degrees of freedom (abbreviated to d.f. or DF) and the sum of squares (s.s. or SS) for each term. The methods for calculating these will not be given here (for an account, see Draper and Smith, 1998, Section 1.3, pp 28–34), but it should be noted that the degrees of freedom for each term represent the number of independent pieces of information to which that term is equivalent. Thus the effect of latitude is a single piece of information, and $DF_{latitude} = 1$. There are 64 houses in the sample, each of which gives a value of $\hat{\varepsilon}_{ij}$, so it might be thought that the 'Residual' term would comprise 64 pieces of information. However, two pieces of information have been 'used up' by the estimation of the intercept and the effect of latitude, so

$$DF_{Residual} = 64 - 2 = 62.$$

This reduction in the residual degrees of freedom is equivalent to that fact that a line of best fit based on only two observations passes exactly through the data points – such a data set provides no information on residual variation.

The mean square for each term (m.s. or MS) is given by SS/DF, and is an estimate of the part of the variance in log(house price) that is accounted for by the term. If there is no real effect of latitude on log(house price), the expected values of $MS_{latitude}$ and $MS_{Residual}$ are the same. Hence on this *null hypothesis* (H_0), the expected value of $MS_{latitude}/MS_{Residual}$ is 1, though the actual value will vary from one data set to another. This ratio, called the variance ratio (abbreviated to v.r.), thus provides a test of H_0. Provided that the residual variation is normally distributed (see Section 1.8), v.r. is also known as the F statistic. If H_0 is true, the distribution of F over an infinite population of samples (data sets) has a definite mathematical form. The precise shape of this distribution depends on the degrees of freedom in the numerator and denominator of the ratio: hence the variable F is referred to more precisely as $F_{DF_{numerator}, DF_{denominator}}$. The distribution in the present case (i.e. the distribution of the variable $F_{1,62}$) is illustrated in Figure 1.3. This curve is interpreted in the same way as the normal distribution illustrated earlier. The area under the curve between any two values of F gives the probability that an observation of F will lie between these values. Hence again the total area under the curve is 1, as any observation of F must have some value between 0 and infinity. Again the variable plotted on the vertical axis is the probability density. This F distribution can be used to determine the probability P of obtaining by chance a value of F larger than that actually observed, as shown in the figure. For example, if $F_{1,62} > 4.00$, then $P < 0.05$, and it is said that the effect under consideration is significant at the 5 % level. Similarly, if $F_{1,62} > 7.06$, then $P < 0.01$, and it is said that the effect is significant at the 1 % level, i.e. highly significant. In the present case $F_{1,62} = 20.55$, and both the anova table and the figure show that P (called F pr. in the table) is less than 0.001: that is, the relationship between latitude and log(house price) in this data set is very highly significant – provided that the model specified is correct. However, there is a diagnostic message which indicates that this may not be the case: GenStat has detected that the residuals do not appear to be random.

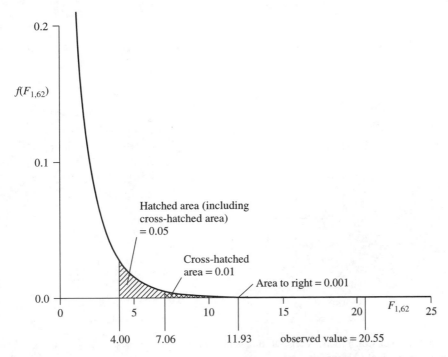

Figure 1.3   Distribution of the variable $F_{1,62}$, showing critical values for significance tests and the corresponding critical regions.

The next item in the output is the parameter estimates, which give the intercept ($\hat{\beta}_0$, the constant term) and slope ($\hat{\beta}_1$) of the line of best fit, with their standard errors (SEs). The negative slope indicates that house prices are higher in the south of England than in the north. The t statistic for the effect of latitude is given by estimate/$SE_{estimate} = -0.0852/0.0188 = -4.53$. Note that $t^2 = (-4.53)^2 = 20.55 = F$, and that for both these statistics the P value (when calculated to a greater degree of precision than is given by the GenStat output) is 0.0000271 – that is, the t test for the significance of the slope is equivalent to the F test in the analysis of variance.

The line of best fit is displayed, together with the data and the mean value for each town, in Figure 1.4. This figure shows that, overall, the regression line fits the data reasonably well. However, observations from the same town generally lie on the same side of the line – that is, the residual values within each town are not mutually independent. For example, as noted in GenStat's diagnostic message, the observations from Ripon, Durham and Carlisle generally lie above the line (GenStat identifies these by the fact that their fitted values of log(house price) lie in the range 5.009 to 5.074), whereas those from Stoke-on-Trent and Crewe all lie below the line (their fitted values lie in the range 5.162 to 5.231).

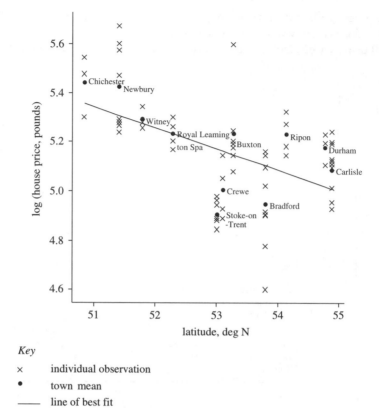

Figure 1.4   Relationship between latitude and house prices in a sample of English towns, showing individual observations, town means and line of best fit.

1.3   Regression analysis on the group means

Because the residual values are not mutually independent, the analysis presented above cannot be relied upon, even though the regression line appears reasonable. In particular, the residual degrees of freedom ($DF_{Residual} = 62$) are an overestimate: we deceive ourselves if we believe that the data comprise 64 independent observations. The simplest way to overcome this problem is to fit the regression model using the mean value of log(house price) for each town. These means are displayed in a spreadsheet in Table 1.2.

The following statements will perform a regression analysis on these mean values:

```
IMPORT \
    'Intro to Mixed Modelling\\Chapter 1\\price, lat, town means.xls'; \
    sheet = 'Sheet1'
MODEL meanlogprice
FIT [FPROB = yes; TPROB = yes] meanlatitude
```

Table 1.2 Mean values of $\log_{10}$(price) of houses in a sample of English towns, and their latitudes.

	A	B	C	D
1	town_unique	n_houses	meanlatitude	meanlogprice
2	Bradford	9	53.7947	4.94745
3	Buxton	8	53.2591	5.23083
4	Carlisle	11	54.8923	5.08552
5	Chichester	3	50.8377	5.44034
6	Crewe	4	53.0998	5.00394
7	Durham	3	54.7762	5.17775
8	Newbury	8	51.4037	5.42456
9	Ripon	4	54.1356	5.23106
10	Royal Leamington Spa	4	52.2876	5.23337
11	Stoke-on-Trent	7	53.0041	4.90642
12	Witney	3	51.7871	5.29207

The output of the FIT statement is as follows:

Regression analysis

Response variate: meanlogprice
Fitted terms: Constant, meanlatitude

Summary of analysis

Source	d.f.	s.s.	m.s.	v.r.	F pr.
Regression	1	0.1160	0.11601	5.19	0.049
Residual	9	0.2010	0.02234		
Total	10	0.3170	0.03170		

Percentage variance accounted for 29.5
Standard error of observations is estimated to be 0.149.

Estimates of parameters

Parameter	estimate	s.e.	t(9)	t pr.
Constant	9.44	1.87	5.04	<.001
meanlatitude	−0.0804	0.0353	−2.28	0.049

The residual degrees of freedom are much fewer ($DF_{Residual} = 9$), reflecting the more reasonable assumption that each town can be considered as an independent observation. Consequently, the relationship between latitude and house price is much less significant: $P = 0.049$, compared with $P < 0.001$ in the previous analysis. However, the estimates of the intercept and slope of the regression line are not much altered.

1.4   A regression model with a term for the groups

The simple method, presented in the previous section, for dealing with the problem of several observations in each town gives no account of the variation within the towns. Nor does it take account of the variation in the number of houses sampled, and hence in the precision of the mean, from one town to another. An alternative approach, which overcomes these deficiencies, is to add a term to the regression model to take account of the variation among towns that is not accounted for by latitude, namely:

$$y_{ij} = \beta_0 + \beta_1 x_i + \tau_i + \varepsilon_{ij} \tag{1.2}$$

where
τ_i = mean deviation from the regression line of observations from the ith town.

The following statements will fit this model (Model 1.2) using the ordinary methods of regression analysis:

```
IMPORT \
    'Intro to Mixed Modelling\\Chapter 1\\house price, latitude.xls'; \
    sheet = 'Sheet1'
CALCULATE logprice = log10(price_pounds)
MODEL logprice
FIT [FPROB = yes; TPROB = yes; \
    PRINT = model, estimates, accumulated] \
    latitude + town
```

The output of the FIT statement is as follows. It is quite voluminous, and each part will be discussed before presenting the next.

Message: term town cannot be fully included in the model because 1 parameter is aliased with terms already in the model.

(town Witney) = 26.80 − (latitude)*0.4981 − (town Buxton)*0.2668 + (town Carlisle)* 0.5467 − (town Chichester)*1.473 − (town Crewe)*0.3461 + (town Durham)*0.4889 −(town Newbury)*1.191 + (town Ripon)*0.1698 − (town Royal Leamington Spa)* 0.7507 − (town Stoke-on-Trent)*0.3938

First comes a message noting that because the variation among the town means is partly accounted for by latitude, the term 'town' cannot be fully included in the regression model. It is said to be *partially aliased* with latitude. (The technical consequence of this partial aliasing is that the effect of one of the towns − Witney, arbitrarily chosen because it comes last when the towns are arranged in alphabetical order − is a function of the effects of the other towns and of latitude, but the numerical details of this relationship, given in the message, need not concern us.)

Next come the statement of the regression model and the estimates of its parameters:

Regression analysis

Response variate: logprice
Fitted terms: Constant + latitude + town

Estimates of parameters

Parameter	estimate	s.e.	t(53)	t pr.
Constant	14.18	2.32	6.12	<.001
latitude	−0.1717	0.0435	−3.95	<.001
town Buxton	0.1914	0.0598	3.20	0.002
town Carlisle	0.3265	0.0885	3.69	<.001
town Chichester	−0.015	0.136	−0.11	0.914
town Crewe	−0.0628	0.0761	−0.83	0.413
town Durham	0.399	0.106	3.75	<.001
town Newbury	0.067	0.102	0.66	0.515
town Ripon	0.3421	0.0840	4.07	<.001
town Royal Leamington Spa	0.0272	0.0874	0.31	0.757
town Stoke-on-Trent	−0.1767	0.0636	−2.78	0.007
town Witney	0	*	*	*

Parameters for factors are differences compared with the reference level:
Factor Reference level
town Bradford

These parameter estimates correctly lead to the sample mean for each town when substituted into the formula

$$\text{mean (log(house price))} = 14.18 - 0.1717 \times \text{latitude} + \text{effect of town}.$$

For example, in Durham,

$$\text{mean (log(house price))} = 14.18 - 0.1717 \times 54.7762 + 0.399 = 5.1739.$$

However, the parameter estimates themselves are arbitrary and uninformative, as illustrated in Figure 1.5. The fitted line is arbitrarily specified to pass through the mean values for Bradford and Witney (the first and last towns in the alphabetic sequence), and the effects of the other towns – their vertical distances from the fitted line – are determined accordingly.

The option setting 'PRINT = accumulated' in the FIT statement specifies that the output should include an accumulated anova, which partitions the variation accounted for by the model between its two terms, namely:

Accumulated analysis of variance

Change	d.f.	s.s.	m.s.	v.r.	F pr.
+ latitude	1	0.70955	0.70955	41.35	<.001
+ town	9	1.23119	0.13680	7.97	<.001
Residual	53	0.90937	0.01716		
Total	63	2.85011	0.04524		

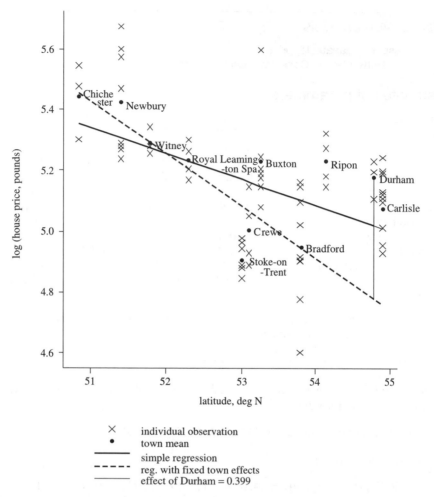

Figure 1.5   Relationship between latitude and house prices in a sample of English towns, comparing the lines of best fit from simple regression analysis and from analysis with town effects treated as fixed.

Despite the arbitrary nature of the parameter estimates, this anova is informative. The term 'latitude' represents the part of the variation in house price that is due to the effect of latitude, and the term 'town' represents the part due to the variation among towns after allowing for latitude; that is, the deviations of the town means from the original line of best fit (not the arbitrary fitted line presented above). Note that the mean square (MS) for latitude is the same as the corresponding value from Model 1.1:

$$MS_{latitude} = MS_{Regression, Model\ 1.1} = 0.70955.$$

That is, the amount of variation explained by latitude is consistent between Models 1.1 and 1.2. Note also that

$$SS_{Total, Model\ 1.1} = SS_{Total, Model\ 1.2} = 2.850,\ allowing\ for\ rounding.$$

Hence Model 1.2 represents a partitioning of the residual term in Model 1.1. Thus

$$DF_{\text{Residual,Model 1.1}} = DF_{\text{town}} + DF_{\text{Residual,Model 1.2}}$$

$$62 = 9 + 53$$

and

$$SS_{\text{Residual,Model 1.1}} = SS_{\text{town}} + SS_{\text{Residual,Model 1.2}}$$

$$2.141 = 1.23119 + 0.90937, \text{ allowing for rounding.}$$

Part of the variation, formerly unexplained, is now attributed to the effects of towns.

1.5 Construction of the appropriate *F* test for the significance of the explanatory variable when groups are present

The accumulated anova can be adapted to provide a realistic assessment of the significance of the effect of latitude. In the GenStat output, both of the mean squares for model terms in this anova are tested against the residual mean square:

- $F_{\text{latitude}} = MS_{\text{latitude}}/MS_{\text{Residual}} = 0.70955/0.01716 = 41.35$
- $F_{\text{town}} = MS_{\text{town}}/MS_{\text{Residual}} = 0.13680/0.01716 = 7.97$.

F_{town} is highly significant ($P < 0.001$), confirming that there is real variation among the towns in addition to that accounted for by latitude. Consequently, F_{latitude} is misleading: it is much larger than the value of 5.19 obtained when the regression is fitted using the town means. In order to obtain the appropriate value from the present analysis, we need to calculate

$$F_{\text{latitude}} = MS_{\text{latitude}}/MS_{\text{town}} = 0.70955/0.13680 = 5.18677.$$

This is almost exactly equivalent to the *F* statistic for latitude in the analysis based on the town means, as shown in Table 1.3. (The GenStat output does not give these

Table 1.3 Comparison between the *F* statistics from regression analyses based on individual houses and on town means.

	Regression based on town means	Regression with term for towns (Model 1.2)
Formula for F_{latitude}	$MS_{\text{Regression}}/MS_{\text{Residual}}$	$MS_{\text{latitude}}/MS_{\text{town}}$
Numerical values	$0.11601/0.02234 = 5.19293$	$0.70955/0.13680 = 5.18677$
$DF_{\text{Numerator}}$	1	1
$DF_{\text{Denominator}}$	9	9
P	0.0487	0.0488

F and P values to sufficient precision to reveal the slight differences between them.) The reason why the two sets of values do not agree exactly will be explained in Section 1.9.

1.6   The decision to regard a model term as random: a mixed model

The F values presented in Table 1.3 are based on a comparison of the variation explained by the regression line with the variation of *the town means* about the line, whereas the much larger value in the accumulated anova ($F_{\text{latitude}} = 41.35$) is based on a comparison with the variation of *individual values of log(house price)* about their respective town means. When we use the variation of the town means as the basis of comparison, we are regarding these means as values of a *random variable* – that is, we are considering the towns in this study as a representative sample from a large population of towns. Formally speaking, we are assuming that the values τ_i are independent values of a variable T, such that

$$T \sim \text{N}(0, \sigma_T^2)$$

– that is, they are very like the values ε_{ij}, except that their variance, σ_T^2, is different. This may seem a radical assumption, but it is necessary to assume that the τ_i are values of a random variable if we are to make any general statement about the relationship between latitude and house prices. (The assumption that this variable has a normal distribution is not a requirement, but other distributions require more advanced modelling methods – see Chapter 9.) If we were to drop one town from the study and replace it with newly chosen town, this would have an effect on the slope of the regression line – a larger effect than would be produced by simply taking a new sample of houses from within the same town. This is because there is more variation among town means than among individual houses from the same town, even after allowing for the effect of latitude: this is made clear by the highly significant value of F_{town}, 7.97. If we insist on regarding our choice of towns as fixed, any inference concerning the relationship between house price and latitude, its magnitude and our confidence that it is real, must be confined to these particular towns. This limitation is reflected by the arbitrary way in which GenStat defines the parameters of Model 1.2. In the language of mixed modelling, in order to obtain a realistic estimate of the effect of latitude on house prices, we must regard the effect of each town as a *random effect*, and town as a *random-effect term* in the regression model. The effect of latitude itself is a non-random or *fixed effect*. Since our model contains effects of both types, it is a *mixed model*.

When deciding whether to regard a factor as random, the essential question to ask is whether the levels studied can be regarded as a representative sample of some large population of levels. In the present case, can the towns sampled be considered representative of the population of English towns? This question of how to determine which model terms should be regarded as random will be discussed more fully, in the context of a wide range of models, in Chapter 6 (Section 6.3).

1.7   Comparison of the tests in a mixed model with a test of lack of fit

The partitioning of the variation around the regression line into two components, one due to the deviations of the town means from the line and the other to the deviations of individual houses from the town means, is similar to the test for lack of fit described by Draper and Smith (1998, Chapter 2, Section 2.1, pp 49–53). Indeed, the calculations performed to obtain the mean squares are identical in the two analyses. However, there is an important difference in the ideas that underlie them, and consequently in the F tests specified, as illustrated in Table 1.4. The angled lines in this table indicate the pairs of mean squares that are compared by the two F tests. The purpose of the test of lack of fit is to determine whether the variation among groups of observations at the same value of X (towns in the present case) is significant, or whether it can be absorbed into the 'Residual' term. In the mixed-model analysis, on the other hand, the reality of the variation among towns is not in doubt. The question is whether the effect of latitude is significant, or whether it can be absorbed into the 'town' term.

Even if the test of lack of fit leads to the conclusion that the variation among towns is significant, the two analyses are not equivalent. Because the test of lack of fit treats only the 'Residual' term as random, it leads to the use of this term as the denominator in all F tests, whereas the mixed model leads to the use of the random-effect term 'town' as the denominator against which to test the significance of the effect of latitude. The full set of tests specified by the two analyses is therefore as shown in Table 1.5. The dotted angled lines in this table indicate the pairs of mean

Table 1.4 Comparison between the F test for lack of fit and the mixed-model F test in the analysis of the effect of latitude on house prices in England.

Source of variation	DF	MS	Test of lack of fit		Mixed-model test	
			F	P	F	P
latitude	1	0.70955			5.19	0.0488
town	9	0.13680	7.97	<0.001		
Residual	53	0.01716				

Table 1.5 Comparison between the F tests conducted in ordinary multiple regression analysis and in mixed-model analysis of the effects of latitude and town on house prices in England.

Source of variation	DF	MS	Analysis with test of lack of fit		Mixed-model analysis	
			F	P	F	P
latitude	1	0.70955	41.35	<0.001	5.19	0.0488
town	9	0.13680	7.97	<0.001	7.97	<0.001
Residual	53	0.01716				

squares that are compared by the additional F tests. In the case considered by Draper and Smith the additional term required (equivalent to 'town' in the present example) is a quadratic one, to allow for curvature in the response to the explanatory variable, and in this situation the test of lack of fit is correct.

1.8   The use of residual maximum likelihood to fit the mixed model

So far we have taken an improvised approach to fitting the mixed model. We have:

- fitted two regression models, one without a term for the effects of towns and the other including such a term;

- taken the estimate of the slope from the first model and the mean squares from the second;

- obtained the F statistic to test the significance of the effect of latitude from the mean squares by hand.

However, the mixed-model analysis can be performed in a more unified manner using the criterion of *residual maximum likelihood* (REML, also known as *restricted maximum likelihood*). The formal meaning of this criterion will be explained in Chapter 10: here, we will simply apply it to the present data. The following GenStat statements specify the mixed model to be fitted:

```
VCOMPONENTS [FIXED = latitude; CADJUST = none] RANDOM = town
REML [PRINT = model, components, Wald, effects] logprice
```

The VCOMPONENTS statement specifies the terms in the model: it is equivalent to the FIT statement in an ordinary regression analysis. The option FIXED specifies the fixed-effect term or terms: in this case, 'latitude'. The constant (the intercept β_0) is also included in the model as a fixed-effect term by default: it does not have to be explicitly specified. The option setting 'CADJUST = none' indicates that no adjustment is to be made to the covariate 'latitude' before analysis: by default, a covariate is *centred* by subtracting its mean value from each of its values (see Chapter 7, Section 7.2). The *parameter* RANDOM specifies the random-effect term(s): in this case, 'town'. The REML statement specifies the response variate whose variation is to be explained by the model: in this case, 'logprice'. It is equivalent to the MODEL statement in an ordinary regression analysis. Note that the VCOMPONENTS and REML statements are given in the opposite order to their equivalents in ordinary regression analysis. The PRINT option indicates what results from the model-fitting process are to be presented in the output.

The output of these statements is as follows:

REML variance components analysis

Response variate:	logprice
Fixed model:	Constant + latitude
Random model:	town
Number of units:	64

Residual term has been added to model

Sparse algorithm with AI optimisation
Covariates not centred

Estimated variance components

Random term	component	s.e.
town	0.01963	0.01081

Residual variance model

Term	Factor	Model(order)	Parameter	Estimate	s.e.
Residual		Identity	Sigma2	0.0171	0.00332

Wald tests for fixed effects

Sequentially adding terms to fixed model

Fixed term	Wald statistic	d.f.	Wald/d.f.	chi pr
latitude	5.04	1	5.04	0.025

Dropping individual terms from full fixed model

Fixed term	Wald statistic	d.f.	Wald/d.f.	chi pr
latitude	5.04	1	5.04	0.025

Message: chi-square distribution for Wald tests is an asymptotic approximation (i.e. for large samples) and underestimates the probabilities in other cases.

Table of effects for Constant

9.497 Standard error: 1.9248

Table of effects for latitude

−0.08147 Standard error: 0.036272

The output begins with a specification of the model fitted, and notes that there are 64 observations in the data analysed. It notes that the residual term in the model

(ε_{ij}), which was not specified in the GenStat statements, has been added by GenStat. GenStat offers two algorithms for fitting mixed models (see Chapter 2, Section 2.5): the output notes that the sparse algorithm with average information (AI) optimisation has been used by default. The details of this algorithm need not concern us here; more will be said about them later (Chapter 10, Section 10.9). Estimates of *variance components* are given for the two random-effect terms, 'town' and the residual term. The concept of a variance component will be explained later (Chapter 3, Sections 3.2 to 3.4). Here, we will simply note that the estimate of residual variance from the REML model ($\hat{\sigma}^2 = 0.0171$) is about the same as that from the accumulated anova including the town term ($\hat{\sigma}^2 = 0.01716$): the two models are almost equivalent in their ability to explain the variation in house prices. The test of significance of the effect of latitude is provided by a measure called the Wald statistic, which has nearly the same value (5.04) as the value of $F_{\text{latitude}}(\approx 5.19)$ obtained earlier (Section 1.5). The P values corresponding to these test statistics are also fairly similar ($P = 0.025$ and 0.049 respectively). GenStat provides separate tests of the effects of adding 'latitude' to, and dropping it from, the model, but in the case of this simple model with only one fixed-effect term (apart from the constant), there is no distinction between these tests. Finally, the output gives the parameter estimates $\hat{\beta}_0$ and $\hat{\beta}_1$ and their SEs.

The estimates and SEs produced by the different methods of analysis are shown in Table 1.6. The three estimates of the effect of latitude agree closely, as do the three estimates of the constant. The SE of the estimated effect of latitude from Model 1.1 is much smaller than those from the regression analysis on town means and Model 1.2. This is because the SE for Model 1.1 is based on the assumption that every house is an independent observation. It overestimates the precision of the line of best fit, just as the F statistic from this model overestimates the significance of the relationship between house price and latitude.

The three fitted lines can be back-transformed to the original units (pounds), using the formula

$$\text{fitted house price} = 10^{\text{fitted log(house price)}}.$$

The resulting curves, together with the town means, back-transformed using the same formula, and the original data, are displayed in Figure 1.6. The three curves agree closely, and all fit the data reasonably well.

Table 1.6   Comparison of the parameter estimates, and their SEs, obtained from different methods of analysis of the effect of latitude on house prices in England.

Term	Regression analysis ignoring towns (Model 1.1)		Regression analysis on town means		Mixed-model analysis (Model 1.2)	
	Estimate	SE$_{\text{Estimate}}$	Estimate	SE$_{\text{Estimate}}$	Estimate	SE$_{\text{Estimate}}$
Constant	9.68	1.00	9.44	1.87	9.497	1.9248
latitude	−0.0852	0.0188	−0.0804	0.0353	−0.08147	0.036272

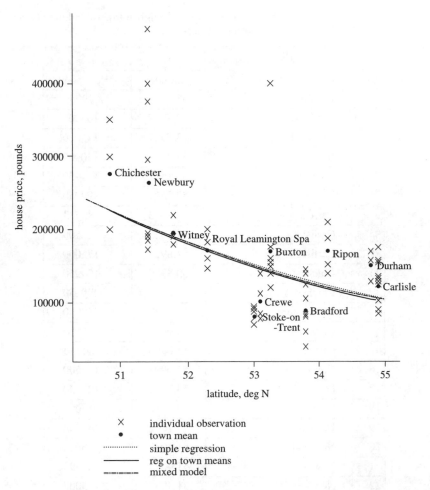

Figure 1.6   Relationship between latitude and house prices in a sample of English towns, comparing the lines of best fit from simple regression analysis, from analysis with town effects treated as fixed and from mixed-model analysis.

1.9   Equivalence of the different analyses when the number of observations per group is constant

The slight discrepancies between the results of the different methods of analysis are due to the unequal numbers of houses observed in different towns. The analysis ignoring towns gives equal weight to each house, whereas the analysis of town means gives equal weight to each town, and hence less weight per house to the houses in towns where a large sample was taken, e.g. Carlisle. The analysis of town means can be adjusted to give more weight to the towns in which more houses were sampled by setting an option in the GenStat MODEL statement, as follows:

```
MODEL [WEIGHTS = n_houses] meanlogprice
FIT [FPROB = yes; TPROB = yes] meanlatitude
```

Table 1.7   Prices of houses in a sample of English towns, and their latitudes, 'trimmed' to three houses per town.

	A	B	C
1	town!	latitude	price_pounds
2	Bradford	53.7947	79950
3	Bradford	53.7947	105000
4	Bradford	53.7947	145000
5	Buxton	53.2591	120000
6	Buxton	53.2591	154950
7	Buxton	53.2591	399950
8	Carlisle	54.8923	85000
9	Carlisle	54.8923	128500
10	Carlisle	54.8923	175000
11	Chichester	50.8377	199950
12	Chichester	50.8377	299250
13	Chichester	50.8377	350000
14	Crewe	53.0998	77500
15	Crewe	53.0998	84950
16	Crewe	53.0998	112500
17	Durham	54.7762	127950
18	Durham	54.7762	157000
19	Durham	54.7762	169950
20	Newbury	51.4037	185000
21	Newbury	51.4037	189995
22	Newbury	51.4037	475000
23	Ripon	54.1356	152000
24	Ripon	54.1356	187950
25	Ripon	54.1356	210000
26	Royal Leamington Spa	52.2876	147000
27	Royal Leamington Spa	52.2876	182500
28	Royal Leamington Spa	52.2876	199950
29	Stoke-on-Trent	53.0041	69950
30	Stoke-on-Trent	53.0041	69950
31	Stoke-on-Trent	53.0041	87950
32	Witney	51.7871	179950
33	Witney	51.7871	189950
34	Witney	51.7871	220000

The mean squares and F value then agree precisely with those from the regression based on individual houses with a term for towns (Model 1.2, Table 1.3)

This source of discrepancy is absent when the number of observations in each group is equal. Therefore, in order to compare the different methods of analysis more closely, we will consider a subset of the data from which randomly chosen houses have been removed, so that there are exactly three houses from each town, as shown in Table 1.7.

The relevant parts of the output produced by the different methods of analysis on this subset of the data are as follows. Ordinary regression analysis ignoring the towns (Model 1.1):

Regression analysis

Response variate: logprice
Fitted terms: Constant, latitude

Summary of analysis

Source	d.f.	s.s.	m.s.	v.r.	F pr.
Regression	1	0.283	0.28349	7.84	0.009
Residual	31	1.121	0.03615		
Total	32	1.404	0.04388		

Percentage variance accounted for 17.6
Standard error of observations is estimated to be 0.190.

Message: the following units have large standardized residuals.

Unit	Response	Residual
6	5.602	2.30

Estimates of parameters

Parameter	estimate	s.e.	t(31)	t pr.
Constant	9.04	1.37	6.57	<.001
latitude	−0.0726	0.0259	−2.80	0.009

Ordinary regression analysis on the town means:

Regression analysis

Response variate: meanlogprice
Fitted terms: Constant, meanlatitude

Summary of analysis

Source	d.f.	s.s.	m.s.	v.r.	F pr.
Regression	1	0.0945	0.09450	3.68	0.087
Residual	9	0.2310	0.02567		
Total	10	0.3255	0.03255		

Percentage variance accounted for 21.1
Standard error of observations is estimated to be 0.160.

Message: the following units have large standardized residuals.

Unit	Response	Residual
10	4.878	−2.04

Estimates of parameters

Parameter	estimate	s.e.	t(9)	t pr.
Constant	9.04	2.01	4.50	0.001
meanlatitude	−0.0726	0.0378	−1.92	0.087

Ordinary regression analysis with town as a term in the model (Model 1.2, town as a fixed-effect term):

Message: term town cannot be fully included in the model because 1 parameter is aliased with terms already in the model.

(town Witney) = 26.80 − (latitude)*0.4981 − (town Buxton)*0.2668 + (town Carlisle)*
0.5467 − (townChichester)*1.473 − (town Crewe)*0.3461 + (town Durham)*0.4889 −
(town Newbury)*1.191 + (town Ripon)*0.1698 − (town Royal Leamington Spa)*
0.7507 − (town Stoke-on-Trent)*0.3938

Regression analysis

Response variate: logprice
Fitted terms: Constant + latitude + town

Accumulated analysis of variance

Change	d.f.	s.s.	m.s.	v.r.	F pr.
+ latitude	1	0.28349	0.28349	14.59	<.001
+ town	9	0.69296	0.07700	3.96	0.004
Residual	22	0.42756	0.01943		
Total	32	1.40402	0.04388		

Mixed-model analysis (Model 1.2, town as a random-effect term):

REML variance components analysis

Response variate:	logprice
Fixed model:	Constant + latitude
Random model:	town
Number of units:	33

Residual term has been added to model

Sparse algorithm with AI optimisation
Covariates not centred

Estimated variance components

Random term	component	s.e.
town	0.01919	0.01226

Residual variance model

Term	Factor	Model(order)	Parameter	Estimate	s.e.
Residual		Identity	Sigma2	0.0194	0.00586

Wald tests for fixed effects

Sequentially adding terms to fixed model

Fixed term	Wald statistic	d.f.	Wald/d.f.	chi pr
latitude	3.68	1	3.68	0.055

Dropping individual terms from full fixed model

Fixed term	Wald statistic	d.f.	Wald/d.f.	chi pr
latitude	3.68	1	3.68	0.055

Message: chi-square distribution for Wald tests is an asymptotic approximation (i.e. for large samples) and underestimates the probabilities in other cases.

Table of effects for Constant

9.038 Standard error: 2.0068

Table of effects for latitude

−0.07260 Standard error: 0.037835

The different analyses are now equivalent in several ways:

- Ordinary regression analysis ignoring the towns, ordinary regression analysis on the town means and mixed-model analysis now all give the same fitted line.

- The regression analysis on the group means and the mixed-model analysis give the same value for the standard error of the slope, $SE_{latitude} = 0.378$.

- the value of the Wald statistic from the regression analysis, 3.68, is now identical to the value of F_{town} value obtained by hand from Model 1.2 with 'town' as a fixed-effect term, i.e. $MS_{latitude}/MS_{town} = 0.28349/0.07700 = 3.68169$. Note that this equivalence only occurs when $DF_{Numerator} = 1$.

This relationship between the F statistic and the Wald statistic deserves further study. Though the values of the two statistics are identical, the corresponding P values, 0.087 for the F statistic and 0.055 for the Wald statistic, are somewhat different. Both these statistics test the null hypothesis that there is no real relationship between latitude and house price. If this hypothesis is true, then the Wald statistic has approximately a χ^2 (chi-square) distribution, in this case with DF $= 1$. Note that the shape of this distribution depends only on the fact that $DF_{\text{latitude}} = 1$. However, the shape of the distribution of the F statistic depends also on the value $DF_{\text{town}} = 9$, which was used to calculate MS_{town}, the denominator of the F statistic. We say that the F statistic is distributed as $F_{DF_{\text{Numerator}}, DF_{\text{Denominator}}}$ (see Section 1.2): in the present case, as $F_{1,9}$. The difference in the shapes of the two distributions is shown in Figure 1.7. It is because of this difference in shape that the two P values differ. The P value is the area under the curve to the right of the observed value of F or the Wald statistic, 3.68. If the χ^2 distribution is used to perform the test, the area to be considered is the cross-hatched area only, giving $P = 0.055$, whereas if the F distribution is used, it is the whole hatched area, giving $P = 0.087$. The χ^2 approximation assumes that, effectively, the degrees of freedom in the denominator are infinite. If this were truly the case, the equivalence between the F statistic and the Wald statistic would be exact: that is, when $F_{1,\infty} = 3.68$, $P = 0.055$. In many mixed-model analyses the value of $DF_{\text{Denominator}}$ that would be obtained if the F statistic were calculated is large, and the assumption that $DF_{\text{Denominator}} = \infty$ gives a good approximation to P. However, when there are few degrees of freedom in the denominator, as in the present case, the approximation is not very close. In this analysis it is straightforward to obtain the equivalent F statistic with the correct degrees of freedom, and hence the correct P

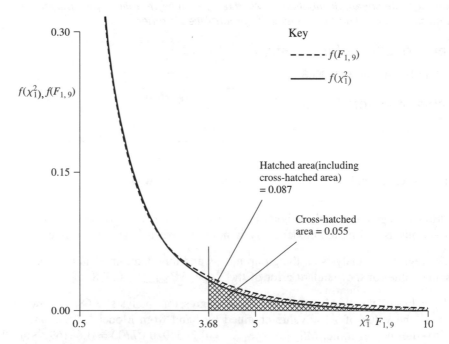

Figure 1.7 Comparison between the distributions of the variables $F_{1,19}$ and χ_1^2.

value. However, we will see later that there are many mixed-model analyses in which the appropriate degrees of freedom are not so easily determined. In this situation, the simplest way to proceed is to use the Wald statistic, but treat the accompanying P value with scepticism. A more formal alternative is described in Chapter 2, Section 2.5.

1.10   Testing the assumptions of the analyses: inspection of the residual values

In order for the significance tests and standard errors in the foregoing analyses to be strictly valid, it is necessary that the ε_{ij} should fulfil the assumption stated in Section 1.2: that is, they should be independent values of a variable E such that

$$E \sim N(0, \sigma^2).$$

GenStat's output has included various warning messages about the distribution of the residual values, and we should examine the residuals from our final model (which are *estimates* of these true residuals) to assess whether they approximately fulfil this assumption. Diagnostic plots of the residuals are produced by the following GenStat statement, executed after the REML statement in Section 1.8:

```
VPLOT [GRAPHICS=high] fittedvalues, normal, halfnormal, histogram
```

These plots are presented in Figure 1.8. If the assumption concerning the distribution of the residuals is correct, these plots are expected to show the following patterns:

- The histogram of residuals should have approximately a normal distribution (i.e. a bell-shaped, symmetrical distribution).

- In the fitted-value plot, the points should lie in a band of nearly constant width, centred near zero over the whole range of the fitted values.

- In the normal and half-normal plots, the points should lie nearly on a straight diagonal line from bottom left to top right.

Overall, the plots fit these expectations reasonably well in the present case. However, there are two possible causes for concern:

- There are two *outliers*, i.e. exceptionally large residual values, one positive and one negative.

- The fitted-value plot shows that the variance of the residuals is larger at some fitted values than at others, i.e. the spread of house prices is wider in some towns than others.

To address the first problem, the outliers might be deleted from the data set, though this is a questionable procedure. To address the second, it would be possible to fit

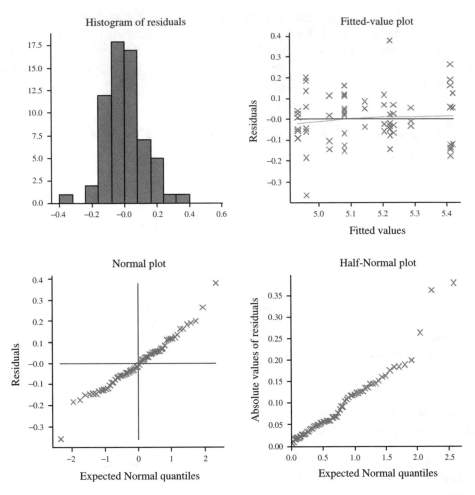

Figure 1.8　Diagnostic plots of residuals from the mixed model relating house prices to latitude in a sample of English towns.

a more elaborate model that allowed for variation of the residual variance among towns. This would give a more accurate estimate of the effect of latitude and of its significance, but lies outside the scope of this chapter.

In order for the test of $F_{latitude}$ and the Wald test to be valid, it is also necessary that

$$T \sim N(0, \sigma_T^2),$$

as discussed in Section 1.5. With a sample of only 11 towns, and only 9 degrees of freedom for variation among them, no very rigorous test of this assumption can be performed. We can, however, inspect the original graph of log(house price) against latitude (Figure 1.4), and note that the town means are reasonably evenly distributed about the fitted line.

1.11   Use of the software R to perform the analyses

All the analyses performed here by GenStat can also be performed by several other statistical software systems. Here we will consider only one of these, namely R, which is reasonably well tested, and has the additional merit of being free. Information on obtaining access to R is given in the preface of this book.

Data are usually presented to R in text files rather than Excel spreadsheets. The data from Sheet1 of file 'Intro to Mixed Modelling\Chapter 1\house price, latitude.xls' are therefore copied to the text file 'house price, latitude.dat' in the same directory, with the following additional modifications:

- the exclamation mark (!) is removed from the end of the heading 'town'

- the town name Royal Leamington Spa is enclosed in double quotes ("), so that R will recognise it as a single value.

Instructions are given to R via the S programming language. (This language is also used by the commercial statistical software S-Plus, so a statistical analysis specified in R can usually also be run in S-Plus with little or no modification.) As in the case of GenStat, only a few key points about these commands will be noted. A full introduction to the R environment and the S language is given by Venables *et al.* (2004).

The following R commands import the data and fit Model 1.1 (Section 1.2), which takes no account of the effects of towns:

```
houseprice_lat <- read.table(
    "Intro to Mixed Modelling\\Chapter 1\\house price, latitude.dat",
    header=TRUE)
attach(houseprice_lat)
logprice <- log(price_pounds, 10)
houseprice.model1 <- lm(logprice ~ latitude)
summary(houseprice.model1)
```

The *function* read.table() specifies that data, arranged in a table (i.e. in rows and columns), are to be read. The *arguments* of the function are between the brackets, and the first of these specifies the file in which the data are to be found. The *assignment symbol* (<-) points to the name, 'houseprice_lat', of the *data frame* that is to hold the data within R. The function attach() makes the *lists* (i.e. columns) of data within this data frame available to subsequent commands. The function log() transforms the house prices to logarithms, and the results are held in the list named 'logprice'. The function lm() indicates that a linear model (i.e. an ordinary regression model) is to be fitted, and the argument of this function, 'logprice ~ latitude', indicates that 'logprice' is the response variable, and that the model to be fitted has the single term 'latitude' (in addition to the constant). The results are stored in the *object* named 'houseprice.model1'. The function summary() extracts a summary of the results from this object, and displays it. The output of these commands is as follows:

```
Call:
lm(formula = logprice ~ latitude)

Residuals:
     Min        1Q     Median        3Q        Max
-0.501150 -0.080188  0.004316  0.109420  0.453718

Coefficients:
             Estimate Std. Error t value Pr(>|t|)
(Intercept)  9.68473    1.00036    9.681 5.14e-14 ***
latitude    -0.08518    0.01879   -4.533 2.71e-05 ***
---
Signif. codes:  0 '***' 0.001 '**' 0.01 '*' 0.05 '.' 0.1 ' ' 1

Residual standard error: 0.1858 on 62 degrees of freedom
Multiple R-Squared: 0.249,      Adjusted R-squared: 0.2368
F-statistic: 20.55 on 1 and 62 DF,  p-value: 2.710e-05
```

The first item in this output is a statement of the model fitted. This is followed by a summary of the distribution of the residual values: their minimum, quartiles and maximum. The three quartiles are the values that cut off the lowest quarter, half and three-quarters of the distribution respectively, i.e. the median is the second quartile. Next come the estimates of the regression coefficients. These are the values referred to in GenStat's output as parameters. The intercept estimated by R is equivalent to the constant estimated by GenStat. Note that the values produced by the two software systems are the same, as they should be. The statistics associated with these estimates are also the same as those from GenStat, with the addition of asterisks indicating the level of significance: *** for $P < 0.001$, ** for $P < 0.01$, etc. Next comes the residual standard error, which is the square root of the value of $MS_{Residual}$ given by GenStat ($0.1858 = \sqrt{0.03453}$), and has the same degrees of freedom, $DF_{Residual} = 62$. The multiple R-squared is a function of the sums of squares given by GenStat ($SS_{Regression}/SS_{Total} = 0.710/2.850 = 0.249$). The F statistic given by R, and the associated degrees of freedom and P value, are the same as those given by GenStat.

The following commands perform the regression analysis on the town means (Section 1.3):

```
houseprice_lat_mean <- read.table(
    "Intro to Mixed Modelling\\Chapter 1\\price, lat, town means.dat",
    header=TRUE)
attach(houseprice_lat_mean)
houseprice.model2 <- lm(meanlogprice ~ meanlatitude)
summary(houseprice.model2)
```

The format of the output from these commands is the same as that of the output presented above, and, again, the results agree numerically with those produced by GenStat.

The following statements fit the ordinary regression model with an additional fixed-effect term for towns (Model 1.2, Section 1.4), and present the accumulated analysis of variance:

```
houseprice_lat <- read.table(
    "Intro to Mixed Modelling\\Chapter 1\\house price, latitude.dat",
    header=TRUE)
attach(houseprice_lat)
logprice <- log(price_pounds, 10)
houseprice.model3 <- lm(logprice ~ latitude + town)
anova(houseprice.model3)
```

Their output is as follows:

```
Analysis of Variance Table

Response: logprice
           Df  Sum Sq Mean Sq F value    Pr(>F)
latitude    1 0.70955 0.70955  41.354 3.710e-08 ***
town        9 1.23119 0.13680   7.973 2.439e-07 ***
Residuals  53 0.90937 0.01716
---
Signif. codes:  0 '***' 0.001 '**' 0.01 '*' 0.05 '.' 0.1 ' ' 1
```

Again, the results agree with those produced by GenStat.

In order to perform the mixed-model analysis of Model 1.2 (Section 1.6) in R, with town as a random-effect term, it is necessary to implement a function, lme() (standing for *l*inear *m*ixed *e*ffects), which is not in the standard set. To do this, a *package* named 'nlme' must be loaded.[1] When this has been done, the following statements will perform the mixed-model analysis and present the results:

```
houseprice.model4 <- lme(logprice ~ latitude, random = ~ 1|town)
summary(houseprice.model4)
anova(houseprice.model4)
```

The first argument of the function lme() specifies that the response variable is logprice and the fixed-effect model comprises the single term 'latitude'. The second argument specifies that the random-effect model comprises the grouping structure 'town'. A formula specifying random effects to be estimated in conjunction with this grouping structure can be specified to the left of the bar (|). However, in the present case all that is required is a single estimated effect for each group, and this is indicated by the minimal model '1'. The output of the function summary() is as follows:

---

[1] The current method of doing this is as follows. Select 'Packages' in the main menu of the R GUI, then, in the Packages sub-menu, select 'Load package...' A window headed 'Select one' opens. In the list of packages in this window, select 'nlme', then click on the button labelled 'OK'. (The package 'nlme' gives access to both non-linear and linear mixed-effect models – hence the 'n' in its name.)

```
Linear mixed-effects model fit by REML
 Data: NULL
        AIC        BIC   logLik
  -42.12122 -33.61268 25.06061

Random effects:
 Formula: ~1 | town
          (Intercept)   Residual
StdDev:    0.1401131 0.1307764

Fixed effects: logprice ~ latitude
               Value Std.Error DF   t-value p-value
(Intercept)  9.496793 1.9247991 53   4.933914  0.0000
latitude    -0.081470 0.0362724  9 -2.246063  0.0513
 Correlation:
          (Intr)
latitude -1

Standardized Within-Group Residuals:
       Min         Q1        Med         Q3        Max
-2.7577350 -0.5397447 -0.1141964  0.4664333  2.8930977

Number of Observations: 64
Number of Groups: 11
```

The estimated values of the fixed effects of Intercept and latitude, and their standard errors, are the same as those given by GenStat. The output of the function anova() is as follows:

```
              numDF   denDF    F-value   p-value
(Intercept)      1      53   12712.430   <.0001
latitude         1       9       5.045   0.0513
```

The value of the F statistic for latitude is the same as that of the Wald statistic produced by GenStat. However, R recognises that $DF_{denominator}$ for this statistic is 9, not infinity, and the corresponding P value produced by R ($= 0.0513$) is therefore somewhat higher that that given by GenStat ($= 0.025$), for the reason explained in Section 1.7.

The following statements produce diagnostic plots of the residuals equivalent to those produced by GenStat, with the exception of the half-normal plot:

```
resmixedlogprice <- residuals(houseprice.model4)
hist(resmixedlogprice)
par(pty = "s")
qqnorm(resmixedlogprice)
qqline(resmixedlogprice)
fitmixedlogprice2 <- fitted(houseprice.model4)
plot(fitmixedlogprice2, resmixedlogprice)
```

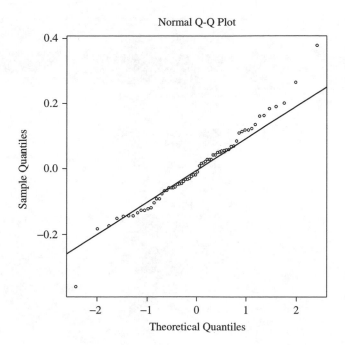

Figure 1.9 Normal plot of residuals from the mixed model relating house prices to latitude in a sample of English towns, produced by R.

The function `residuals()` extracts the residual values obtained by fitting the model 'houseprice.model4', and the function `hist()` produces a histogram of these values. Next come the commands that produce a normal plot from these residuals. The function `par()` is used here to ensure that this plot will be displayed in a square region. The function `qqnorm()` produces the normal plot (also known as a *quantile–quantile* or *Q–Q* plot, as it is produced by plotting the quantiles of the observed distribution against the quantiles of the corresponding normal distribution), and the function `qqline()` adds a line connecting the first and third quartiles on this plot. The function `fitted()` extracts the fitted values from the model 'houseprice.model4', and the function `plot()` plots the residual values against these. The normal plot is the only one of these diagnostic plots that has important differences from its GenStat equivalent: it is therefore presented in Figure 1.9.

1.12 Fitting a mixed model using GenStat's GUI

Most of the mixed models considered in this book can be fitted not only by the execution of statements written in the GenStat language, but alternatively by using GenStat's GUI. We will examine the use of this method for specifying the mixed model fitted above.

The appearance of the GUI when GenStat is first opened is illustrated in Figure 1.10. (Some details of the appearance differ from user to user.)

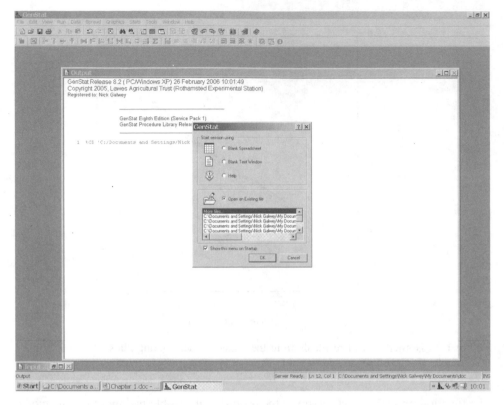

Figure 1.10 The GenStat GUI when first opened.

In order to specify the model, the user should proceed as follows. In the small start-dialogue window headed 'GenStat' in the middle of the GUI, click on the button labelled 'Cancel'. Next, in order to open the file holding the data, select 'File' in the GenStat main menu across the top of the screen, then, in the 'File' sub-menu, select 'Open'. A window opens headed 'Select Input file...'. In the box in this window labelled 'Files of type:' select 'Other spreadsheet files', then navigate to the directory holding the data file required. Select the file – in this case, 'house price, latitude for GUI.xls'. The appearance of the GUI is now as shown in Figure 1.11.

Click on the button labelled 'Open'. A window opens headed 'Select Excel Worksheet for Import'. Within this window, click on 'S: Sheet1', then click on the button labelled 'Finish'. The specified sheet of the Excel workbook opens, and the appearance of the GUI is as shown in Figure 1.12. Note that the variate 'logprice' has been added to the data prior to import.

In order to specify the mixed model, select 'Stats' in the main menu, then in the 'Stats' sub-menu select 'Mixed Models (REML)', then in the 'Mixed Models (REML)' sub-menu select 'Linear Mixed Models...'. The appearance of the GUI immediately before and immediately after making this last selection is as shown in Figure 1.13.

In the box labelled 'Y-variate:' within the window headed 'Linear Mixed Models', enter 'logprice'. In the box labelled 'Fixed Model:' enter 'latitude'. In the box labelled

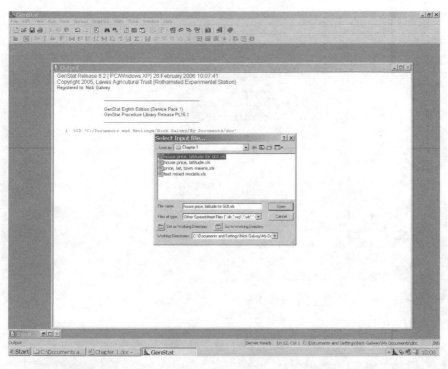

Figure 1.11 The GenStat GUI during selection of the data file to be analysed.

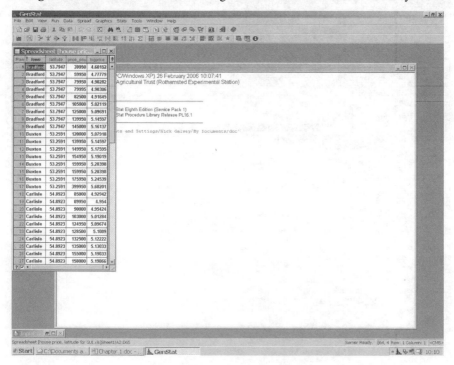

Figure 1.12 The GenStat GUI after opening the data file.

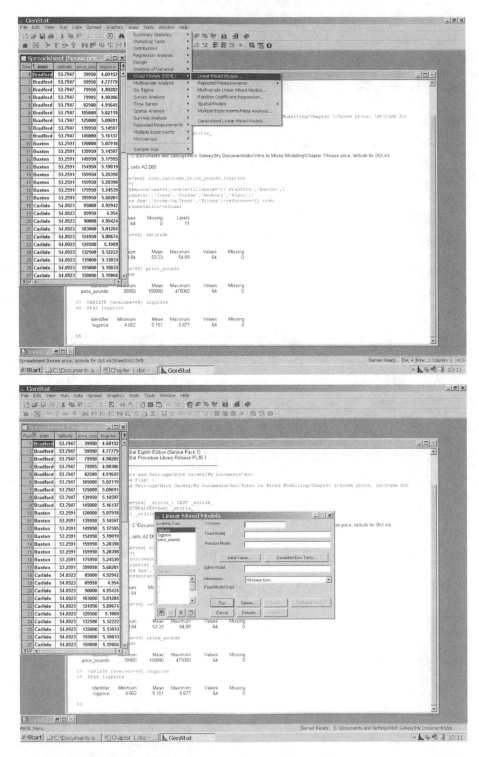

Figure 1.13   The GenStat GUI before and after selection of the menu item 'Linear Mixed Models'.

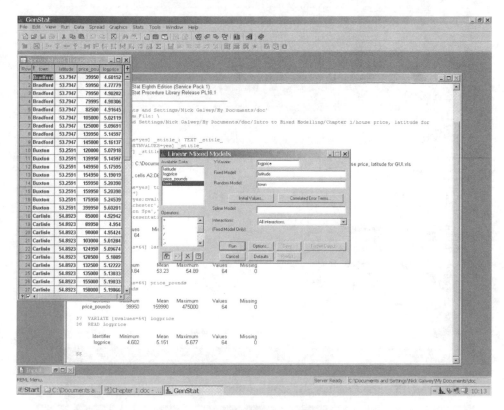

Figure 1.14 The GenStat GUI after specification of the model terms.

'Random Model:', enter 'town'. The appearance of the GUI is then as shown in Figure 1.14.

In the window headed 'Linear Mixed Models', click on the button labelled 'Run'. The model is fitted, and the results are sent to the output window. When this is brought to the front and appropriately sized, the appearance of the GUI is as shown in Figure 1.15.

Note that the output includes the GenStat commands that have been generated by the GUI. These provide an important record of the analysis specified. The results agree with those presented in Section 1.8, but they do not include the estimates of the fixed effects. To obtain these, return to the window headed 'Linear Mixed Models' (currently hidden behind the Output window), and click on the button labelled 'Further Output'. A window opens headed 'Linear Mixed Model Further Output'. In this window, tick the box labelled 'Estimated Effects'. The appearance of the GUI is then as shown in Figure 1.16.

Click on the button labelled 'Run'. The estimates of the fixed effects are then added to the output, as shown in Figure 1.17.

The specification of mixed models via the GUI may be easier than writing commands in the GenStat language, at least initially, but it is much more voluminous to illustrate. Subsequent examples in this book will therefore be illustrated using the command language only.

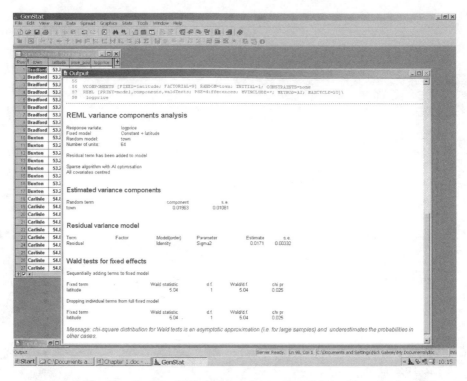

Figure 1.15   The GenStat GUI showing the results of fitting the mixed model.

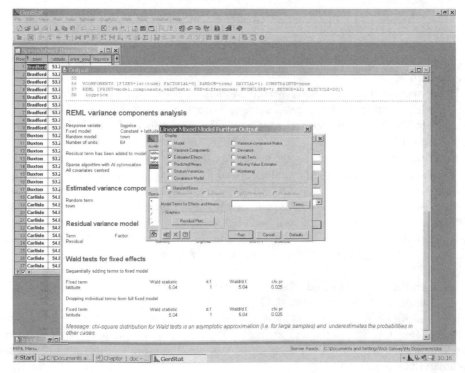

Figure 1.16   The GenStat GUI when the presentation of estimated effects has been specified.

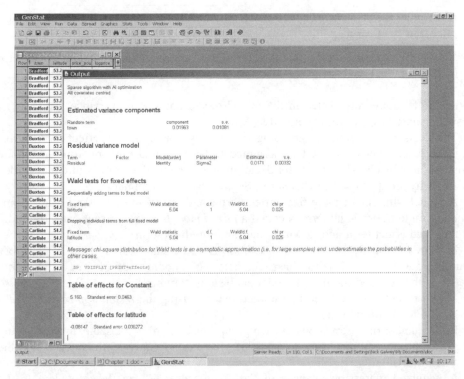

Figure 1.17 The GenStat GUI showing estimates of fixed effects from the mixed model.

1.13 Summary

When several observations of a response variable Y are taken at each value of an explanatory variable X, we may be able to recognise two sources of random variation, namely:

- variation among groups

- variation among observations within each group.

In this situation, a regression model with more than one random-effect term is required.

This situation is illustrated by a data set in which the price of houses in a sample of English towns is related to their latitude. The houses sampled in each town form a natural group, and houses in the same town are taken to lie at the same latitude.

If these data are analysed as if houses within the same town were mutually independent, the following results will be overestimated:

- the residual degrees of freedom

- the significance of the effect of latitude (i.e. P will be smaller than it should be).

Two methods are shown for taking into account the non-independence of houses in the same town:

- performing the analysis using the mean value of the response variable ($\log_{10}$(house price)) in each town;

- adding the grouping factor 'town' to the regression model as a random-effect term (a mixed-model analysis).

When F tests are constructed in the mixed-model analysis of variance, $MS_{latitude}$ is tested against MS_{town}, not against $MS_{Residual}$.

The decision to regard the grouping factor 'town' as a random-effect term means that the levels (towns) studied are assumed to be chosen at random from an infinite population of levels. The criteria for deciding whether a model term should be regarded in this way are discussed in Chapter 6, Section 6.3 and Summary.

A model with one or more fixed-effect terms and one or more random-effect terms in addition to the residual term is called a *mixed model*. In the present case, 'latitude' is the fixed-effect term and 'town' is the random-effect term.

In regression analysis, repeated observations at the same value of the explanatory variable are often used as the basis for a test of goodness of fit. The relationship between this test and the mixed-model analysis is explored.

Mixed models can be fitted in a unified manner using the *residual maximum likelihood* (REML) criterion.

If the number of observations in each group (each town) is the same, analysis using the town means and mixed-model analysis give exactly equivalent results. If not, the results differ slightly.

A weighted regression analysis of the mean values is exactly equivalent to the mixed-model analysis even if the number of observations varies from group to group. More weight is given to those means that are based on larger samples of observations.

The significance of each fixed-effect term in a mixed model can be tested by a Wald statistic. This is a χ^2 statistic, closely related to the F statistic from anova. However, it somewhat overestimates the significance (i.e. the value of P is smaller than it should be), as it is effectively based on the assumption that $DF_{Denominator\ of\ F statistic} = \infty$.

The validity of the significance tests used in mixed modelling depends on the assumption that every random-effect term, including the residual term, represents a normally distributed variable. Diagnostic plots of the residuals are used, and other random effects are inspected, to determine whether this assumption is approximately valid.

The use of the software systems GenStat and R to fit a mixed model is described. In GenStat, mixed models can be specified either using a command language or *via* a graphical user interface (GUI).

1.14 Exercises

1. The spreadsheet in Table 1.8 gives data on the greyhounds that ran in the Kanyana Stake (2) in Western Australia in December 2005. (Data reproduced by kind permission of David Shortte, Western Australian Greyhound Racing Association.)

   (a) Calculate the average speed of each animal in each of its recent races. Plot the speeds against the age of each animal.

Table 1.8 Data on greyhounds that ran in the Kanyana Stake (2) in December 2005. box = box from which the animal started this race; time = time taken to complete recent races (seconds); distance = distance run in recent races (metres).

	A	B	C	D	E
1	name	box	birthdate	time	distance
2	Leprechaun Kate	1	Jan-04		515
3	Leprechaun Kate	1	Jan-04		525
4	Leprechaun Kate	1	Jan-04	30.89	530
5	Leprechaun Kate	1	Jan-04	31.31	530
6	Leprechaun Kate	1	Jan-04	31.23	530
7	Mystifier	2	Dec-03		430
8	Mystifier	2	Dec-03		450
9	Mystifier	2	Dec-03		457
10	Mystifier	2	Dec-03		457
11	Mystifier	2	Dec-03		430
12	Mystifier	2	Dec-03	24.43	410
13	Mystifier	2	Dec-03	24.35	410
14	Mystifier	2	Dec-03	24.33	410
15	Proudly Agro	3	Feb-03	31.72	530
16	Proudly Agro	3	Feb-03	31.23	530
17	Proudly Agro	3	Feb-03	31.72	530
18	Proudly Agro	3	Feb-03	31.30	530
19	Proudly Agro	3	Feb-03	31.55	530
20	Proudly Agro	3	Feb-03	31.65	530
21	Proudly Agro	3	Feb-03	31.15	530
22	Proudly Agro	3	Feb-03	31.37	530
23	Desperado Lover	4	Jan-02	30.50	509
24	Desperado Lover	4	Jan-02	30.80	509
25	Desperado Lover	4	Jan-02	30.92	509
26	Desperado Lover	4	Jan-02	30.84	509
27	Desperado Lover	4	Jan-02	31.39	509
28	Desperado Lover	4	Jan-02	31.40	530
29	Desperado Lover	4	Jan-02	31.68	530
30	Desperado Lover	4	Jan-02	31.81	530
31	Squeaky Cheeks	5	Nov-03	34.82	530
32	Squeaky Cheeks	5	Nov-03	31.72	530
33	Squeaky Cheeks	5	Nov-03	32.34	530
34	Squeaky Cheeks	5	Nov-03	31.32	530
35	Squeaky Cheeks	5	Nov-03	31.85	530

(Continued overleaf)

Table 1.8   (*continued*)

	A	B	C	D	E
36	Squeaky Cheeks	5	Nov-03	31.22	530
37	Squeaky Cheeks	5	Nov-03	31.55	530
38	Squeaky Cheeks	5	Nov-03	31.52	530
39	Beyond the Sea	6	Apr-04	31.37	530
40	Beyond the Sea	6	Apr-04	30.75	530
41	Keith Kaos	7	Feb-03	30.56	509
42	Keith Kaos	7	Feb-03	31.31	530
43	Keith Kaos	7	Feb-03	31.81	530
44	Keith Kaos	7	Feb-03	31.71	530
45	Keith Kaos	7	Feb-03	31.58	530
46	Keith Kaos	7	Feb-03	31.04	530
47	Keith Kaos	7	Feb-03	31.61	530
48	Keith Kaos	7	Feb-03	31.59	530
49	Elza Prince	8	Nov-02		530
50	Elza Prince	8	Nov-02	31.30	530
51	Elza Prince	8	Nov-02	31.55	530
52	Elza Prince	8	Nov-02	31.92	530
53	Elza Prince	8	Nov-02	31.26	530
54	Elza Prince	8	Nov-02	31.73	530
55	Elza Prince	8	Nov-02	31.86	530
56	Elza Prince	8	Nov-02	31.52	530
57	Jarnat Boy	9	Apr-03	31.30	530
58	Jarnat Boy	9	Apr-03	31.90	530
59	Jarnat Boy	9	Apr-03	31.59	530
60	Jarnat Boy	9	Apr-03	31.55	530
61	Jarnat Boy	9	Apr-03	31.28	530
62	Jarnat Boy	9	Apr-03	31.12	530
63	Jarnat Boy	9	Apr-03	32.44	530
64	Jarnat Boy	9	Apr-03	31.64	530
65	Shilo Mist	10	Apr-03	31.71	530
66	Shilo Mist	10	Apr-03	32.30	530
67	Shilo Mist	10	Apr-03	31.49	530
68	Shilo Mist	10	Apr-03	32.17	530
69	Shilo Mist	10	Apr-03	32.14	530
70	Shilo Mist	10	Apr-03	31.74	530
71	Shilo Mist	10	Apr-03	31.98	530
72	Shilo Mist	10	Apr-03	31.82	530

Table 1.9   Levels of available chlorine in batches of a chemical product manufactured at two-week intervals, after a period of storage.

Length of time since production (weeks)	Available chlorine			
8	0.49	0.49		
10	0.48	0.47	0.48	0.47
12	0.46	0.46	0.45	0.43
14	0.45	0.43	0.43	
16	0.44	0.43	0.43	
18	0.46	0.45		
20	0.42	0.42	0.43	
22	0.41	0.41	0.40	
24	0.42	0.40	0.40	
26	0.41	0.40	0.41	
28	0.41	0.40		
30	0.40	0.40	0.38	
32	0.41	0.40		
34	0.40			
36	0.41	0.38		
38	0.40	0.40		
40	0.39			
42	0.39			

(b) The first value of speed for 'Squeaky Cheeks' (in row 31 of the spreadsheet) is an outlier: it is much lower than the other speeds achieved by this animal. Consider the arguments for and against excluding this value from the analysis of the data.

For the remainder of this exercise, exclude this outlier from the data.

(c) Perform a regression analysis with speed as the response variable and age as the explanatory variable, treating each observation as independent. Obtain the equation of the line of best fit, and draw the line on your plot of the data.

(d) Specify a more appropriate regression model for these data, making use of the fact that a group of observations was made on each animal. Fit your model to the data by the ordinary methods of regression analysis. Obtain the accumulated analysis of variance from your analysis.

(e) Which is the appropriate term against which to test the significance of the effect of age:

   (i) if 'name' is regarded as a fixed-effect term?

   (ii) if 'name' is regarded as a random-effect term?

Obtain the F statistic for age using both approaches, and obtain the corresponding P values. Note which test gives the higher level of significance, and explain why.

(f) Reanalyse the data by mixed modelling, fitting a model with the same terms but regarding 'name' as a random-effect term. Use the Wald statistic to test the significance of the effect of age.

(g) Obtain the equation of the line of best fit from your mixed-model analysis. Draw the line on your plot of the data, and compare it with that obtained when every observation was treated as independent.

(h) Obtain a subset of the data comprising only the last two observations on each animal. Repeat your analysis on this subset, and confirm that the Wald statistic for the effect of age now has the same value as the corresponding F statistic.

2. A chemical product was manufactured at two-week intervals, and after a period of storage the level of available chlorine, which is known to decline with time, was measured in cartons of the product (Draper and Smith, 1998, Chapter 24, Section 24.3, pp 518–519). The results obtained are presented in Table 1.9. (Data reproduced by permission of John Wiley & Sons, Inc.)

(a) Arrange these data for analysis by GenStat or R.

(b) Plot the values of available chlorine against the times since production. Obtain the line of best fit relating these two variables, treating each observation as independent, and display it on your plot.

(c) Consider whether observations made at the same time form a natural group. If so, consider whether it is reasonable to consider the variation of the group means around a fitted line as a random variable.

These data will be analysed further in Chapter 7, Exercise 1.

2

The need for more than one random-effect term in a designed experiment

2.1 The split plot design: a design with more than one random-effect term

In Chapter 1 we examined the idea that more than one random-effect term may be necessary in regression analysis, but this idea also arises in the analysis of designed experiments. In the simplest experimental designs, the fully randomised design leading to the one-way anova and the randomised block design leading to the two-way anova, it is sufficient to recognise a single source of random variation, the residual term. In designs with several treatment factors, a single random-effect term is still sufficient. But in designs with more elaborate block structures, it is necessary to recognise the block effects as random effects, either explicitly by the specification of a mixed model, or implicitly by using an anova protocol specific to the design in question. Such designs are known as incomplete block designs, because each block does not contain a complete set of treatments. One of the simplest designs of this type is the split plot design.

The theory of the split plot design was worked out in the 1930s in the context of field experiments on crops, and such experiments provide a simple context in which to illustrate it. (The literature from this period is not very accessible, but some account of it is given by Cochran and Cox (1957, Chapter 7, pp 293–316).) If two *treatment factors*, such as the choice of crop variety and the level of application of nitrogen fertiliser, are to be studied in all combinations, the experimenter may decide to apply each variety.nitrogen combination to several replicate field plots in a randomised block design. Alternatively, he or she may decide to sow each variety on a relatively large area of land, referred to a *main plot*, but to apply each level of nitrogen to a small *sub-plot* within each main plot. This might be done if comparisons between the levels

Introduction to Mixed Modelling: Beyond Regression and Analysis of Variance   N. W. Galwey
© 2006 John Wiley & Sons, Ltd

of nitrogen were of more interest than comparisons between the varieties: the former will be based on comparisons between the sub-plots, which are expected to be more precise than comparisons between the main plots. More often, the use of the split plot design is imposed by a practical constraint – for example, if the seed drill used is too wide to sow more than one variety in each main plot. For a fuller account of split plot designs, see, for example, Mead (1988, Chapter 14, pp 382–421). For other types of incomplete block design, see Chapter 7, Section 7.2, pp 134–142 and Chapter 15, pp 422–469 in the same book.

The split plot design is applicable to many other disciplines besides agricultural field experiments – for example, industrial experimentation, pharmacology, physiology, animal nutrition and food science. Here we will illustrate the design with data from an evaluation of four commercial brands of ravioli by nine trained assessors. (Data reproduced by kind permission of Guillermo Hough, DESA-ISETA, Argentina.) The purpose of the study was to identify differences in taste and texture between the brands. Knowledge of such differences is of great commercial importance to food manufacturers, but difficult to obtain: these sensory characteristics must ultimately be assessed by the subjective impressions of a human observer, which vary among individuals, and over occasions in the same individual. However, if the subjective assessment of some aspect of taste or texture (such as saltiness or gumminess) is consistent, for a particular brand, among individuals and over occasions – that is, if the perceived differences between brands are statistically significant – it is safe to conclude that these differences are real. Differences among assessors are of less interest. Different individuals may simply be using different parts of the assessment scale to describe the same sensations: who can say whether food tastes saltier to you than it does to me? However, if there are significant *interactions* between brand and assessor – for example, if the assessor ANA consistently perceives Brand A as saltier than Brand B, whereas GUI consistently ranks these brands in the opposite order – this is of interest to the investigator.

The brands of ravioli were cooked, served into small dishes and presented hot to the assessors. Three replicate evaluations were made, each being completed on a single day; hence each day comprised a block. There may have been uncontrolled and unobserved variation from day to day in the cooking and serving conditions – for example, the temperature of the room may have changed. On each day, the order in which the four brands were presented to the assessors was randomised. However, on any given day, all the assessors received the brands in the same order: for this type of product it is complicated to randomise the order of presentation among assessors. Hence each presentation of a brand comprised a main plot: the brand varied only between presentations, but the whole set of assessors received the brand within each presentation. Each serving, in a single dish, comprised a sub-plot. During each presentation, the servings were shuffled before being taken to the assessors; thus the assessors were informally randomised over the sub-plots within each main plot. (It would have been cumbersome to follow a formal randomisation at this stage: it was more important to get the servings to the assessors while they were still hot.) This experimental design is illustrated in Figure 2.1, detail being shown only for Day 1.

Each assessor gave the serving presented to him or her a numerical score for saltiness. The data obtained are displayed in a spreadsheet in Table 2.1. The allocation of

Day 1 Presentation 1 Brand B	Serving	1	2	3	4	5	6	7	8	9
	Assessor	PER	FAB	HER	MJS	ANA	GUI	ALV	MOI	NOR

Presentation 2 Brand A	Serving	1	2	3	4	5	6	7	8	9
	Assessor	MJS	ANA	GUI	MOI	FAB	NOR	HER	ALV	PER

Presentation 3 Brand C	Serving	1	2	3	4	5	6	7	8	9
	Assessor	NOR	FAB	GUI	ALV	MJS	ANA	HER	PER	MOI

Presentation 4 Brand D	Serving	1	2	3	4	5	6	7	8	9
	Assessor	ALV	HER	MOI	MJS	GUI	PER	ANA	NOR	FAB

Day 2

.

.

.

Day 3

.

.

.

Figure 2.1   The arrangement, in a split plot design, of an experiment to compare the perception of aspects of taste and texture of four commercial brands of ravioli by nine trained assessors.

assessors to servings within each presentation is an example of a randomisation that *might have* occurred, as the *actual* allocation produced by shuffling is unknown.

2.2   The analysis of variance of the split plot design: a random-effect term for the main plots

The algebraic model for the effects in this experimental design is

$$y_{ijk} = \mu + \delta_i + \pi_{ij} + \varepsilon_{ijk} + \alpha_{l|ij} + \beta_{m|ijk} + (\alpha\beta)_{lm|ijk} \qquad (2.1)$$

where

y_{ijk} = the evaluation of saltiness of the kth serving in the jth presentation on the ith day,

μ = the grand mean (overall mean) value of saltiness,

δ_i = the effect of the ith day, i.e. the departure of the mean of the ith day from the grand mean,

π_{ij} = the effect of the jth presentation on the ith day, i.e. the deviation of the mean of this presentation from the value predicted on the basis of μ, δ_i and $\alpha_{l|ij}$,

ε_{ijk} = the residual effect, i.e. the effect of the kth serving in the ijth day.presentation combination, which is the deviation of y_{ijk} from the value predicted on the basis of μ, δ_i, π_{ij}, $\alpha_{l|ij}$, $\beta_{m|ijk}$ and $(\alpha\beta)_{lm|ijk}$,

Table 2.1 Perceived saltiness of four commercial brands of ravioli by nine trained assessors, investigated in an experiment with a split plot design.
The conventions used in this spreadsheet are the same as those used in Table 1.1, Chapter 1.

	A	B	C	D	E	F
1	day!	presentation!	serving!	brand!	assessor!	saltiness
2	1	1	1	B	PER	15.59
3	1	1	2	B	FAB	28.95
4	1	1	3	B	HER	8.91
5	1	1	4	B	MJS	6.68
6	1	1	5	B	ANA	0.00
7	1	1	6	B	GUI	24.50
8	1	1	7	B	ALV	33.41
9	1	1	8	B	MOI	11.14
10	1	1	9	B	NOR	24.50
11	1	2	1	A	MJS	40.09
12	1	2	2	A	ANA	11.14
13	1	2	3	A	GUI	22.27
14	1	2	4	A	MOI	13.36
15	1	2	5	A	FAB	44.55
16	1	2	6	A	NOR	22.27
17	1	2	7	A	HER	31.18
18	1	2	8	A	ALV	26.73
19	1	2	9	A	PER	17.82
20	1	3	1	C	NOR	66.82
21	1	3	2	C	FAB	75.73
22	1	3	3	C	GUI	64.59
23	1	3	4	C	ALV	49.00
24	1	3	5	C	MJS	62.36
25	1	3	6	C	ANA	13.36
26	1	3	7	C	HER	53.45
27	1	3	8	C	PER	37.86
28	1	3	9	C	MOI	40.09
29	1	4	1	D	ALV	40.09
30	1	4	2	D	HER	35.64
31	1	4	3	D	MOI	33.41
32	1	4	4	D	MJS	55.68
33	1	4	5	D	GUI	46.77
34	1	4	6	D	PER	6.68
35	1	4	7	D	ANA	0.00

Table 2.1 (*continued*)

	A	B	C	D	E	F
36	1	4	8	D	NOR	55.68
37	1	4	9	D	FAB	60.14
38	2	1	1	C	MOI	37.86
39	2	1	2	C	PER	13.36
40	2	1	3	C	HER	22.27
41	2	1	4	C	ALV	35.64
42	2	1	5	C	NOR	33.41
43	2	1	6	C	MJS	57.91
44	2	1	7	C	GUI	62.36
45	2	1	8	C	ANA	11.14
46	2	1	9	C	FAB	89.09
47	2	2	1	D	GUI	31.18
48	2	2	2	D	ALV	31.18
49	2	2	3	D	ANA	8.91
50	2	2	4	D	NOR	17.82
51	2	2	5	D	HER	17.82
52	2	2	6	D	MOI	28.95
53	2	2	7	D	MJS	42.32
54	2	2	8	D	PER	33.41
55	2	2	9	D	FAB	60.14
56	2	3	1	B	NOR	24.50
57	2	3	2	B	ALV	26.73
58	2	3	3	B	GUI	35.64
59	2	3	4	B	HER	26.73
60	2	3	5	B	ANA	0.00
61	2	3	6	B	MOI	15.59
62	2	3	7	B	PER	28.95
63	2	3	8	B	FAB	13.36
64	2	3	9	B	MJS	20.05
65	2	4	1	A	FAB	60.14
66	2	4	2	A	GUI	24.50
67	2	4	3	A	ALV	35.64
68	2	4	4	A	MOI	20.05
69	2	4	5	A	MJS	13.36
70	2	4	6	A	ANA	2.23
71	2	4	7	A	NOR	11.14
72	2	4	8	A	PER	6.68

(*Continued overleaf*)

Table 2.1 (*continued*)

	A	B	C	D	E	F
73	2	4	9	A	HER	28.95
74	3	1	1	A	MJS	20.05
75	3	1	2	A	PER	15.59
76	3	1	3	A	GUI	22.27
77	3	1	4	A	MOI	13.36
78	3	1	5	A	FAB	17.82
79	3	1	6	A	ANA	0.00
80	3	1	7	A	HER	24.50
81	3	1	8	A	ALV	28.95
82	3	1	9	A	NOR	13.36
83	3	2	1	C	FAB	60.14
84	3	2	2	C	NOR	62.36
85	3	2	3	C	HER	44.55
86	3	2	4	C	ALV	33.41
87	3	2	5	C	ANA	11.14
88	3	2	6	C	GUI	53.45
89	3	2	7	C	MJS	51.23
90	3	2	8	C	PER	11.14
91	3	2	9	C	MOI	13.36
92	3	3	1	B	HER	17.82
93	3	3	2	B	GUI	31.18
94	3	3	3	B	ALV	37.86
95	3	3	4	B	FAB	13.36
96	3	3	5	B	PER	2.23
97	3	3	6	B	NOR	51.23
98	3	3	7	B	MOI	13.36
99	3	3	8	B	MJS	6.68
100	3	3	9	B	ANA	0.00
101	3	4	1	D	HER	37.86
102	3	4	2	D	PER	26.73
103	3	4	3	D	FAB	46.77
104	3	4	4	D	MOI	17.82
105	3	4	5	D	GUI	35.64
106	3	4	6	D	ALV	44.55
107	3	4	7	D	NOR	17.82
108	3	4	8	D	ANA	0.00
109	3	4	9	D	MJS	55.68

$\alpha_{l|ij}$ = the main effect of the lth brand, being the brand served in the ijth day.presentation combination,

$\beta_{m|ijk}$ = the main effect of the mth assessor, being the assessor who evaluated the ijkth day.presentation.serving combination,

$(\alpha\beta)_{lm|ijk}$ = the interaction effect between the lth brand and the mth assessor, being the brand.assessor combination used in the ijkth day.presentation.serving combination.

The *main effect* of a particular brand is its average effect, over all assessors, on the response variable – that is, the average difference between observations on this brand and the grand mean. Similarly, the main effect of a particular assessor is his or her average effect, over all brands. The interaction between brand and assessor is the effect that is specific to that particular brand.assessor combination, but shared by all observations from that combination. It is the departure of the mean for the brand.assessor combination from the value expected on the basis of the two main effects. For a fuller explanation of main effects and interaction effects, see Mead (1988, Chapter 3, Section 3.3, pp 33–36).

The terms on the right hand side of Equation 2.1, μ, δ_i, α_l, etc., are parameters to be estimated: their estimates are designated $\hat{\mu}$, $\hat{\delta}_i$, $\hat{\alpha}_l$, etc. Note that the observation on the response variable (y_{ijk}) is identified only by the suffixes i, j and k, not by l and m – that is, it is identified only in terms of the day, presentation and serving to which it relates, not in terms of the experimental treatments (combinations of brand and assessor). The same is true of the residual effect. The effects of day, presentation and serving are grouped together towards the beginning of the model. These terms represent the structure of the *natural variation* in the experiment, independent of the experimental treatments that were imposed; they are sufficient to identify every observation in the experiment. The treatment terms, $\alpha_{l|ij}$, $\beta_{m|ijk}$ and $(\alpha\beta)_{lm|ijk}$, are grouped together at the end of the model. The suffixes of these terms indicate not only the treatment-factor level or combination of levels applied, specified before the bar (|), but also the observation or set of observations to which they were applied, specified after the bar.

As in the models introduced in Chapter 1, it is assumed that the *true* values ε_{ijk} are independent values of a variable E, such that

$$E \sim N(0, \sigma^2).$$

The other terms that represent the structure of the natural variation are also assumed to be random-effect terms. That is, it is further assumed that the δ_i are independent values of a random variable Δ, such that

$$\Delta \sim N(0, \sigma_\Delta^2),$$

and the π_{ij} are independent values of a random variable Π, such that

$$\Pi \sim N(0, \sigma_\Pi^2).$$

The justification for these assumptions will be considered later (Section 2.6). These sources of natural variation are not of intrinsic interest: the aim of the experiment is to compare the treatments applied – the combinations of brand and assessor. The sources of natural variation are *nuisance variables*, which reduce the precision of such comparisons.

Although the concepts underlying Model 2.1 are the same as those underlying the regression models presented in Chapter 1, there is a set of specialised techniques and customs for the analysis of designed, balanced experiments like this. Therefore the model to be fitted is specified not in MODEL and FIT statements like those used previously, but in more specialised statements, as follows:

```
IMPORT 'Intro to Mixed Modelling\\Chapter 2\\ravioli.xls'; \
    SHEET = 'sheet1'
BLOCKSTRUCTURE day / presentation / serving
TREATMENTSTRUCTURE brand * assessor
ANOVA [FPROB = yes] saltiness
```

The part of the model that relates to the nuisance variables is specified in the BLOCK-STRUCTURE statement, and that which relates to the variables under investigation is specified in the TREATMENTSTRUCTURE statement. (In experimental design, a group of observations that share a common value of a nuisance variable is referred to as a block.) Together, these two statements are analogous to the FIT statement. The composite model terms 'brand * assessor' and 'day / presentation / serving' will be explained below. The ANOVA statement specifies the response variable, in this case saltiness. It is analogous to the MODEL statement.

The output of these statements is as follows. It is quite voluminous, and the first part will be discussed before presenting the rest.

Analysis of variance

Variate: saltiness

Source of variation	d.f.	s.s.	m.s.	v.r.	F pr.
day stratum	2	743.4	371.7	2.67	
day.presentation stratum					
brand	3	9859.9	3286.6	23.59	0.001
Residual	6	835.9	139.3	1.39	
day.presentation.serving stratum					
assessor	8	15474.5	1934.3	19.29	<.001
brand.assessor	24	6075.3	253.1	2.52	0.002
Residual	64	6417.5	100.3		
Total	107	39406.5			

Message: the following units have large residuals.

day 1 presentation 1	−5.65	s.e.	2.78
day 1 presentation 4 serving 6	−19.80	s.e.	7.71
day 1 presentation 4 serving 8	21.03	s.e.	7.71

Before we consider the numerical values in this anova table, we need to examine the model notation used, which is that of Wilkinson and Rogers (1973). In this notation the composite model term 'brand * assessor' can be expanded to 'brand + assessor + brand.assessor', and each term in this expanded model appears in the column headed 'Source of variation'. The terms 'brand' and 'assessor' represent the main effects of these factors – that is, the effects represented by $\alpha_{l|ij}$ and $\beta_{m|ijk}$ in Equation 2.1. The term 'brand.assessor' represents the effects of interaction between these factors, represented by $(\alpha\beta)_{lm|ijk}$ in Equation 2.1. The composite model term 'day / presentation / serving' can also be expanded, to 'day + day.presentation + day.presentation.serving', and each term in this expanded model also appears in the column headed 'Source of variation'. The term 'day' represents the main effects of days, but there is no corresponding term for the main effects of presentations. This is because Presentation 1 on Day 1 is not the same presentation as Presentation 1 on Day 2, and it would therefore not be meaningful to obtain the mean for Presentation 1 over days. Because no main effects of presentation are estimated, the term 'day.presentation' does *not* represent day × presentation interaction effects. Instead, it represents the effects of presentations *within* each day. These effects are the variation among presentations which, it is estimated, would have been present even if a single brand had been used throughout the experiment. Similarly, Serving 1 in a particular presentation on a particular day is not the same serving as Serving 1 in any other presentation. Hence no main effects of serving are estimated, and the term 'day.presentation.serving' does not represent day × presentation × serving interaction effects, but the effects of servings within each presentation.

Thus the model in the BLOCKSTRUCTURE statement partitions the experiment into three strata, each representing a different source of natural variation. The meaning of $MS_{Residual}$ in each stratum is shown in Table 2.2. Each of the treatment terms, 'brand', 'assessor' and 'brand.assessor', is tested in the appropriate stratum. Each treatment occurs once in each day, so comparisons between the day means give no information about treatment effects, and no treatment term is tested in the 'day' stratum. Comparisons between brands are based on presentation means, so the main effect of brand is tested in the 'day.presentation' stratum: that is, the F value for brand is given by

$$F_{brand} = MS_{brand}/MS_{Residual,day.presentation\ stratum} = 3286.6/139.3 = 23.59.$$

Comparisons between assessors, and between brand × assessor interaction effects, are based on individual servings within each presentation, so these effects are tested in the 'day.presentation.serving' stratum: that is,

$$F_{assessor} = MS_{assessor}/MS_{Residual,day.presentation.serving\ stratum} = 1934.3/100.3 = 19.29,$$

Table 2.2 The strata of a split plot experiment to compare the perception of saltiness of commercial brands of ravioli by trained assessors.

Stratum	Meaning of $MS_{Residual}$
day	Natural variation among days
day.presentation	Natural variation among presentations on the same day
day.presentation.serving	Natural variation among servings in the same presentation

and

$$F_{\text{brand.assessor}} = MS_{\text{brand.assessor}}/MS_{\text{Residual,day.presentation.serving stratum}}$$

$$= 253.1/100.3 = 2.52.$$

Note that the value of MS_{Residual} is greater in higher strata than in lower strata: that is,

$$MS_{\text{Residual, day stratum}} > MS_{\text{Residual, day.presentation stratum}}$$

$$> MS_{\text{Residual, day.presentation.serving stratum}}$$

$$371.7 > 139.3 > 100.3.$$

This is to be expected: there is more natural variation among days than among presentations on the same day, and more natural variation among presentations than among servings in the same presentation. These sources of variation will be considered in more detail in Chapter 3, Section 3.17.

Below the anova table, GenStat gives a warning that two observations have large residual values. We will examine the residual values from this analysis later (Section 2.7).

The rest of the output of these BLOCKSTRUCTURE, TREATMENTSTRUCTURE and ANOVA statements is as follows:

Tables of means

Variate: saltiness

Grand mean 29.28

brand		A	B	C	D			
		21.78	19.22	43.23	32.91			

assessor		ALV	ANA	FAB	GUI	HER	MJS	MOI
		35.27	4.83	47.52	37.86	29.14	36.01	21.53

assessor		NOR	PER
		33.41	18.00

brand	assessor	ALV	ANA	FAB	GUI	HER	MJS
A		30.44	4.46	40.84	23.01	28.21	24.50
B		32.67	0.00	18.56	30.44	17.82	11.14
C		39.35	11.88	74.99	60.13	40.09	57.17
D		38.61	2.97	55.68	37.86	30.44	51.23

brand	assessor	MOI	NOR	PER
A		15.59	15.59	13.36
B		13.36	33.41	15.59
C		30.44	54.20	20.79
D		26.73	30.44	22.27

Standard errors of differences of means

Table	brand	assessor	brand assessor
rep.	27	12	3
s.e.d.	3.213	4.088	8.351
d.f.	6	64	66.70

Except when comparing means with the same level(s) of brand	
brand	8.176
d.f.	64

The main effects of brand are highly significant: that is, the P value associated with F_{brand} is less than 0.001, indicating that we may confidently reject the null hypotheses that the main effects of all brands are zero. It is therefore legitimate to interpret the one-way table of brand means, because at least some of the differences between these means are almost certainly real. Similarly, the main effects of assessor are highly significant, indicating that it is legitimate to interpret the one-way table of assessor means. The brand × assessor interaction term is also highly significant, so the two-way table of means for brand.assessor combinations may also be interpreted. The one-way tables show that Brand C was perceived as considerably saltier than the other brands, and that ANA's mean assessment score is strikingly lower than that of the other assessors. The perceived saltiness of Brand C may be important to manufacturers, retailers or consumers. However, ANA's low scores are not necessarily important: as suggested earlier (Section 2.1), they may simply reflect this assessor's use of the assessment scale. The two-way table shows that ANA's scores are also less variable than those of the other assessors – they range only from 0.00 to 11.88, whereas GUI's scores range from 23.01 to 60.13. (Such variation among assessors in the use of the assessment scale can be taken into account by more sophisticated methods of data analysis than the split plot anova presented here, e.g. generalised Procrustes analysis (Gower, 1975).) There is also some variation in the ranking of the brands between the assessors: for example, ANA perceived Brand B as the least salty, whereas GUI perceived Brand A as the least salty. Such crossover effects may be important: they suggest an obstacle to designing a brand that will please all consumers. However, all assessors perceived Brand C as the saltiest. The table of standard errors of differences between means will be considered in Chapter 4, Section 4.3.

2.3 Consequences of failure to recognise the main plots when analysing the split plot design

In order to understand why the variation among presentations within the same day must be specified as a term in the model for the analysis of this experiment, it may be helpful to look at what happens if this term is omitted. The model fitted (Model 2.2) is then

$$y_{ij} = \mu + \delta_i + \varepsilon_{ij} + \alpha_{l|ij} + \beta_{m|ij} + (\alpha\beta)_{lm|ij} \tag{2.2}$$

where

$\qquad y_{ij} =$ the evaluation of saltiness of the jth serving on the ith day,

$\qquad \mu =$ the grand mean (overall mean) value of saltiness,

$\qquad \delta_i =$ the effect of the ith day,

$\qquad \varepsilon_{ij} =$ the residual effect, i.e. the effect of the jth serving on the ith day, which is the deviation of y_{ij} from the value predicted on the basis of μ, δ_i, $\alpha_{l|ij}$, $\beta_{m|ij}$, $(\alpha\beta)_{lm|ij}$,

$\alpha_{l|ij}$ = the main effect of the lth brand, being the brand served in the jth serving on the ith day,

$\beta_{m|ij}$ = the main effect of the mth assessor, being the assessor who evaluated the jth serving on the ith day,

$(\alpha\beta)_{lm|ij}$ = the interaction effect between the lth brand and the mth assessor, being the brand.assessor combination used in the jth serving on the ith day.

The following statements specify and fit this model:

```
BLOCKSTRUCTURE day
TREATMENTSTRUCTURE brand * assessor
ANOVA [FPROB = yes; PRINT = aovtable] saltiness
```

Note that the model in the BLOCKSTRUCTURE statement has been reduced from 'day / presentation / serving' to 'day': that is, not only has the term 'day.presentation' been omitted from the model, but so has the term 'day.presentation.serving'. The latter term represents residual variation, and therefore need not be specified explicitly.

The output of the ANOVA statement is as follows:

Analysis of variance

Variate: saltiness

Source of variation	d.f.	s.s.	m.s.	v.r.	F pr.
day stratum	2	743.4	371.7	3.59	
day.*Units* stratum					
brand	3	9859.9	3286.6	31.72	<.001
assessor	8	15474.5	1934.3	18.67	<.001
brand.assessor	24	6075.3	253.1	2.44	0.002
Residual	70	7253.5	103.6		
Total	107	39406.5			

The 'Residual' term in this anova is obtained by pooling the terms 'day.presentation' and 'day.presentation.serving' in the anova from Model 2.1 presented in the previous section: that is,

$$DF_{\text{Residual, Model 2.2}} = DF_{\text{Residual, day.presentation stratum, Model 2.1}}$$
$$+ DF_{\text{Residual, day.presentation.serving stratum, Model 2.1}}$$
$$70 = 6 + 64$$

and

$$SS_{\text{Residual, Model 2.2}} = SS_{\text{Residual, day.presentation stratum, Model 2.1}}$$
$$+ SS_{\text{Residual, day.presentation.serving stratum, Model 2.1}}$$
$$7253.5 = 835.9 + 6417.5, \text{ allowing for rounding.}$$

For each model term in the TREATMENTSTRUCTURE statement, an F value is calculated, using $MS_{Residual}$ as the denominator in every case. For example, the F value for brand is given by $MS_{brand}/MS_{Residual} = 3286.6/103.6 = 31.72$, and that for assessor by $MS_{assessor}/MS_{Residual} = 1934.3/103.6 = 18.67$.

This method of constructing the F tests is clearly seen to be incorrect when the variation among presentations is considered. A single brand is used in each presentation, and any comparison between brands must be made between presentations: hence if the main effects of brand only were to be studied, the analysis could be performed on the presentation means. At this level, the experiment has a randomised block design, each brand being allocated to a random presentation within each day. Hence in this part of the analysis, $DF_{Residual}$ should be

$$\text{no. of presentations} - 1 - DF_{day} - DF_{brand} = 12 - 1 - 2 - 3 = 6,$$

not 70. The variation within a presentation is irrelevant to the comparisons between brands. Hence it cannot be right to include this variation in the calculation of the $MS_{Residual}$ against which MS_{brand} is tested. We deceive ourselves if we imagine that the 24 servings on each day represent 24 independent observations. The consequences of this mistake are seen when the F statistics for the main effects of brand obtained by the two methods are compared:

- correct analysis gives $F_{3,6} = 23.59$, $P = 0.00101$;

- analysis ignoring presentations gives $F_{3,70} = 31.72$, $P \approx 0$.

Because the natural variation among servings in the same presentation is less than that among servings in different presentations, pooling these terms underestimates the $MS_{Residual}$ against which MS_{brand} is tested, and this inflates the F statistic. The excessive number of degrees of freedom in the denominator further exaggerates its significance.

Whereas any comparison between brands must be made between presentations, any comparison between assessors can be made within a single presentation. Hence it cannot be right to include the variation among presentations in the calculation of the $MS_{Residual}$ against which $MS_{assessor}$ is tested. The analysis ignoring presentations will tend to deflate the value of $F_{assessor}$, and underestimate its significance. As for the brand.assessor interaction effects, by definition they sum to zero within each presentation, because all nine assessors are represented in the presentation. Hence variation among presentations is irrelevant to comparisons between these effects also. To summarise: only variation among presentations is relevant to comparisons between brands, whereas only variation within presentations is relevant to comparisons between assessors, and between brand.assessor interaction effects.

2.4 The use of mixed modelling to analyse the split plot design

The foregoing account has emphasised the need to recognise random-effect model terms additional to the residual term when analysing a split plot design. Because the model used (Model 2.1) contains such terms, it is a mixed model, and although a split plot design is usually analysed with specialised tools, such as GenStat's BLOCK-STRUCTURE, TREATMENTSTRUCTURE and ANOVA directives, it can also be fitted by more general tools for mixed-model analysis, such as GenStat's VCOMPONENTS and REML directives. Comparison of the results with the relatively familiar split plot anova (Section 2.2) will help to elucidate the concepts of mixed modelling.

The mixed-model analysis of Model 2.1 can be specified as follows:

```
VCOMPONENTS [FIXED = brand * assessor] \
   RANDOM = day / presentation / serving
REML [PRINT = model, components, Wald, means] saltiness
```

The model formerly specified in the TREATMENTSTRUCTURE statement, 'brand * assessor', is now specified in the option FIXED of the VCOMPONENTS statement, while the model formerly specified in the BLOCKSTRUCTURE statement, 'day / presentation / serving', is now specified in the parameter RANDOM of this statement. The response variable, saltiness, formerly specified in the ANOVA statement, is now specified in the REML statement.

The first part of the output of the REML statement is as follows. (The remainder will be presented in Chapter 4, Section 4.5.)

REML variance components analysis

Response variate: saltiness
Fixed model: Constant + brand + assessor + brand.assessor
Random model: day + day.presentation + day.presentation.serving
Number of units: 108

day.presentation.serving used as residual term

Sparse algorithm with AI optimisation

Estimated variance components

Random term	component	s.e.
day	6.5	10.6
day.presentation	4.3	9.2

Residual variance model

Term	Factor	Model(order)	Parameter	Estimate	s.e.
Ł day.presentation.serving		Identity	Sigma2	100.3	17.7

Wald tests for fixed effects

Sequentially adding terms to fixed model

Fixed term	Wald statistic	d.f.	Wald/d.f.	chi pr
brand	70.77	3	23.59	<0.001
assessor	154.32	8	19.29	<0.001
brand.assessor	60.59	24	2.52	<0.001

Dropping individual terms from full fixed model				
Fixed term	Wald statistic	d.f.	Wald/d.f.	chi pr
brand.assessor	60.59	24	2.52	<0.001

Message: chi-square distribution for Wald tests is an asymptotic approximation (i.e. for large samples) and underestimates the probabilities in other cases.

As in the example in Chapter 1, Section 1.8, the output begins with a specification of the model fitted, and notes the number of units (observations) in the data analysed. GenStat detects that there is only one observation of each day.presentation.serving combination, and hence that this term should be used as the residual term. Next come estimates of variance components for the terms in the random-effect model: these will be considered later (Chapter 3, Section 3.17).

Next come Wald tests for the terms in the fixed-effect model, which correspond to the F tests for the treatment terms in the corresponding anova (Section 2.3). For each treatment term,

$$DF_{\text{Wald statistic}} = DF_{\text{Numerator of } F\text{statistic}},$$

and

$$\text{Wald statistic}/DF_{\text{Wald statistic}} = F\text{statistic}. \tag{2.3}$$

For example,

$$\text{Wald statistic}_{\text{brand}}/DF_{\text{brand}} = F_{\text{brand}}$$

$$70.77/3 = 23.59.$$

As in the mixed-model analysis of the effect of latitude on house prices in Chapter 1, Section 1.9, the P values associated with the corresponding Wald and F statistics are similar, but not identical. These values, given with greater precision than in the GenStat output, are shown in Table 2.3. As in the previous case, the P values associated with the Wald statistic are smaller, because when this statistic is used it is effectively assumed that $DF_{\text{Denominator of } F\text{statistic}} = \infty$. GenStat provides separate tests of the effects of adding and dropping terms to and from the model. However, in a balanced experimental design in which each treatment term is orthogonal to (independent of) all the others, as in the present case, there is no distinction between these two tests.

Table 2.3   Comparison of the P values associated with Wald statistics and F statistics in a split plot experiment to compare the perception of saltiness of commercial brands of ravioli by trained assessors.

Term	P	
	for Wald statistic	for F
brand	<0.00001	0.00101
assessor	<0.00001	<0.00001
brand.assessor	0.00005	0.00175

The means for the treatment terms given by the mixed-modelling analysis are the same as those given by the corresponding anova.

2.5   A more conservative alternative to the Wald statistic

It has been noted earlier (Chapter 1, Section 1.9) that the Wald statistic overestimates the significance of the fixed effect tested (i.e. underestimates the P value), because it effectively assumes that $DF_{Denominator} = \infty$ in the corresponding F test. A more conservative alternative is provided by the likelihood ratio test statistic proposed by Welham and Thompson (1997): this test tends to underestimate the significance of the effect (i.e. to overestimate the P value). Like the F statistic in an anova, the likelihood ratio statistic compares the full fixed-effect model with a submodel (reduced model) in which one or more terms are omitted. For example, the use of this test to determine the significance of the brand.assessor interaction can be specified as follows:

```
VCOMPONENTS [FIXED = brand * assessor] \
    RANDOM = day / presentation / serving
REML [PRINT = deviance; METHOD = Fisher; \
    SUBMODEL = brand + assessor] saltiness
```

Recalling that the fixed-effect model 'brand * assessor' can be expanded to 'brand + assessor + brand.assessor', we see that the SUBMODEL option in the REML statement specifies this model with the brand.assessor term omitted. When the SUBMODEL option is used, it is necessary to fit the model by the *Fisher-scoring algorithm*. This is indicated by the METHOD option: in the analysis reported in Chapter 1, Section 1.8, the default *Average Information* (AI) algorithm was used. These algorithms will be compared in more detail in Chapter 10, Section 10.9. The output of these statements is as follows:

Deviance: -2*Log-Likelihood		
Submodel:	Constant + brand + assessor	
Full fixed model:	Constant + brand + assessor + brand.assessor	
Source	deviance	d.f.
Submodel	490.54	93
Full model	447.91	69
Change	42.63	24

The change in deviance between the submodel and the full model is interpreted as a χ^2 statistic, and the difference between the degrees of freedom of the two deviances gives the degrees of freedom for this χ^2 value. In the present case,

$$P(\chi^2_{24} > 42.63) = 0.01096.$$

As expected, this value is larger (less significant, more conservative) than those given by either the Wald statistic ($P = 0.00005$) or the F statistic ($P = 0.00175$).

Having established that the 'brand.assessor' interaction is significant, it would not be appropriate to test the significance of the main effects 'brand and 'assessor', as these are marginal to the interaction (see Chapter 7, Section 7.2).

2.6   Justification for regarding block effects as random

In order to obtain the correct analysis of the split plot design using the mixed-modelling directives, we had to specify the block terms as random-effect terms, and it is reasonable to question whether we may truly regard block effects as random. In order to do so, we must regard the blocks of each type (day, presentation within day and serving within presentation) as a small sample from an infinite population. What is this infinite population? To recognise it, we must regard the days on which the experiment *was* performed as a random sample from an effectively infinite population of days on which it *might have been* performed. Likewise, within each day, the presentations that were made must be regarded as a random sample of the presentations that might have been made. Finally, the servings within each presentation must be regarded as a sample of the servings that might have been served.

In addition to regarding the population of potential blocks as infinite, we must regard the blocks studied as a sample chosen from this population randomly, and independently of each other. Is this reasonable? Can the presentations within a day be considered random, when they were made at successive times? Can they be considered independent, when there may be a trend over time – perhaps towards a decreased perception of saltiness, as palates become jaded? The justification for considering block effects as random comes from the randomisation *of the treatments over blocks*. Even if there is a tendency for the last presentation on each day to produce a lower estimate of saltiness, this will not bias the estimate for any individual brand – *provided that* no individual brand is more likely than another to be allocated to this presentation. If the brands are randomised over the presentations, then the presentation effects that *do* occur in combination with Brand A will be a random sample from the presentation effects that are *liable to* occur. The same argument applies to the servings within each presentation: the assessors must be randomised over the servings.

It is usual to treat all block terms as random: in particular, in the present case 'day' is also placed in the random-effect model. However, because each treatment occurs once on each day, no treatment term is tested in the day stratum. Hence the decision to treat 'day' as a random-effect term has no effect on the values of the F statistics or the Wald statistics, or on the values of $SE_{Difference}$. Perhaps because complete blocks (such as days in the present example) have no effect on these statistics, the decision to regard complete-block terms as random or fixed is not always considered to be of central importance. It does, however, have an effect on the SEs of treatment means, which will be discussed later (Chapter 4, Section 4.6).

As for the assumption that the block effects are normally distributed, both the anova and mixed-model analysis are fairly robust in the presence of departures from this assumption. This is because, strictly speaking, it is the estimated means of the levels in each treatment term (i.e. brand means, assessor means and means of brand.assessor combinations) that must be normally distributed, not the individual values. There is a mathematical theorem, the central limit theorem, which states that the mean of a large number of independent observations from the same distribution is approximately normally distributed, regardless of the distribution of the individual observations, and this turns out to be a fairly good approximation even when the mean is based on only a small number of observations. (For an account of the central limit theorem, which is frequently used by statisticians to derive distributions for test statistics, see, for example, Bulmer, 1979, Chapter 7, pp 115–120.) It is also assumed that the observations are independent, and that their variance is constant: it is much more important that these assumptions should be approximately true. Indeed, the purpose of adding the term 'town' to Model 1.2, or including 'presentation' in Model 2.2, is to recognise that houses in the same town, or servings in the same presentation, are not independent. As we have seen, failure to do so results in P values that are very unreliable.

2.7 Testing the assumptions of the analyses: inspection of the residual values

As in the case of the analyses presented in Chapter 1, it is important to check that the assumptions concerning the distributions of random effects are at least approximately fulfilled. Diagnostic plots of the residual values $\hat{\varepsilon}_{ijk}$ are produced by the statement

```
VPLOT [GRAPHICS=high] fittedvalues, normal, halfnormal, histogram
```

and are presented in Figure 2.2. Although GenStat identified two observations with large residual values, one positive and one negative (Section 2.2), these plots indicate that the distribution of the residuals fits the assumptions of the analysis very well, according to the criteria introduced in Chapter 1, Section 1.10.

As there are only 12 day.presentation combinations, with only 6 degrees of freedom, diagnostic plots of the values of $\hat{\tau}_{ij}$ are not as informative of those of $\hat{\varepsilon}_{ijk}$. However, it is worth inspecting a histogram of the values of $\hat{\tau}_{ij}$. The following statement obtains these values:

```
REML [PRINT = effects; PTERMS = day.presentation; \
    METHOD = Fisher] saltiness
```

The output of this statement is as follows:

Table of effects for day.presentation

presentation	1	2	3	4
day				
1	−1.2154	0.4262	1.6977	0.5648
2	−0.6865	−0.6402	0.7009	0.3312
3	−0.7574	−1.0113	0.5145	0.0754

Standard errors of differences

Average:	2.680
Maximum:	2.727
Minimum:	2.574

Average variance of differences: 7.186

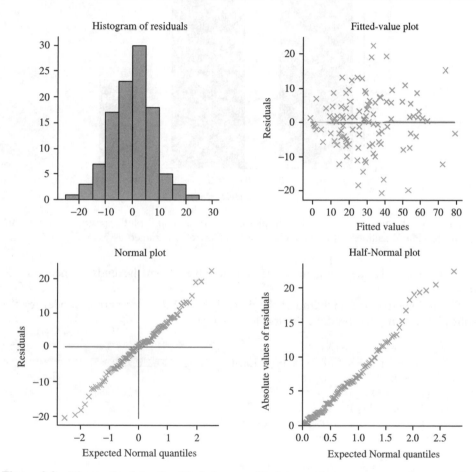

Figure 2.2 Diagnostic plots of residuals from a split plot experiment to compare the perception of saltiness of commercial brands of ravioli by trained assessors.

The histogram of these values (Figure 2.3) does at least show that there are no extreme outliers among the presentation effects, as is expected if the assumptions underlying the analysis are correct.

2.8 Use of R to perform the analyses

To prepare the data for analysis by R, they are transferred from the Excel workbook 'Intro to Mixed Modelling\Chapter 2\ravioli.xls' to the text file 'ravioli.dat' in the

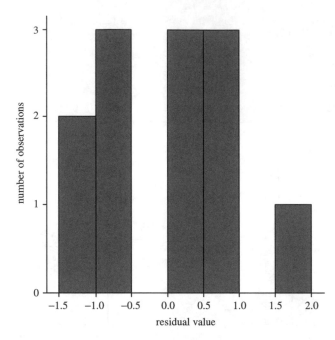

Figure 2.3   Histogram of residual main plot effects from a split plot experiment to compare the perception of saltiness of commercial brands of ravioli by trained assessors.

same directory. Exclamation marks (!) are removed from the ends of headings in this file.

The following commands import the data, fit Model 2.1 (the correct model for the split plot design), and produce the resulting anova:

```
ravioli <- read.table(
   "Intro to Mixed Modelling\\Chapter 2\\ravioli.dat",
   header=TRUE)
attach(ravioli)
day <- factor(day)
presentation <- factor(presentation)
serving <- factor(serving)
ravioli.model1 <-
   aov(saltiness ~ brand * assessor +
   Error(day / presentation / serving))
summary(ravioli.model1)
model.tables(ravioli.model1, type = "means", se = TRUE)
```

The lists 'day', 'presentation' and 'serving' are not automatically recognised by R as factors, because their values are numerical. The function factor() is therefore used to convert them to factors, and each factor is then assigned to the same name as before. The function aov() is used to perform analysis of variance on balanced experimental designs. It is not to be confused with the function anova(), which is used to retrieve an accumulated anova from a regression analysis. The model to be fitted is specified as an argument of aov() in the same way as in functions lm() and lme() (Chapter 1, Section 1.11), but with an additional feature: within the model specification that comprises the argument of this function, the function Error() specifies the block terms. The output of the function summary() is as follows:

```
Error: day
              Df Sum Sq Mean Sq F value Pr(>F)
Residuals      2 743.38  371.69

Error: day:presentation
              Df Sum Sq Mean Sq F value    Pr(>F)
brand          3 9859.4  3286.5  23.588 0.001013 **
Residuals      6  836.0   139.3
---
Signif. codes:  0 '***' 0.001 '**' 0.01 '*' 0.05 '.' 0.1 ' ' 1

Error: day:presentation:serving
               Df   Sum Sq Mean Sq F value    Pr(>F)
assessor        8 15474.9  1934.4 19.2912 2.121e-14 ***
brand:assessor 24  6075.3   253.1  2.5245  0.001715 **
Residuals      64  6417.4   100.3
---
Signif. codes:  0 '***' 0.001 '**' 0.01 '*' 0.05 '.' 0.1 ' ' 1
```

The MS values, the F tests performed and the F and P values are the same as those given by GenStat (Section 2.2). The output of the function model.tables() is as follows:

```
Tables of means
Grand mean

29.28451

 brand
           A     B     C     D
       21.78 19.22 43.23 32.91
rep    27.00 27.00 27.00 27.00

 assessor
          ALV    ANA   FAB   GUI   HER   MJS   MOI   NOR   PER
        35.27  4.826 47.52 37.86 29.14 36.01 21.53 33.41 18.00
rep     12.00 12.000 12.00 12.00 12.00 12.00 12.00 12.00 12.00

 brand:assessor
        assessor
brand ALV    ANA   FAB   GUI   HER   MJS   MOI   NOR   PER
  A   30.44  4.45 40.83 23.02 28.21 24.50 15.59 15.59 13.36
  rep  3.00  3.00  3.00  3.00  3.00  3.00  3.00  3.00  3.00
  B   32.67  0.00 18.56 30.44 17.82 11.14 13.36 33.41 15.59
```

```
rep   3.00   3.00   3.00   3.00   3.00   3.00   3.00   3.00   3.00
C     39.35 11.88 74.98 60.14 40.09 57.17 30.44 54.20 20.79
rep   3.00   3.00   3.00   3.00   3.00   3.00   3.00   3.00   3.00
D     38.61  2.97 55.68 37.86 30.44 51.23 26.73 30.44 22.27
rep   3.00   3.00   3.00   3.00   3.00   3.00   3.00   3.00   3.00
Warning message:
SEs for type  means  are not yet implemented in:
model.tables.aovlist(ravioli.model1, type = "means", se = TRUE)
```

The mean values are the same as those given by GenStat. However, note that R is not able to give SEs for these means, though a message indicates that this may become possible in future.

The following commands fit Model 2.2, which takes no account of the effects of presentations:

```
ravioli.model2 <- aov(saltiness ~ brand * assessor + Error(day))
summary(ravioli.model2)
```

The values in the resulting anova table are the same as those given by GenStat (Section 2.3), allowing for rounding.

The following commands use REML to fit Model 2.1, and summarise the results:

```
ravioli.model3 <- lme(saltiness ~ brand * assessor,
    random = ~ 1|day / presentation / serving)
anova(ravioli.model3)
```

Their output is as follows:

	numDF	denDF	F-value	p-value
(Intercept)	1	64	249.18923	<.0001
brand	3	6	23.58671	0.0010
assessor	8	64	19.29129	<.0001
brand:assessor	24	64	2.52455	0.0017

Note that the term represented as 'brand.assessor' in the GenStat output is represented as 'brand:assessor' in the output of R. Similarly, 'day.presentation' is represented as 'day:presentation', and 'day.presentation.serving' by 'day:presentation:serving'. Note also that R adds an F test of the null hypothesis that the constant term in the model (which R fits as an intercept, not as the grand mean) is zero. The F values and the degrees of freedom are the same as those obtained by GenStat.

The following statements produce diagnostic plots of the residuals equivalent to those produced by GenStat, with the exception of the half-normal plot:

```
resmixedsaltiness <- residuals(ravioli.model3)
hist(resmixedsaltiness)
par(pty = "s")
qqnorm(resmixedsaltiness)
qqline(resmixedsaltiness)
```

```
fitmixedsaltiness <- fitted(ravioli.model3)
plot(fitmixedsaltiness, resmixedsaltiness)
```

Details of the functions used to produced diagnostic plots are given in Chapter 1, Section 1.11.

2.9 Summary

The need for mixed models occurs not only in regression analysis, but also in the analysis of designed experiments.

In a fully randomised design (leading to one-way anova) or a randomised block design (leading to two-way anova), it is sufficient to recognise the residuals as a random-effect term.

However, in designs with more elaborate block structures, it is necessary to recognise the block effects as random effects. One of the simplest designs of this type is the split plot design.

The application of mixed modelling to the analysis of designed experiments is illustrated by a split plot experiment to assess the sensory characteristics of ravioli. The treatment factors are brand and assessor. The block factors are day (complete block), presentation (main plot) and serving (sub-plot). The brand varies only among main plots (not within each main plot), but the assessor varies among sub-plots within each main plot.

If the main plots are not recognised when the experiment is analysed, the following results will be wrongly specified:

- the value of $MS_{Residual}$ against which MS_{brand} is tested will be underestimated;

- the value of $DF_{Residual}$ for the test statistic F_{brand} will be exaggerated;

- the significance of 'brand' will be overestimated (i.e. P will be smaller than it should be).

Mixed modelling can be used, as an alternative to anova, to analyse the split plot design. The relationship between these two methods of specifying the analysis is shown in Table 2.4.

The Wald statistic overestimates the significance of a fixed-effect term (see also Chapter 1, Section 1.9 and Summary). A more conservative test is provided by the likelihood ratio statistic proposed by Welham and Thompson (1997).

Table 2.4 The relationship between the specification of the analysis of a split plot design by anova and by mixed modelling

Terms		Method of specification of terms	
in the general case	in the example presented	in analysis of variance	in mixed model
$A * B$[1]	brand*assessor	treatment structure	fixed-effect model
block / mainplot	day / presentation	block structure	random-effect model
/subplot	/serving		

[1] A and B are the treatment factors.

An argument justifying the decision to regard all block terms as random is presented. This argument is based on the fact that the treatments are randomised over the blocks.

2.10   Exercises

1. A field experiment to investigate the effect of four levels of nitrogen fertiliser on the yield of three varieties of oats was laid out in a split plot design (Yates, 1937). The experiment comprised six blocks of land, each divided into three main plots. Each main plot was sown with a single variety of oats. Each variety was sown on one main plot within each block; within the block, the allocation of varieties to main plots was randomised. Each main plot was divided into four sub-plots. A different level of nitrogen fertiliser was applied to each of the four sub-plots within each main plot; within the main plot, the allocation of fertiliser levels to sub-plots was randomised. The results obtained are presented in Table 2.5.

   (a) Make a sketch the field layout of Block 1. Note that it is not possible to determine the orientation of the sub-plots relative to the main plots from the information provided: make a sensible assumption about this.

   (b) Analyse the experiment by the methods of analysis of variance. Determine whether each of the following treatment terms is significant according to the F test:

      (i) variety

      (ii) nitrogen level

      (iii) variety × nitrogen level interaction.

   (c) Analyse the experiment by mixed modelling. Confirm that the Wald statistic for each treatment term has the expected relationship to the corresponding F statistic. Explain the difference in the P values between each F statistic and the corresponding Wald statistic.

2. An experiment was conducted to compare the effects of three types of oil on the amount of wear suffered by piston rings in an engine (Bennett and Franklin, 1954, Chapter 8, Section 8.62, pp 542–543). Five piston rings in the engine were weighed at the beginning and end of each 12-hour test run, and the weight loss of each ring was calculated. The engine was charged with a new oil at the beginning of each test run. The experiment comprised five replications, so there were 15 test runs in total. The results obtained are presented in Table 2.6. (Data reproduced by kind permission of John Wiley & Sons, Inc). The oil type and the rings are to be regarded as treatment factors.

   (a) The experiment has a split plot design. Identify the block, main plot and sub-plot factors. Identify the treatment factor that varies only between main plots, and the treatment factor that varies between sub-plots within each main plot.

   (b) What steps should have been taken during the planning of this experiment to ensure that treatment effects are not confounded with any other trends?

Table 2.5 Results of an experiment to determine the effect of nitrogen fertiliser on the yield of oat varieties, laid out in a split plot design.

The conventions used in this spreadsheet are the same as those used in Table 1.1, Chapter 1.

	A	B	C	B	D	E
1	block	mainplot	subplot	variety	nitrogen	yield
2	1	1	1	Marvellous	0.6 cwt	156
3	1	1	2	Marvellous	0.4 cwt	118
4	1	1	3	Marvellous	0.2 cwt	140
5	1	1	4	Marvellous	0 cwt	105
6	1	2	1	Victory	0 cwt	111
7	1	2	2	Victory	0.2 cwt	130
8	1	2	3	Victory	0.6 cwt	174
9	1	2	4	Victory	0.4 cwt	157
10	1	3	1	Golden rain	0 cwt	117
11	1	3	2	Golden rain	0.2 cwt	114
12	1	3	3	Golden rain	0.4 cwt	161
13	1	3	4	Golden rain	0.6 cwt	141
14	2	1	1	Marvellous	0.4 cwt	104
15	2	1	2	Marvellous	0 cwt	70
16	2	1	3	Marvellous	0.2 cwt	89
17	2	1	4	Marvellous	0.6 cwt	117
18	2	2	1	Victory	0.6 cwt	122
19	2	2	2	Victory	0 cwt	74
20	2	2	3	Victory	0.2 cwt	89
21	2	2	4	Victory	0.4 cwt	81
22	2	3	1	Golden rain	0.2 cwt	103
23	2	3	2	Golden rain	0 cwt	64
24	2	3	3	Golden rain	0.4 cwt	132
25	2	3	4	Golden rain	0.6 cwt	133
26	3	1	1	Golden rain	0.2 cwt	108
27	3	1	2	Golden rain	0.4 cwt	126
28	3	1	3	Golden rain	0.6 cwt	149
29	3	1	4	Golden rain	0 cwt	70
30	3	2	1	Marvellous	0.6 cwt	144
31	3	2	2	Marvellous	0.2 cwt	124
32	3	2	3	Marvellous	0.4 cwt	121
33	3	2	4	Marvellous	0 cwt	96
34	3	3	1	Victory	0 cwt	61
35	3	3	2	Victory	0.6 cwt	100

(Continued overleaf)

Table 2.5   (*continued*)

	A	B	C	B	D	E
36	3	3	3	Victory	0.2 cwt	91
37	3	3	4	Victory	0.4 cwt	97
38	4	1	1	Marvellous	0.4 cwt	109
39	4	1	2	Marvellous	0.6 cwt	99
40	4	1	3	Marvellous	0 cwt	63
41	4	1	4	Marvellous	0.2 cwt	70
42	4	2	1	Golden rain	0 cwt	80
43	4	2	2	Golden rain	0.4 cwt	94
44	4	2	3	Golden rain	0.6 cwt	126
45	4	2	4	Golden rain	0.2 cwt	82
46	4	3	1	Victory	0.2 cwt	90
47	4	3	2	Victory	0.4 cwt	100
48	4	3	3	Victory	0.6 cwt	116
49	4	3	4	Victory	0 cwt	62
50	5	1	1	Golden rain	0.6 cwt	96
51	5	1	2	Golden rain	0 cwt	60
52	5	1	3	Golden rain	0.4 cwt	89
53	5	1	4	Golden rain	0.2 cwt	102
54	5	2	1	Victory	0.4 cwt	112
55	5	2	2	Victory	0.6 cwt	86
56	5	2	3	Victory	0 cwt	68
57	5	2	4	Victory	0.2 cwt	64
58	5	3	1	Marvellous	0.4 cwt	132
59	5	3	2	Marvellous	0.6 cwt	124
60	5	3	3	Marvellous	0.2 cwt	129
61	5	3	4	Marvellous	0 cwt	89
62	6	1	1	Victory	0.4 cwt	118
63	6	1	2	Victory	0 cwt	53
64	6	1	3	Victory	0.6 cwt	113
65	6	1	4	Victory	0.2 cwt	74
66	6	2	1	Golden rain	0.6 cwt	104
67	6	2	2	Golden rain	0.4 cwt	86
68	6	2	3	Golden rain	0 cwt	89
69	6	2	4	Golden rain	0.2 cwt	82
70	6	3	1	Marvellous	0 cwt	97
71	6	3	2	Marvellous	0.2 cwt	99
72	6	3	3	Marvellous	0.4 cwt	119
73	6	3	4	Marvellous	0.6 cwt	121

Table 2.6 Weight loss of piston rings during test runs of an engine in an experiment to compare the effects of three types of oil. The values presented are logarithms of loss of weight in gramms ×100.

Replicate	1			2			3			4			5		
Oil Ring	A	B	C	A	B	C	A	B	C	A	B	C	A	B	C
1	1.782	1.568	1.507	1.642	1.539	1.562	1.682	1.616	1.630	1.654	1.680	1.740	1.496	1.626	1.558
2	1.306	1.223	1.240	1.346	1.064	1.334	1.322	1.369	1.428	1.532	1.452	1.408	1.354	1.466	1.478
3	1.149	1.029	1.068	1.090	0.778	1.136	1.176	1.053	1.202	1.233	1.193	1.228	1.038	1.167	1.330
4	1.025	0.919	0.982	1.012	0.690	1.021	0.930	0.935	1.057	0.992	0.973	1.093	0.924	0.974	0.996
oil ring	1.110	1.093	1.094	1.000	0.733	0.987	0.892	0.845	1.029	0.940	0.786	1.060	0.863	0.881	0.968

(c) Arrange the data for analysis by GenStat or R.

(d) Analyse the experiment by the methods of analysis of variance. Identify the three terms that belong to the treatment structure, and determine whether each is significant according to the F test.

(e) Analyse the experiment by mixed modelling. Confirm that the Wald statistic for each term in the treatment structure has the expected relationship to the corresponding F statistic.

(f) The mean log-transformed weight loss is smallest during test runs with Oil B. Is it safe to conclude that this type of oil causes less wear than the other two?

Note that the sums of squares etc. obtained in the analysis of these data to not agree exactly with those presented by Bennett and Franklin.

3

Estimation of the variances of random-effect terms

3.1 The need to estimate variance components

In Chapters 1 and 2 we recognised random-effect terms, and took them into account when fitting models to data, in order to interpret the fixed-effect terms more reliably. In Chapter 1 the emphasis was on the effect of latitude (a fixed-effect term) on house prices; the additional variation among towns (a random-effect term) was treated as a nuisance variable that reduced the precision with which the fixed effect was estimated, and that must be taken into account in order to obtain a realistic value of $SE_{latitude}$. In Chapter 2 the emphasis was on the effects of brand and assessor (fixed effects) on the perceived saltiness of ravioli, and the additional variation among days, presentations and servings (random-effect terms) was treated as a set of nuisance variables. However, a random-effect term may be of interest in its own right. Sometimes the means of individual levels of a random-effect factor are of interest, but even if this is not the case, it may still be useful to estimate the variance of the effects among the population of levels from which the sample studied was drawn. This chapter examines the methods for estimating the variance due to each random-effect term and the interpretation of the results.

3.2 A hierarchical random-effect model for a three-stage assay process

In an assay process, several sources of random variation commonly occur, and the relative magnitude of these has an important bearing on the cost and accuracy of the process. We will examine this problem in a data set obtained from delivery batches of a chemical paste (Davies and Goldsmith, 1984, Chapter 6, Section 6.5, pp 135–141). Three casks from each batch were sampled, and two analytical tests were performed on the contents of each cask to determine the 'strength' (per cent) of the paste. The data are displayed in the spreadsheet in Table 3.1. (Data reproduced by kind permission of Pearson Education Ltd).

Introduction to Mixed Modelling: Beyond Regression and Analysis of Variance N. W. Galwey
© 2006 John Wiley & Sons, Ltd

Table 3.1 'Strength' (per cent) of a chemical paste, assessed by two tests on each of three casks in each of 10 delivery batches.
The conventions used in this spreadsheet are the same as those used in Table 1.1, Chapter 1.

	A	B	C	D
1	delivery!	cask!	test!	strength
2	1	1	1	62.8
3	1	1	2	62.6
4	1	2	1	60.1
5	1	2	2	62.3
6	1	3	1	62.7
7	1	3	2	63.1
8	2	1	1	60.0
9	2	1	2	61.4
10	2	2	1	57.5
11	2	2	2	56.9
12	2	3	1	61.1
13	2	3	2	58.9
14	3	1	1	58.7
15	3	1	2	57.5
16	3	2	1	63.9
17	3	2	2	63.1
18	3	3	1	65.4
19	3	3	2	63.7
20	4	1	1	57.1
21	4	1	2	56.4
22	4	2	1	56.9
23	4	2	2	58.6
24	4	3	1	64.7
25	4	3	2	64.5
26	5	1	1	55.1
27	5	1	2	55.1
28	5	2	1	54.7
29	5	2	2	54.2
30	5	3	1	58.8
31	5	3	2	57.5
32	6	1	1	63.4
33	6	1	2	64.9
34	6	2	1	59.3

Table 3.1   (*continued*)

	A	B	C	D
35	6	2	2	58.1
36	6	3	1	60.5
37	6	3	2	60.0
38	7	1	1	62.5
39	7	1	2	62.6
40	7	2	1	61.0
41	7	2	2	58.7
42	7	3	1	56.9
43	7	3	2	57.7
44	8	1	1	59.2
45	8	1	2	59.4
46	8	2	1	65.2
47	8	2	2	66.0
48	8	3	1	64.8
49	8	3	2	64.1
50	9	1	1	54.8
51	9	1	2	54.8
52	9	2	1	64.0
53	9	2	2	64.0
54	9	3	1	57.7
55	9	3	2	56.8
56	10	1	1	58.3
57	10	1	2	59.3
58	10	2	1	59.2
59	10	2	2	59.2
60	10	3	1	58.9
61	10	3	2	56.6

A natural model for these data (Model 3.1) is

$$y_{ijk} = \mu + \delta_i + \gamma_{ij} + \varepsilon_{ijk} \qquad (3.1)$$

where
y_{ijk} = the value of strength from the kth test on the jth cask from the ith delivery,
μ = the grand mean (overall mean) value of strength,
δ_i = the effect of the ith delivery,
γ_{ij} = the effect of the jth cask from the ith delivery,
ε_{ijk} = the effect of the kth test in the ijth delivery.cask combination.

It is natural to treat delivery, cask within delivery and test within cask all as random effect terms – that is, to make the following assumptions:

- that the δ_i are independent values of a random variable Δ, such that

$$\Delta \sim N(0, \sigma_\Delta^2);$$

- that the γ_{ij} are independent values of a random variable Γ, such that

$$\Gamma \sim N(0, \sigma_\Gamma^2);$$

and

- that the ε_{ijk} are independent values of a random variable E, such that

$$E \sim N(0, \sigma_E^2).$$

The following GenStat statements read the data and analyse them according to Model 3.1:

```
IMPORT 'Intro to Mixed Modelling\\Chapter 3\\paste strength.xls';\
    SHEET = 'sheet1'
BLOCKSTRUCTURE delivery / cask / test
ANOVA [FPROB = yes] strength
```

Note that as all the terms in this model are random-effect terms, they are all specified in the BLOCKSTRUCTURE statement: there is no TREATMENTSTRUCTURE statement. The output of the ANOVA statement is as follows:

Analysis of variance

Variate: strength

Source of variation	d.f.	s.s.	m.s.	v.r.	F pr.
delivery stratum	9	247.4027	27.4892	1.57	
delivery.cask stratum	20	350.9067	17.5453	25.88	
delivery.cask.test stratum					
	30	20.3400	0.6780		
Total	59	618.6493			

Message: the following units have large residuals.

delivery 5	−4.15	s.e.	2.03
delivery 4 cask 3	4.90	s.e.	2.42
delivery 9 cask 2	5.32	s.e.	2.42

Tables of means

Variate: strength

Grand mean 60.05

As in the analysis of the split plot design (Model 2.1, Chapter 2, Section 2.2), each random-effect term defines a stratum in the anova table. Like the random-effect part of the model for the split plot design, Model 3.1 is *hierarchical*: its terms are *nested*, one within another. This hierarchical structure is illustrated diagrammatically in Figure 3.1. (For clarity, the detail of the lower strata is shown only for the first five deliveries.)

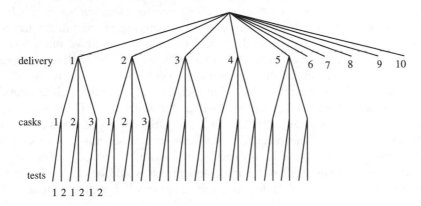

Figure 3.1   The hierarchical structure of the experiment to assess the effects of delivery batches, casks and tests on the strength of a chemical paste.

Each of the variables Δ, Γ and E contributes to the total variation in paste strength, and the variances σ_Δ^2, σ_Γ^2 and σ_E^2 are referred to as *components* of the total variance. Each term in the model is tested against the term nested within it, i.e. the term below it in the hierarchy, and hence in the anova table. Thus:

- $F_{\text{delivery}} = \text{MS}_{\text{delivery}}/\text{MS}_{\text{delivery.cask}} = 27.4892/17.5453 = 1.57,$

- $F_{\text{delivery.cask}} = \text{MS}_{\text{delivery.cask}}/\text{MS}_{\text{delivery.cask.test}} = 17.5453/0.6780 = 25.88.$

As all the model terms are included in the GenStat BLOCKSTRUCTURE statement, they are all assumed by GenStat to represent nuisance variables. Hence the P values corresponding to these F statistics (see Chapter 1, Section 1.2) are not given, and no tables of means are presented (except the grand mean). However, the P values can be looked up by many standard pieces of software, including GenStat. These P values, together with the information required to obtain them, are presented in Table 3.2. There is a suggestion that the paste strength varies between deliveries, but this falls short of significance. However, there is highly significant variation among casks within a delivery, relative to the variation between tests on the same cask.

Table 3.2   Precise P values from the anova of the strength of a chemical paste in a hierarchical design.

Term	$\text{DF}_{\text{Numerator}}$	$\text{DF}_{\text{Denominator}}$	F	P
delivery	9	20	1.57	0.19
delivery.cask	20	30	25.88	>0.0001

3.3   The relationship between variance components and stratum mean squares

It is natural to enquire not only into the significance of the different sources of variation, but also into their magnitude. This may be of importance for various purposes – for example, to decide whether an allowance for variation between deliveries should be written into the specification of a manufacturing process or the terms of a contract, or to decide how many casks should be sampled, and how many tests conducted on each cask, in future deliveries. Broadly speaking, the 'delivery' stratum in the preceding anova represents the variation among deliveries, the 'delivery.cask' stratum represents variation among casks within each delivery, and the 'delivery.cask.test' stratum represents variation among tests within the same cask. However, $MS_{delivery}$ cannot be used directly as an estimate of σ_Δ^2, nor $MS_{delivery.cask}$ as an estimate of σ_Γ^2: the relationship between mean squares and variance components is somewhat more elaborate. In order to explore it, we need to introduce the concept of the *expected mean square* – the mean value of the mean square that would be obtained over a long run of similar experiments, sampled from the same underlying distributions.

Let

$$n_\Gamma = \text{the number of casks within each delivery} = 3 \qquad (3.2)$$

$$n_E = \text{the number of tests within each cask} = 2. \qquad (3.3)$$

The relationships between the true values of the variance components and the expected mean squares in the present experiment are then as shown in Table 3.3.

The simple formulae in this table are applicable only in the case of a balanced experimental design, where n_Γ is constant over deliveries, and n_E is constant over casks within each delivery and over deliveries. We will not derive them formally here, but we will note some intuitively reasonable properties that they have. These will be reviewed first in relation to the comparison between $MS_{delivery.cask}$ and $MS_{delivery.cask.test}$:

- The variation among tests in the same cask ($MS_{delivery.cask.test}$) is not influenced by the variation among casks (σ_Γ^2) or among deliveries (σ_Δ^2).

- However, the variation among casks in the same delivery ($MS_{delivery.cask}$) is influenced by the variation among tests within each cask σ_E^2. This means that even if there were no real variation among casks within each delivery (i.e. if σ_Γ^2 were zero), there would still be some variation among the observed cask means, due to the random effects of tests.

Table 3.3 Expected mean squares of the terms in the anova of the strength of a chemical paste in a hierarchical design.

Source of variation	Expected MS
delivery	$n_\Gamma n_E \sigma_\Delta^2 + n_E \sigma_\Gamma^2 + \sigma_E^2$
delivery.cask	$n_E \sigma_\Gamma^2 + \sigma_E^2$
delivery.cask.test	σ_E^2

- If n_E is large, then the influence of σ_E^2 on $MS_{delivery.cask}$ is small relative to that of σ_Γ^2. In the limiting case, if an infinite number of tests were performed on each cask, then their mean would be the true mean value for the cask, and $MS_{delivery.cask}$, which is closely related to the variance among the cask means, would be determined only by σ_Γ^2.

- These expected mean squares justify the F test of $MS_{delivery.cask}$ against $MS_{delivery.cask.test}$. If the hypothesis

$$H_0 : \sigma_\Gamma^2 = 0$$

is true, then

$$\text{expected } MS_{delivery.cask} = \text{expected } MS_{delivery.cask.test}$$

and

$$\text{expected value of } F = (\text{expected } MS_{delivery.cask})/(\text{expected } MS_{delivery.cask.test}) = 1.$$

The F statistic then follows the F distribution with the appropriate degrees of freedom.

- The greater the value of n_E, the greater the power of the F test. For a given value of σ_Γ^2, a greater value of n_E gives a greater expected value of F, and a greater probability that H_0 will be rejected.

Similar considerations apply to the comparison between $MS_{delivery}$ and $MS_{delivery.cask}$:

- The variation among casks in the same delivery ($MS_{delivery.cask}$) is not influenced by the variation among deliveries (σ_Δ^2).

- However, the variation among deliveries ($MS_{delivery}$) is influenced by the variation among casks within each delivery (σ_Γ^2) and among tests within each cask (σ_E^2). This means that even if there were no real variation among deliveries (i.e. if σ_Δ^2 was zero), there would still be some variation among the observed delivery means, due to the random effects of casks and/or tests.

- If n_Γ is large, then the influence of σ_Γ^2 on $MS_{delivery}$ is small relative to that of σ_Δ^2. In the limiting case, if an infinite number of casks were sampled in each delivery, then their mean would be the true mean value for the delivery, and $MS_{delivery}$, which is closely related to the variance among the delivery means, would be determined only by σ_Δ^2.

- These expected mean squares justify the F test of $MS_{delivery}$ against $MS_{delivery}$. If the hypothesis

$$H_0 : \sigma_\Delta^2 = 0$$

is true, then

$$\text{expected } MS_{delivery} = \text{expected } MS_{delivery.cask}$$

and

$$\text{expected value of } F = (\text{expected } MS_{delivery})/(\text{expected } MS_{delivery.cask}) = 1.$$

The F statistic then follows the F distribution with the appropriate degrees of freedom.

- The greater the value of n_Γ, the greater the power of the F test. For a given value of σ_Δ^2, a greater value of n_Γ gives a greater expected value of F, and a greater probability that H_0 will be rejected.

3.4   Estimation of the variance components in the hierarchical random-effect model

The formulae for the expected mean squares, together with the observed mean squares from the anova table, can be used to obtain estimates, $\hat{\sigma}_\Delta^2, \hat{\sigma}_\Gamma^2$ and $\hat{\sigma}_E^2$, of the true variances $\sigma_\Delta^2, \sigma_\Gamma^2$ and σ_E^2 respectively. Thus

$$\hat{\sigma}_E^2 = \text{MS}_{\text{delivery.cask.test}} = 0.6780. \tag{3.4}$$

Similarly,

$$n_E \hat{\sigma}_\Gamma^2 + \hat{\sigma}_E^2 = \text{MS}_{\text{delivery.cask}} = 17.5453. \tag{3.5}$$

Rearranging Equation 3.5, and substituting the numerical value of n_E and of the estimate $\hat{\sigma}_E^2$ from Equations 3.3 and 3.4 respectively, we obtain

$$\hat{\sigma}_\Gamma^2 = \frac{17.5453 - 0.6780}{2} = 8.43365. \tag{3.6}$$

Similarly,

$$n_\Gamma n_E \hat{\sigma}_\Delta^2 + n_E \hat{\sigma}_\Gamma^2 + \hat{\sigma}_E^2 = \text{MS}_{\text{delivery}} = 27.4892. \tag{3.7}$$

Rearranging Equation 3.7, and substituting the numerical values of n_Γ and n_E and of the estimates $\hat{\sigma}_E^2$ and $\hat{\sigma}_\Gamma^2$ from Equations 3.2, 3.3, 3.4 and 3.6 respectively, we obtain

$$\hat{\sigma}_\Delta^2 = \frac{27.4892 - 2 \times 8.43365 - 0.6780}{6} = 1.657317.$$

These variance-component estimates can be produced directly by the anova directives BLOCKSTRUCTURE and ANOVA. The statement required is as follows:

```
ANOVA [PRINT = stratumvariances] strength
```

The output of this statement is given below:

Estimated stratum variances

Variate: strength

Stratum	variance	effective d.f.	variance component
delivery	27.489	9.000	1.657
delivery.cask	17.545	20.000	8.434
delivery.cask.test	0.678	30.000	0.678

The variance-component estimates can also be produced by the mixed-modelling directives. The statements required are as follows:

```
VCOMPONENTS RANDOM = delivery / cask / test
REML [PRINT = model, components] strength
```

Note that as the model contains no fixed-effect terms, the option FIXED in the VCOMPONENTS statement is not set. Similarly, as the Wald test is applicable only to fixed-effect terms, the setting 'Wald' is omitted from the PRINT option in the REML statement. The output of these statements is given below:

REML variance components analysis

Response variate:	strength
Fixed model:	Constant
Random model:	delivery + delivery.cask + delivery.cask.test
Number of units:	60

delivery.cask.test used as residual term

Sparse algorithm with AI optimisation

Estimated variance components

Random term	component	s.e.
delivery	1.6573	2.3494
delivery.cask	8.4337	2.7755

Residual variance model

Term	Factor	Model(order)	Parameter	Estimate	s.e.
delivery.cask.test		Identity	Sigma2	0.678	0.1751

The estimates of the variance components for delivery and delivery.cask agree with the values derived from the mean squares in the anova. GenStat identifies delivery.cask.test as the residual term (because there is only one observation for each combination of these factors – see Chapter 2, Section 2.4), and the estimated variance component for this term is presented as the *residual variance*. It agrees with $MS_{\text{delivery.cask.test}}$ in the anova. A standard error is presented with each estimated variance component. However, whereas estimated means are normally distributed around the true value (provided that the assumptions underlying the analysis are true – see Chapter 1, Section 1.10), estimated variances are not: a variance cannot be less than zero, and the distribution of estimated variances must therefore be asymmetrical. Consequently the SE of a variance is not as informative as that of a mean. Nevertheless, these values do give a rough indication of the precision with which the variance components have been estimated. For example, we can say that $\hat{\sigma}_{\Delta}^2 = 1.6573 \pm 2.3494$. We note that in this case the SE of the estimate is larger that the estimate itself, which is consistent with the finding that the F value for the effect of deliveries is non-significant. We can also say that $\hat{\sigma}_\Gamma^2 = 8.4337 \pm 2.7755$, and can be fairly confident that the true variance among casks within each delivery lies somewhere in the range indicated.

3.5   Design of an optimum strategy for future sampling

Knowledge of the magnitude of each variance component can be used to design an optimum strategy for future sampling. This will be of importance if, for example, it is desired to estimate the mean paste strength of future deliveries to a certain degree of precision, without incurring unnecessary cost. The investigator can choose the number of casks to sample from each delivery (n_Γ), and the number of tests to perform on each cask (n_E). Should he or she perform more tests on fewer casks, or fewer tests on more casks, in order to deploy a given amount of effort to best effect?

In order to answer this question, the cost of sampling is expressed as a function of three components, which it is assumed can be treated as constants. The first is the fixed costs, which are the same regardless of the number of casks sampled or the number of tests performed on each cask, and which do not concern us further. The remaining two components are:

- c_Γ, the average cost of obtaining samples from each cask, after the fixed costs have been met, regardless of the number of tests performed on the cask; and

- c_E, the average cost of each test on a cask, after the cost of obtaining samples from it has been met.

It is assumed that $c_\Gamma \geqslant 0$ and $c_E \geqslant 0$. The total cost of sampling is then given by

$$k = c_\Gamma n_\Gamma + c_E n_\Gamma n_E = n_\Gamma (c_\Gamma + c_E n_E). \tag{3.8}$$

The true mean paste strength in the ith delivery is

$$\mu_i = \mu + \delta_i,$$

and an estimate of this is given by

$$\hat{\mu}_i = \frac{\sum_{j=1}^{n_\Gamma} \sum_{k=1}^{n_E} y_{ijk}}{n_\Gamma n_E}. \tag{3.9}$$

Note that this is the ordinary mean of the observed values, not the 'shrunk estimate' introduced in Chapter 5, Section 5.2. The SE of this estimate is given by

$$SE_{\hat{\mu}_i} = \sqrt{\frac{\sigma_\Gamma^2}{n_\Gamma} + \frac{\sigma_E^2}{n_\Gamma n_E}} = \sqrt{\frac{1}{n_\Gamma} \left(\sigma_\Gamma^2 + \frac{\sigma_E^2}{n_E} \right)}. \tag{3.10}$$

More will be said about the calculation of the SE of a parameter estimate, and the meaning of this statistic, in Chapter 4. For the moment, it is sufficient to note that $SE_{\hat{\mu}_i}$ indicates the precision with which $\hat{\mu}_i$ estimates μ_i. $SE_{\hat{\mu}_i}$ is small, and the precision of $\hat{\mu}_i$ is high, when:

- the amount of variation among individual observations, determined by the variance components σ_Γ^2 and σ_E^2, is small; and

- the sample size, indicated by n_Γ and n_E, is large.

From Equation 3.8,

$$n_\Gamma = \frac{k}{c_\Gamma + c_E n_E}. \tag{3.11}$$

That is, if we stipulate that the total cost k is to be kept constant, the investigator is not able to vary and n_E independently: any choice of n_E enforces a corresponding value of n_Γ. Substituting the expression for n_Γ from Equation 3.11 into Equation 3.10, we obtain

$$SE_{\hat{\mu}_i} = \sqrt{\frac{c_\Gamma + c_E n_E}{k} \left(\sigma_\Gamma^2 + \frac{\sigma_E^2}{n_E} \right)}, \tag{3.12}$$

and n_E is the only term in this expression that can be manipulated by the investigator. The problem then is to find the value of n_E that minimises the value of $SE_{\hat{\mu}_i}$, and it can be shown that this is given by

$$n_E = \sqrt{\frac{c_\Gamma \sigma_E^2}{c_E \sigma_\Gamma^2}} \tag{3.13}$$

(Snedecor and Cochran, 1989, Chapter 21, Section 21.10, pp 447–450). That is, the number of tests made on each cask should be large, and the number of casks sampled should be correspondingly small, if:

- the cost of sampling each cask (c_Γ) is high;

- the cost of each test on a cask already sampled (c_E) is low;

- the variation among tests within each cask (σ_E^2) is large;

- the variation among casks within each delivery (σ_Γ^2) is small.

For example, suppose the cost of sampling each cask is small, say $0.3 \times$ the cost of each test within a cask already sampled. This cost might represent a few moments of the sampler's time to open the cask. Then

$$\frac{c_\Gamma}{c_E} = 0.3.$$

Substituting this value and the estimates $\hat{\sigma}_E^2$ and $\hat{\sigma}_\Gamma^2$ obtained in Equations 3.4 and 3.6 respectively into Equation 3.13, we obtain

$$n_E = \sqrt{\frac{0.3 \times 0.678}{8.4337}} = 0.0241.$$

That is, there is no justification for performing more than one test per cask. On the other hand, if the cost of sampling each cask is large (e.g. if the whole contents of the cask are rendered unusable by sampling), say

$$\frac{c_\Gamma}{c_E} = 30,$$

then

$$n_E = \sqrt{\frac{30 \times 0.678}{8.4337}} = 2.41,$$

and two or three tests should be performed on each cask.

This approach can be extended to assay processes with more than two stages (Snedecor and Cochran, 1967, Chapter 17, Section 17.12, pp 533–534). For example, suppose that the mean strength of the paste in a future series of deliveries is to be estimated. The investigator must now decide not only how many tests to perform on each cask, but also how many casks to sample within each delivery. There is now an additional component to the cost of sampling, namely:

- c_Δ, the cost of obtaining samples from each delivery.

It is assumed that $c_\Delta \geqslant 0$. The total cost of sampling is then given by

$$k = c_\Delta n_\Delta + c_\Gamma n_\Delta n_\Gamma + c_E n_\Delta n_\Gamma n_E = n_\Delta(c_\Delta + (c_\Gamma + c_E n_E)n_\Gamma), \qquad (3.14)$$

where
n_Δ = number of deliveries sampled,

which is an equation of the same form as Equation 3.8. Let

$$\hat{\mu} = \frac{\sum_{i=1}^{n_\Delta} \sum_{j=1}^{n_\Gamma} \sum_{k=1}^{n_E} y_{ijk}}{n_\Delta n_\Gamma n_E}$$

$$= \text{estimated grand mean strength over deliveries, casks and tests.} \qquad (3.15)$$

Then

$$\mathrm{SE}_{\hat{\mu}} = \sqrt{\frac{\sigma_\Delta^2}{n_\Delta} + \frac{\sigma_\Gamma^2}{n_\Delta n_\Gamma} + \frac{\sigma_E^2}{n_\Delta n_\Gamma n_E}} = \sqrt{\frac{1}{n_\Delta}\left(\sigma_\Delta^2 + \frac{\left(\sigma_\Gamma^2 + \frac{\sigma_E^2}{n_E}\right)}{n_\Gamma}\right)}, \qquad (3.16)$$

which is an equation of the same form as Equation 3.10, and the value of $\mathrm{SE}_{\hat{\mu}}$ is minimised when

$$n_\Gamma = \sqrt{\frac{c_\Delta\left(\sigma_\Gamma^2 + \frac{\sigma_E^2}{n_E}\right)}{(c_\Gamma + c_E n_E)\sigma_\Delta^2}}, \qquad (3.17)$$

which is an equation of the same form as Equation 3.13. The argument concerning the number of tests per cask is unchanged, and when the value of n_E has been decided as described above, it can be substituted from Equation 3.11 into Equation 3.17. This equation shows that the number of casks to be sampled in each delivery should be large, and the number of deliveries to be sampled correspondingly small, if:

- the cost of sampling each delivery (c_Δ) is high;
- the variation among deliveries (σ_Δ^2) is low.

For example, suppose that the cost of sampling each cask is 0.3× the cost of an additional test within a cask already sampled, as discussed above. In practice, the number of tests performed within each cask should then be set to the lowest value physically possible,

$$n_E = 1.$$

Suppose further that c_Δ, the cost of sampling each delivery, is 3.5× the cost of each test. This might reflect the cost of sending a sampler to attend the delivery. Substituting these values, and the other values obtained earlier, into Equation 3.17 gives

$$
n_\Gamma = \sqrt{\frac{3.5 \times \left(17.5453 + \dfrac{0.6780}{1} \right)}{(0.3 + 1 \times 1) \times 27.4892}} = 1.7848,
$$

indicating that two casks should be sampled from each delivery.

3.6 Use of R to analyse the hierarchical three-stage assay process

To prepare the paste strength data for analysis by R, they are transferred from the Excel workbook 'Intro to Mixed Modelling\Chapter 3\paste strength.xls' to the text file 'paste strength.dat' in the same directory. Exclamation marks (!) are removed from the ends of headings in this file.

The following commands import the data, fit Model 3.1, and produce the resulting anova:

```
pastestrength <- read.table(
    "Intro to Mixed Modelling\\Chapter 3\\paste strength.dat",
    header=TRUE)
attach(pastestrength)
delivery <- factor(delivery)
cask <- factor(cask)
test <- factor(test)
pastestrength.model1aov <- aov(strength ~
    Error(delivery / cask / test))
summary(pastestrength.model1aov)
```

The output from these commands is as follows:

```
Error: delivery
          Df  Sum Sq Mean Sq F value Pr(>F)
Residuals  9 247.403  27.489

Error: delivery:cask
          Df Sum Sq Mean Sq F value Pr(>F)
Residuals 20 350.91   17.55

Error: delivery:cask:test
          Df Sum Sq Mean Sq F value Pr(>F)
Residuals 30 20.340   0.678
```

The degrees of freedom, sum of squares and mean square for each term are the same as those produced by GenStat, and the F tests comparing the mean square for each term with that for the term below can be conducted by hand.

The following commands perform a mixed-model analysis on the same model:

```
pastestrength.model1lme <- lme(strength ~ 1,
    random = ~ 1| delivery / cask / test)
summary(pastestrength.model1lme)
```

The output of these statements is as follows:

```
Linear mixed-effects model fit by REML
 Data: NULL
        AIC       BIC    logLik
 256.9907 267.3784 -123.4954

Random effects:
 Formula: ~1 | delivery
         (Intercept)
StdDev:    1.287327

 Formula: ~1 | cask %in% delivery
         (Intercept)
StdDev:    2.904076

 Formula: ~1 | test %in% cask %in% delivery
         (Intercept)   Residual
StdDev:    0.652641 0.5020602

Fixed effects: strength ~ 1
               Value Std.Error DF  t-value p-value
(Intercept) 60.05333 0.6768627 30 88.72307       0

Standardised Within-Group Residuals:
        Min            Q1           Med            Q3           Max
-0.902265501 -0.314365731  0.005791316  0.287789749  0.847339795

Number of Observations: 60
Number of Groups:
                        delivery          cask %in% delivery
                              10                          30
     test %in% cask %in% delivery
                              60
```

The standard deviation (abbreviated to StdDev or SD) given by R for the model component '$\sim 1|$ delivery' is the square root of the variance component for the term 'delivery'. Thus

$$1.287327^2 = 1.6572 = \hat{\sigma}_\Delta^2$$

as given by GenStat (Section 3.4), allowing for rounding error. Similarly, the SD for model component '$\sim 1|$ cask %in% delivery' is the square root of the variance component for the term 'delivery.cask'. Thus

$$2.904076^2 = 8.4337 = \hat{\sigma}_{\Gamma}^2.$$

The two SD values given in connection with the model component '$\sim 1|$ test %in% cask %in% delivery' must be combined to obtain an estimate of the variance component for the term 'delivery.cask.test', as follows:

$$0.652641^2 + 0.5020602^2 = 0.678 = \hat{\sigma}_E^2.$$

This agrees with the value given by GenStat.

3.7   Genetic variation: a crop field trial with an unbalanced design

There are many other situations, besides the optimisation of assay procedures, in which it is useful to estimate the magnitude of the different variance components that contribute to a response variable. In particular, variance components are routinely estimated in studies related to plant and animal breeding. The amount of genetic variation among the individuals of a species of crop plant or domesticated animal can be compared with the amount of variation due to non-genetic causes in a ratio called the *heritability*. Estimates of these two variance components can also be used to give a prediction of the potential for genetic improvement.

We will explore this issue in data from a field trial of barley breeding lines (reproduced by kind permission of Reg Lance, Department of Agriculture, Western Australia). The lines studied were derived from a cross between two parent varieties, 'Chebec' and 'Harrington'. They were 'doubled haploid' lines, which means they were obtained by a laboratory technique that ensures that all plants within the same breeding line are genetically identical, so that the line will breed true. This feature improves the precision with which genetic variation among the lines can be estimated. The trial considered here, conducted in Western Australia in 1995, was arranged in two randomised blocks. Within each block, each line occupied a single rectangular field plot. All lines were present in Block 1, but due to limited seed stocks, some were absent from Block 2. The grain yield (g/m^2) was measured in each field plot. The data obtained are displayed in the spreadsheet in Table 3.4.

A natural model for these data (Model 3.18) is

$$y_{ij} = \mu + \delta_i + \varepsilon_{ij} + \phi_{k|ij} \tag{3.18}$$

where
  y_{ij} = the grain yield of the jth plot in the ith block,
  μ = the grand mean value of grain yield,
  δ_i = the effect of the ith block,
  ε_{ij} = the effect of the jth plot in the ith block,
  $\phi_{k|ij}$ = the effect of the kth breeding line, being the line sown in the ijth block.plot combination.

Table 3.4　Yields of doubled haploid breeding lines of barley, assessed in a field trial with an unbalanced randomised block design.

The conventions used in this spreadsheet are the same as those used in Table 1.1, Chapter 1.

	A	B	C
1	block!	line!	yield_g_m2
2	1	2	483.33
3	1	39	145.84
4	1	41	321.84
5	1	4	719.14
6	1	79	317.63
7	1	76	344.48
8	1	78	260.02
9	1	35	374.28
10	1	25	428.61
11	1	67	407.25
12	1	17	551.84
13	1	30	353.29
14	1	73	355.30
15	1	64	647.92
16	1	72	165.76
17	1	44	517.52
18	1	18	366.24
19	1	82	251.30
20	1	29	606.37
21	1	23	605.75
22	1	38	641.42
23	1	31	166.75
24	1	40	410.87
25	1	57	181.97
26	1	63	562.90
27	1	83	280.44
28	1	36	800.35
29	1	66	687.92
30	1	10	764.88
31	1	13	541.15
32	1	62	730.48
33	1	81	315.63
34	1	43	678.46

Table 3.4   (*continued*)

	A	B	C
35	1	26	580.22
36	1	45	519.88
37	1	12	436.59
38	1	24	671.22
39	1	51	692.55
40	1	6	849.66
41	1	14	910.76
42	1	80	487.86
43	1	46	724.01
44	1	42	793.43
45	1	47	192.43
46	1	37	895.30
47	1	27	731.87
48	1	54	809.41
49	1	32	669.16
50	1	70	996.19
51	1	9	774.84
52	1	8	636.45
53	1	77	357.94
54	1	22	340.65
55	1	11	644.83
56	1	58	521.67
57	1	19	622.72
58	1	49	830.57
59	1	28	679.92
60	1	50	721.13
61	1	74	489.31
62	1	7	907.38
63	1	68	325.96
64	1	59	553.46
65	1	16	210.71
66	1	33	770.23
67	1	53	559.14
68	1	75	617.33
69	1	20	632.46
70	1	61	611.52

(*Continued overleaf*)

Table 3.4   (*continued*)

	A	B	C
71	1	1	717.78
72	1	71	595.86
73	1	55	555.29
74	1	69	467.24
75	1	60	572.90
76	1	21	514.62
77	1	65	818.74
78	1	34	673.43
79	1	5	798.99
80	1	48	786.06
81	1	56	522.61
82	1	3	873.04
83	1	15	600.06
84	1	52	603.04
85	2	64	681.64
86	2	27	762.42
87	2	29	932.33
88	2	68	385.47
89	2	16	240.00
90	2	33	846.85
91	2	67	702.58
92	2	51	746.11
93	2	66	846.05
94	2	50	885.67
95	2	36	1054.70
96	2	80	478.12
97	2	17	959.25
98	2	26	639.39
99	2	63	755.90
100	2	74	551.41
101	2	56	435.62
102	2	81	303.72
103	2	43	836.82
104	2	79	439.17
105	2	14	934.72
106	2	32	836.95

Table 3.4 (*continued*)

	A	B	C
107	2	70	904.90
108	2	41	538.00
109	2	31	226.12
110	2	71	569.61
111	2	65	713.43
112	2	7	820.08
113	2	58	435.34
114	2	69	378.89
115	2	34	639.11
116	2	11	516.84
117	2	44	873.18
118	2	49	823.25
119	2	54	859.36
120	2	72	258.59
121	2	9	587.07
122	2	23	817.51
123	2	48	645.10
124	2	59	634.58
125	2	78	260.26
126	2	61	472.44
127	2	45	575.76
128	2	18	265.37
129	2	19	423.76
130	2	35	554.69
131	2	6	755.05
132	2	40	568.31
133	2	8	299.92
134	2	1	591.19
135	2	10	756.63
136	2	42	552.53
137	2	46	627.25
138	2	37	552.70
139	2	60	284.72
140	2	38	540.68
141	2	13	475.10
142	2	24	463.22
143	2	39	212.66

It is also natural to treat the block and residual terms as random-effect terms: that is, to assume that the δ_i are independent values of a random variable Δ, such that

$$\Delta \sim N(0, \sigma_\Delta^2),$$

and the ε_{ij} are independent values of a random variable E, such that

$$E \sim N(0, \sigma_E^2).$$

The following statements might be used to read these data and to attempt to analyse them according to Model 3.18:

```
IMPORT \
    'Intro to Mixed Modelling\\Chapter 3\\barley progeny.xls'; \
    SHEET = 'sheet1'
BLOCKSTRUCTURE block
TREATMENTSTRUCTURE line
ANOVA [FPROBABILITY = yes; PRINT = aovtable, information] \
    yield_g_m2
```

However, the ANOVA statement produces the following output:

> *Fault 11, code AN 1, statement 1 on line 6*
>
> Command: ANOVA [FPROBABILITY = yes; PRINT = aovtable, information] yield_g_m2
> Design unbalanced - cannot be analysed by ANOVA
> Model term line (non-orthogonal to term block) is unbalanced, in the block.*Units* stratum.

Because some breeding lines are present in both blocks but others only in Block 1, the design is unbalanced, and analysis of variance cannot proceed.

3.8 Production of a balanced experimental design by 'padding' with missing values

Such an unbalanced design can be analysed by mixed modelling (see Section 3.10). However, a less rigorous method of overcoming the problem of imbalance – or at least disguising it – is to 'pad' the spreadsheet with rows representing the missing breeding lines in Block 2. A missing value (represented by an asterisk (*)) is inserted for 'yield_g_m 2' in each row added. To facilitate this, the spreadsheet is de-randomised, and sorted by the identifying numbers of the breeding lines. The padded, de-randomised spreadsheet is as shown in Table 3.5.

Table 3.5 Spreadsheet holding yields of doubled haploid breeding lines of barley, 'padded' to produce a balanced randomised block design.
The conventions used in this spreadsheet are the same as those used in Table 1.1, Chapter 1

	A	B	C		A	B	C
1	block!	line!	yield_g_m2	46	1	45	519.88
2	1	1	717.78	47	1	46	724.01
3	1	2	483.33	48	1	47	192.43
4	1	3	873.04	49	1	48	786.06
5	1	4	719.14	50	1	49	830.57
6	1	5	798.99	51	1	50	721.13
7	1	6	849.66	52	1	51	692.55
8	1	7	907.38	53	1	52	603.04
9	1	8	636.45	54	1	53	559.14
10	1	9	774.84	55	1	54	809.41
11	1	10	764.88	56	1	55	555.29
12	1	11	644.83	57	1	56	522.61
13	1	12	436.59	58	1	57	181.97
14	1	13	541.15	59	1	58	521.67
15	1	14	910.76	60	1	59	553.46
16	1	15	600.06	61	1	60	572.90
17	1	16	210.71	62	1	61	611.52
18	1	17	551.84	63	1	62	730.48
19	1	18	366.24	64	1	63	562.90
20	1	19	622.72	65	1	64	647.92
21	1	20	632.46	66	1	65	818.74
22	1	21	514.62	67	1	66	687.92
23	1	22	340.65	68	1	67	407.25
24	1	23	605.75	69	1	68	325.96
25	1	24	671.22	70	1	69	467.24
26	1	25	428.61	71	1	70	996.19
27	1	26	580.22	72	1	71	595.86
28	1	27	731.87	73	1	72	165.76
29	1	28	679.92	74	1	73	355.30
30	1	29	606.37	75	1	74	489.31
31	1	30	353.29	76	1	75	617.33
32	1	31	166.75	77	1	76	344.48
33	1	32	669.16	78	1	77	357.94
34	1	33	770.23	79	1	78	260.02
35	1	34	673.43	80	1	79	317.63
36	1	35	374.28	81	1	80	487.86
37	1	36	800.35	82	1	81	315.63
38	1	37	895.30	83	1	82	251.30
39	1	38	641.42	84	1	83	280.44
40	1	39	145.84	85	2	1	591.19
41	1	40	410.87	86	2	2	*
42	1	41	321.84	87	2	3	*
43	1	42	793.43	88	2	4	*
44	1	43	678.46	89	2	5	*
45	1	44	517.52	90	2	6	755.05

(Continued overleaf)

Table 3.5   (*continued*)

	A	B	C		A	B	C
91	2	7	820.08	130	2	46	627.25
92	2	8	299.92	131	2	47	*
93	2	9	587.07	132	2	48	645.10
94	2	10	756.63	133	2	49	823.25
95	2	11	516.84	134	2	50	885.67
96	2	12	*	135	2	51	746.11
97	2	13	475.10	136	2	52	*
98	2	14	934.72	137	2	53	*
99	2	15	*	138	2	54	859.36
100	2	16	240.00	139	2	55	*
101	2	17	959.25	140	2	56	435.62
102	2	18	265.37	141	2	57	*
103	2	19	423.76	142	2	58	435.34
104	2	20	*	143	2	59	634.58
105	2	21	*	144	2	60	284.72
106	2	22	*	145	2	61	472.44
107	2	23	817.51	146	2	62	*
108	2	24	463.22	147	2	63	755.90
109	2	25	*	148	2	64	681.64
110	2	26	639.39	149	2	65	713.43
111	2	27	762.42	150	2	66	846.05
112	2	28	*	151	2	67	702.58
113	2	29	932.33	152	2	68	385.47
114	2	30	*	153	2	69	378.89
115	2	31	226.12	154	2	70	904.90
116	2	32	836.95	155	2	71	569.61
117	2	33	846.85	156	2	72	258.59
118	2	34	639.11	157	2	73	*
119	2	35	554.69	158	2	74	551.41
120	2	36	1054.70	159	2	75	*
121	2	37	552.70	160	2	76	*
122	2	38	540.68	161	2	77	*
123	2	39	212.66	162	2	78	260.26
124	2	40	568.31	163	2	79	439.17
125	2	41	538.00	164	2	80	478.12
126	2	42	552.53	165	2	81	303.72
127	2	43	836.82	166	2	82	*
128	2	.44	873.18	167	2	83	*
129	2	45	575.76				

This spreadsheet is stored as 'sheet 2' in the same workbook as the previous one. The same statements are used to analyse the padded data, with the substitution of 'sheet 2' for 'sheet 1' in the IMPORT statement. The output of the ANOVA statement is now as follows:

Analysis of variance

Variate: yield_g_m2

Source of variation	d.f.	(m.v.)	s.s.	m.s.	v.r.	F pr.
block stratum	1		7015.	7015.	0.52	
block.*Units* stratum						
line	82		6351920.	77462.	5.78	<.001
Residual	58	(24)	777747.	13409.		
Total	141	(24)	6179665.			

Message: the following units have large residuals.

block 1 *units* 8	174.8	s.e.	68.4
block 1 *units* 17	−197.2	s.e.	68.4
block 1 *units* 37	177.8	s.e.	68.4
block 1 *units* 44	−171.3	s.e.	68.4
block 2 *units* 8	−174.8	s.e.	68.4
block 2 *units* 17	197.2	s.e.	68.4
block 2 *units* 37	−177.8	s.e.	68.4
block 2 *units* 44	171.3	s.e.	68.4

The option setting 'PRINT = aovtable, information' in the ANOVA statement specifies that only the anova table, and summary information such as details of any large residuals, are to be presented. In particular, the tables of treatment means are not to be included in the output: they are voluminous, and are not needed here.

This anova, like that of the split plot design presented in Chapter 2 (Section 2.2), contains a stratum for each random-effect term. However, a slight difference in the way these strata have been specified in the present case should be noted. The term 'block' was explicitly mentioned in the BLOCKSTRUCTURE statement (Section 3.7), but the term 'block.*Units*' was not. It relates to the variation among individual values (referred to as *units*) within each block. There was no implicitly defined term like this in the anova of the split plot design, because in the specification of that analysis, the units (servings) were explicitly mentioned in the BLOCKSTRUCTURE statement. In the present case, a factor named 'plot' might be defined to number the plots within each block. The model in the BLOCKSTRUCTURE statement could then be modified to 'block / plot', and the 'block.*Units*' stratum would be relabelled as 'block.plot'. The anova would be numerically unchanged.

The column holding the degrees of freedom in the anova table contains information on the number of missing values (abbreviated to m.v.). The data set comprises 83 breeding lines in two blocks, so if there were no missing values, the total degrees of freedom would be given by

$$DF_{total} = 83 \times 2 - 1 = 165.$$

However, there are 24 missing values, as noted in brackets in the 'Total' row of the anova, so

$$DF_{total} = 165 - 24 = 141.$$

How should the missing degrees of freedom be distributed between the terms 'line' and 'Residual'? There is no breeding line for which all observations are missing. Therefore DF_{line} is unaffected by the missing values, and the residual degrees of freedom are reduced from

$$DF_{Residual} = \text{No. of units} - 1 - DF_{line} - DF_{block} = 166 - 1 - 82 - 1 = 82$$

to

$$DF_{Residual} = \text{No. of units} - 1 - DF_{line} - DF_{block} - \text{No. of missing values}$$
$$= 166 - 1 - 82 - 1 - 24 = 58.$$

Estimates of these missing values are obtained in order to proceed with the analysis, and the resulting anova is an approximation. This is revealed by the fact that the sums of squares for the various terms do not add up to the total sum of squares:

$$7015 + 6351920 + 777747 = 7136682 \neq 6179665.$$

On the null hypothesis that there is no real variation in grain yield among the breeding lines, the expected value of the F statistic (variance ratio, v.r. – see Chapter 1, Section 1.2) for the term 'line' is 1. The observed value, 5.78, is considerably larger than this, and the associated P value is small (<0.001), indicating that real variation is present – provided that the assumptions that underlie the analysis of variance are correct. Some observations are noted as having large residual values, which may cast doubt on these assumptions; however, we will postpone a fuller exploration of the residuals until we have arrived at an analysis that can be applied to the original, unbalanced data set. We will, however, take a closer look at this F statistic, as we will shortly be comparing it with other significance tests of the same null hypothesis (Sections 3.10–3.12). Values of F that provide significant evidence against the null hypothesis lie in the hatched region of Figure 3.2 – the upper tail of the F distribution. Values of F well below 1 are also extreme and improbable, lying in the lower tail of the distribution. However, they do not provide evidence against the null hypothesis: any variation among the breeding lines will tend to *increase* the value of F.

3.9   Regarding a treatment term as a random-effect term. The use of mixed-model analysis to analyse an unbalanced data set

The cross Chebec × Harrington could produce many other progeny lines besides those studied here, and the lines in this field trial may reasonably be considered as a random

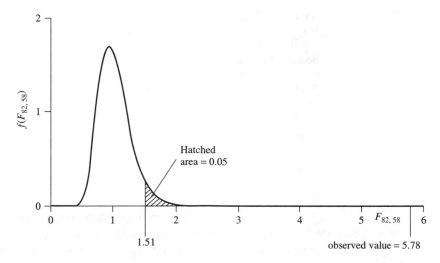

Figure 3.2 Distribution of the variable $F_{82,58}$, showing the critical value and critical region for significance at the 5 % level.

sample from this population of potential lines. Thus it is reasonable to consider 'line' as a random-effect term – that is, to assume that in Model 3.18, the ϕ_k are independent values of a random variable Φ, such that

$$\Phi \sim N(0, \sigma_\phi^2).$$

The variance component σ_ϕ^2 can be estimated by considering the expected MS for each term in the anova. These are as shown in Table 3.6. Thus (following the method presented in Section 3.4),

$$83\hat{\sigma}_\Delta^2 + \hat{\sigma}_E^2 = 7015, \quad 2\hat{\sigma}_\phi^2 + \hat{\sigma}_E^2 = 77462 \quad \text{and} \quad \hat{\sigma}_E^2 = 13409.$$

Rearranging these expressions, and substituting for $\hat{\sigma}_E^2$, we obtain

$$\hat{\sigma}_\Delta^2 = \frac{7015 - 13409}{83} = -77.0361$$

$$\hat{\sigma}_\phi^2 = \frac{77462 - 13409}{2} = 32026.5.$$

The estimate of the variance component due to blocks is negative, and small compared with the other components. We may decide that the best estimate of this component is zero, i.e. that there is no real difference between the mean grain yield values of the two blocks. That is, we may place a *constraint* upon the variance-component estimate. The estimate of the variance due to breeding lines is about double the residual variance.

So far, the random-effect terms that we have encountered have been treated as block terms: when data have been analysed using GenStat's anova directives, they have been included in the model in the BLOCKSTRUCTURE statement. However, it is not appropriate to treat 'line' as a block term. The variable Φ is not a nuisance

Table 3.6 Expected mean squares of the terms in the anova of yields of breeding lines of barley in a balanced randomised block design.

Source of variation	Expected MS
block	$83\sigma_\Delta^2 + \sigma_E^2$
line	$2\sigma_\Phi^2 + \sigma_E^2$
Residual	σ_E^2

variable: we are interested in knowing whether the variation among breeding lines is real (i.e. we are interested in testing its significance) and in estimating its variance component σ_Φ^2. The proper place for 'line' is among the treatment terms. In due course (Chapter 5, Section 5.1) we will want to estimate the mean yield of each breeding line. GenStat's analysis of variance directives require that random-effect terms be treated as block terms, but the mixed-modelling directives are not subject to this constraint. Moreover, unlike the anova directives, they do not require a balanced experimental design; hence they can be used on the original, 'unpadded' data in the spreadsheet stored as 'sheet1'. Thus the following statements will perform the mixed-model analysis that corresponds to the anova in Section 3.8, except that the unpadded data are used and 'line' is regarded as a random-effect term:

```
IMPORT \
    'Intro to Mixed Modelling\\Chapter 3\\barley progeny.xls'; \
    SHEET = 'sheet1'
VCOMPONENTS RANDOM = block + line
REML [PRINT = model, components, deviance, means; \
    PTERMS = 'constant'] yield_g_m2
```

The option setting 'PTERMS = 'constant'' specifies that the estimated value of the constant (the overall mean) is to be printed, but not the means for levels of 'block' or 'line'.

The output of the REML statement above is as follows:

REML variance components analysis

Response variate: yield_g_m2
Fixed model: Constant
Random model: block + line
Number of units: 142
Residual term has been added to model

Sparse algorithm with AI optimisation

Estimated variance components

Random term	component	s.e.
block	15.	325.
line	30645.	6242.

Residual variance model

Term	Factor	Model(order)	Parameter	Estimate	s.e.
Residual		Identity	Sigma2	13222.	2431.

Deviance: -2*Log-Likelihood

Deviance	d.f.
1613.24	138

Note: deviance omits constants which depend on fixed model fitted.

Table of predicted means for Constant

572.6 Standard error: 21.84

We will compare these results with those from the anova on the balanced, 'padded' experimental design. The estimates of the three variance components, $\hat{\sigma}_A^2 = -77.0361$, $\hat{\sigma}_\phi^2 = 32026.5$ and $\hat{\sigma}_E^2 = 13409$, are similar to the corresponding values obtained from the anova, but not identical. As the values from the mixed-model analysis are based on the true, unbalanced experimental design, they are the more reliable.

3.10   Comparison of a variance-component estimate with its standard error

As before (Section 3.4), each variance component in the output of the REML statement is accompanied by a standard error: for example, the variation among breeding lines is estimated to be $\hat{\sigma}_\phi^2 = 30646 \pm 6251$. This SE provides an indication of the precision with which the variance component is estimated. It may also provide a tentative indication of whether any variation is really accounted for by the term under consideration. If the null hypothesis

$$H_0 : \sigma_\phi^2 = 0$$

is true, the ratio

$$z = \hat{\sigma}_\phi^2 / \text{SE}(\hat{\sigma}_\phi^2) \tag{3.19}$$

is expected to have a value of about ± 1, on average. In the present case,

$$z = 30646/6251 = 4.90.$$

This is considerably larger than the expected value, suggesting that there is real variation among the breeding lines.

If $\hat{\sigma}_\phi^2$ were an observation of a normally distributed variable, this ratio could be made the basis for a formal significance test. z would then be approximately an observation of a *standard normal variable*: that is, of a variable Z such that

$$Z \sim N(0, 1).$$

(For an account of the normal distribution, see Chapter 1, Section 1.2.) Figure 3.3 shows that a value of Z greater than 1.645 provides significant evidence, at the 5 % level, against H_0. The observed value z is well above this value, and, if the assumptions underlying this test were fulfilled, it would provide highly significant evidence of variation among the breeding lines. However, $\hat{\sigma}_\phi^2$ is *not* an observation of a normally distributed variable. As noted in Section 3.4, estimates of a variance component do not follow a normal distribution at all reliably, and they should not be used as the basis of a formal significance test. It is nevertheless worth comparing each variance-component estimate with its SE informally. In the case of the variation among blocks, the ratio is

$$z = 15/236 = 0.0636.$$

As this is well below 1, we might consider dropping the term 'block' from the model fitted, and proceed to conduct a formal significance test to decide whether this was justified (see Sections 3.11–3.12). However, many statisticians would argue that if a term represents part of the design of the experiment – as in the present case – it should be retained, regardless of its significance or the magnitude of its variance component.

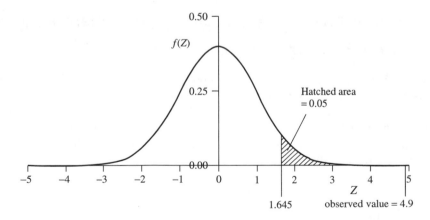

Figure 3.3   The standard normal distribution, showing the critical value and critical region for significance at the 5 % level.

Although a SE is presented for the residual variance component, $\hat{\sigma}_E^2$, it is not appropriate to consider this in relation to the H_0 that $\sigma_E^2 = 0$, even informally. There is some variation among the observations that is not accounted for by the effects of breeding lines or blocks, so this null hypothesis is known with certainty to be false.

3.11 An alternative significance test for variance components

There is an alternative method for testing the significance of variance components, based on the *deviance* presented in the output of the REML statement in Section 3.9. Deviance is a measure of the goodness of fit of the model to the data: the better the fit, the smaller the deviance. By comparing the deviance in the output in Section 3.9 with that obtained when the term under consideration is omitted from the random-effect model, the contribution made by this term to the fit of the model can be assessed. The model fitted above is referred to as the *full model*. The deviance from the *reduced model*, from which the term 'line' is omitted, is obtained by the following statements:

```
VCOMPONENTS RANDOM = block
REML [PRINT = deviance] yield_g_m2
```

The output of the REML statement is as follows:

Deviance: -2*Log-Likelihood

Deviance	d.f.
1652.92	139

Note: deviance omits constants which depend on fixed model fitted.

The output notes that certain constants are omitted from the calculation of the deviance. These constants cancel out when the difference between two deviances is calculated, *provided that* the deviances were obtained from models with the same fixed-effect terms. This condition is met in the present case, giving

$$\text{deviance}_{\text{reduced model}} - \text{deviance}_{\text{full model}} = 1652.92 - 1613.24 = 39.68. \quad (3.20)$$

It is important to note that if the fixed-effect terms differ between the two models, the constants do not cancel out and the deviances cannot be compared. Because the full and reduced models differ by a single random-effect term, the degrees of freedom of these two deviances differ by 1:

$$\text{DF}(\text{deviance}_{\text{reduced model}}) - \text{DF}(\text{deviance}_{\text{full model}}) = 139 - 138 = 1.$$

This is also a requirement if two models are to be compared by the method described here.

The interpretation of this difference between the deviances requires some explanation. It can be treated approximately as a χ^2 statistic with 1 degree of freedom. A large value of this statistic will be obtained from a data set in which the estimate of the variance component due to lines, $\hat{\sigma}_\phi^2$, is large and positive. Such a data set will give an F statistic that lies in the upper tail of the distribution presented in Section 3.8, providing evidence against the null hypothesis

$$H_0 : \sigma_\phi^2 = 0.$$

However, a data set in which $\hat{\sigma}_\phi^2$ is large *and negative* will also give a large *positive* difference between the deviances. Such a data set will give an F statistic that lies in the lower tail of the distribution, and does *not* provide evidence against the null hypothesis, unless we are willing to consider the possibility that the true value of the variance component is negative. An approximate adjustment for this can be made by halving the probability value associated with the χ^2 statistic. Formal arguments concerning the validity of this adjustment are given by Stram and Lee (1994) and by Crainiceanu and Ruppert (2004).

In the present case, we obtain

$$P(\chi_1^2 > 39.68) = 2.99 \times 10^{-10}$$

and hence, approximately,

$$P(\text{deviance}_{\text{reduced model}} - \text{deviance}_{\text{full model}} > 39.68) = \frac{1}{2} \times 2.99 \times 10^{-10}$$
$$= 1.50 \times 10^{-10}.$$

As before, the variation among breeding lines is found to be highly significant.

The significance of the variation among blocks can also be tested in this way. The following statements obtain the difference between the deviances obtained from the models including and excluding the term 'block' (including the term 'line' in both cases):

```
VCOMPONENTS RANDOM = line
REML [PRINT = deviance] yield_g_m2
```

The output of the REML statement is as follows:

Deviance: -2*Log-Likelihood

Deviance	d.f.
1613.24	139

Note: deviance omits constants which depend on fixed model fitted.

There is no difference between this value and that given by the full model, to the degree of precision presented in the GenStat output. However, the deviance from the reduced model can be saved, and printed with more precision, by the following statements:

```
VKEEP [DEVIANCE = devreduced]
PRINT devreduced; decimals = 6
```

The output of the PRINT statement is as follows:

```
devreduced
  1613.243236
```

When the precise deviance from the full model is obtained by the same method, we obtain the result

$$\text{deviance}_{\text{reduced model}} - \text{deviance}_{\text{full model}} = 1613.243236 - 1613.241029 = 0.002207,$$

and

$$P(\chi_1^2 > 0.002207) = 0.9625.$$

Hence

$$P((\text{deviance}_{\text{reduced model}} - \text{deviance}_{\text{full model}}) > 0.002207) = \frac{1}{2} \times 0.9625 = 0.4813.$$

As before, the difference between the blocks is found to be non-significant.

3.12   Comparison among significance tests for variance components

The two types of significance test for random effect introduced above (Sections 3.8 and 3.11), and the informal comparison of a variance-component estimate with its SE (Sections 3.10), are summarised and compared in Table 3.7. The P values for all these tests depend on distributional assumptions and should not be taken too literally. This is particularly true for very small values in the upper tail of the distribution, where the relative values given by the numerical approximations used are not very precise. However, these P values do illustrate some important general comparisons among the tests, as follows:

- Of the three tests compared, the F test from the anova makes the fullest use of the available information and is the most accurate *in the circumstances in which it is valid*. This is reflected by the fact that it gives the highest level of significance (i.e. the smallest value of P) for the effect of 'line'. However, in the present case the data had to be 'padded' in order to conduct this test, which means that it is not strictly valid, and there are many other experimental designs and types of data set for which the F test is unavailable.

- The significance of the deviance-based test does not depend on the degrees of freedom of the model strata concerned, which means that it does not fully take account of the consequences of limited sample size. This makes it less reliable than the F test. Nevertheless, it is a good substitute for the F test when this is not available.

Table 3.7 Comparison of different significance tests in the statistical analysis of yields of breeding lines of barley in a randomised block design.

Term	F from anova on 'padded' data		$\chi^2 = \text{deviance}_{\text{reducedmodel}} - \text{deviance}_{\text{fullmodel}}$		$z = \hat{\sigma}^2_{\text{term}}/\text{SE}(\hat{\sigma}^2_{\text{term}})$
	Test statistic	P	Test statistic	P	
block	$F_{1,58} = 0.52$	0.4737	0.002207	0.4813	0.0636
line	$F_{82,58} = 5.78$	3.43×10^{-11}	39.68	1.50×10^{-10}	4.90

- The ratio z does not provide a reliable significance test, due to the non-normality of the distribution of the estimated variance component. It is nevertheless a useful tool. It does not require a reduced model to be fitted – an important consideration when there are several random-effect terms, any of which might be dropped from the model, giving a large number of models to be explored and compared. If a variance-component estimate is much smaller than its standard error, as in the case of the 'block' term in the present analysis, its significance can be tested formally, and the possibility of dropping it from the model can be considered.

3.13   Inspection of the residual values

The validity of the F test presented in Section 3.8, and of the alternative significance test introduced in Section 3.11, depends on the assumption that the random variable E is normally distributed. This assumption can be explored by obtaining diagnostic plots of residuals, using the following statement:

```
VPLOT [GRAPHICS=high] fittedvalues, normal, halfnormal, histogram
```

The output of this statement is as shown in Figure 3.4. These plots show slight indications of departure from the assumptions of the analysis. The histogram of residuals is slightly skewed, and its tails are slightly compressed relative to those of a normal distribution: that is, there are *fewer* extreme residual values than would be expected by chance. There is a definite trend in the fitted-value plot, larger fitted values more commonly being associated with positive residuals. The normal and half-normal plots conform well to the assumptions, except for a slight flattening at both ends of the former and the upper end of the latter, again indicating a slight deficit of extreme values. A fuller analysis of this data set would include a search for the cause of the trend in the fitted-value plot, perhaps in the spatial pattern of variation of yield values over the trial site. The trend could then be eliminated by the addition of a term representing such spatial variation to the mixed model. (See Chapter 9, Section 9.7, for a fuller account of such methods.)

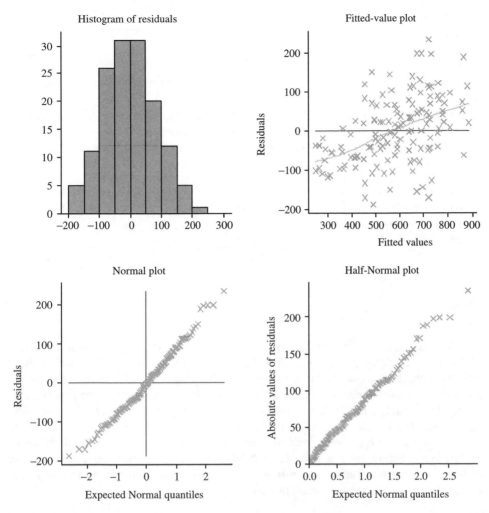

Figure 3.4   Diagnostic plots of residuals from the mixed-model analysis of yields of breeding lines of barley in an unbalanced randomised block design.

3.14   Heritability. The prediction of genetic advance under selection

It is often of interest to compare the magnitude of the different components of variance that contribute to the total variance of a response variable. For example, in the genetic improvement of plants and animals, the genetic (i.e. heritable) and environmental (i.e. non-heritable) sources of variation are routinely compared by means of a quantity called the *heritability*. The method for calculating heritability depends on the genetic structure of the population considered – inbred, random-mating, back-crossed, and so on. Here we will demonstrate a method of calculation presented by Allard (1960, pp 94–98) which is appropriate to the barley field trial under consideration. The heritability of a trait is defined as

$$h^2 = \frac{\hat{\sigma}_G^2}{\hat{\sigma}_P^2} \tag{3.21}$$

where
$\hat{\sigma}_G^2$ = genetic component of variance, i.e. the part of the variation in the organism's *phenotype* (its observable traits) that is due to genetic effects,
$\hat{\sigma}_P^2$ = phenotypic variance, i.e. the variance due to the combined effects of genotype and environment.

Note that h^2, not h, stands for heritability. The variance-component estimates given by the mixed-model analysis can be substituted into the variance estimates in this formula, as follows:

$$h^2 = \frac{\hat{\sigma}_\phi^2}{\hat{\sigma}_\phi^2 + \dfrac{\hat{\sigma}_E^2}{n_\Delta}} \tag{3.22}$$

where
n_Δ = the number of replications per line.

The presence of n_Δ in the formula for heritability perhaps requires explanation, as this coefficient is a feature of the experimental design, not of the sources of variation under investigation. The heritability is a measure of the proportion of variance *among the estimated breeding-line means* that is genetic in origin. The more replicate observations are made on each line (i.e. the higher the value of n_Δ), the more reliable are these means, and the higher the heritability. Conversely, the variance component for the block term, $\hat{\sigma}_P^2 = 15$, is absent from the formula for heritability, although this is one of the sources of variation under investigation. This is because the block effects contribute equally to each estimated breeding-line mean, and therefore do not contribute to the variation among them. In the present case,

$$h^2 = \frac{30646}{30646 + \dfrac{13222}{1.63}} = \frac{30646}{38758} = 0.791 \text{ or } 79.1\,\%.$$

The heritability can be used to calculate the *expected genetic advance under selection* in a plant or animal breeding programme. This is given by the formula

$$G_s = i\sigma_P h^2 \tag{3.23}$$

where
i = an index of the intensity of selection.

The index i is defined in relation to the standard normal distribution: that is, the distribution of a variable Z, such that

$$Z \sim N(0, 1)$$

(see Section 3.11). It is the value of Z that corresponds to the fraction (k) of the population that is to be selected. For example, suppose that the highest-yielding 5 % of breeding lines are to be selected from the present field trial. The hatched area of Figure 3.5 is the corresponding part of a standard normal distribution – that is, an area of 0.05 at the upper end of the distribution.

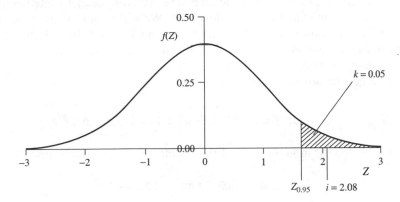

Figure 3.5 The standard normal distribution, showing the mean value of a selected part of the distribution.

The mean of *the specified part of* the distribution can be obtained by numerical evaluation of the function

$$i = \int_{Z_{1-k}}^{\infty} Z f(Z) \, dZ \tag{3.24}$$

where
Z_{1-k} = the $(1 - k)$th quantile of the distribution of Z, i.e. the value that cuts off a fraction k at the upper end of the distribution.

This integral has been tabulated (e.g. by Becker, 1992), but for values of k in the range 0.1 to 0.001, a reasonable approximation is given by

$$i = 1.13 + 0.73 \log_{10}(1/k). \tag{3.25}$$

Substitution of $k = 0.05$ into Equation 3.25 gives $i = 2.08$, and substituting this value into Equation 3.23 we obtain the following expression for the expected genetic advance under selection:

$$G_s = 2.08 \times \sqrt{38758 \, \text{g}^2/\text{m}^4} \times 0.791 = 323.9 \, \text{g/m}^2.$$

The estimated grand mean grain yield of all the breeding lines, given by the estimate of the constant in the output from the REML statement presented in Section 3.9, is $572.6 \, \text{g/m}^2$. Hence the expected mean grain yield that would be obtained if the selected fraction of the breeding lines were sown in a new field trial is

$$572.6 \, \text{g/m}^2 + 323.9 \, \text{g/m}^2 = 896.5 \, \text{g/m}^2.$$

This is an advance of

$$\frac{323.9}{572.6} \times 100 = 56.6\%.$$

The mean grain yield of the highest-yielding 5 % of the lines in the present field trial is 920.1 g/m². The expected mean in a future trial falls short of this value because the presence of environmental variation interferes with the crop breeder's attempt to select on the genetic variation among the breeding lines. We will return to this 'shrinkage' of predicted values, relative to our naïve expectation, in Chapter 5. In fact, the shrinkage due to residual variation is greater than this simple comparison of the expected future means and the present means suggests. If the heritability of grain yield were 100 %, the expected genetic advance would be

$$G_s = 2.08 \times \sqrt{38758 \text{ g}^2/\text{m}^4} \times 1 = 409.5 \text{ g/m}^2,$$

giving an expected mean grain yield of the selected lines in a future trial of

$$572.6 \text{ g/m}^2 + 409.5 \text{ g/m}^2 = 982.1 \text{ g/m}^2.$$

If the distribution of the breeding-line means were exactly normal, this would be the same as the mean value of the selected lines in the present trial. The distribution of the breeding-line means in this field trial is superimposed on the normal distribution with the same mean and variance in Figure 3.6. The upper tail of this distribution is

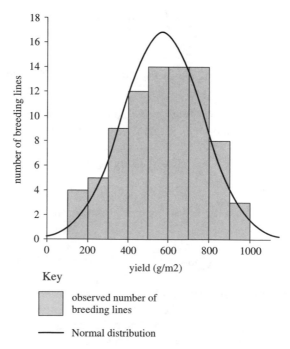

Key

☐ observed number of breeding lines

── Normal distribution

Figure 3.6 Distribution of the mean yields of breeding lines of barley, and the normal distribution with the same mean and variance.

somewhat shorter than that of the normal distribution: the normal distribution leads us to expect one or two breeding lines with a mean yield above $1,000 \, g/m^2$, whereas none are observed. This accounts for the low mean yield of the breeding lines in this tail.

Just as the formula for heritability does not include the variance due to blocks, so it does not take into account the difference in growing conditions between the present location and season and the environment in which a future trial might be conducted. Thus it is somewhat misleading to add the expected genetic advance to the mean of the present trial and obtain an expected mean for a future trial. All lines might yield more or less in the new environment, but this would not interfere with the validity of the breeder's selection. It is more precise to use G_s as an indication of the *difference* expected between the mean of the selected breeding lines and that which would be obtained from the full set of lines studied, if the full set were grown again. Even this interpretation takes no account of genotype $\times$ environment interaction. It may be that the response of the selected lines to the new environment will be different from that of the unselected lines, so that the gap between the two groups will be wider or narrower than in the present field trial. Thus any calculation of the heritability and the expected genetic advance under selection in a substantial crop breeding programme has to take account of components of variance due to genotype $\times$ year, genotype $\times$ location and genotype $\times$ year $\times$ location interaction effects, as well as the effects considered in this simple example.

3.15   Use of R to analyse the unbalanced field trial

To prepare the original, unbalanced barley breeding-line data set for analysis by R, it is transferred from 'sheet 1' in the Excel workbook 'Intro to Mixed Modelling\ Chapter 3\barley progeny.xls' to the text file 'barley progeny.dat' in the same directory. Exclamation marks (!) are removed from the ends of column headings in this file.

The following commands import the data, fit Model 3.18, and produce the resulting anova:

```
barleyprogeny.unbalanced <- read.table(
    "Intro to Mixed Modelling\\Chapter 3\\barley progeny.dat",
    header=TRUE)
attach(barleyprogeny.unbalanced)
fline <- factor(line)
fblock <- factor(block)
barleyprogeny.model1aov <- aov(yield_g_m2 ~ fline +
    Error(fblock))
summary(barleyprogeny.model1aov)
```

The output of these commands is as follows:

```
Error: fblock
        Df Sum Sq Mean Sq
fline    1  58204   58204
```

```
Error: Within
           Df  Sum Sq Mean Sq F value    Pr(>F)
fline      82 5343031   65159  4.8605 1.322e-09 ***
Residuals  58  777542   13406
---
Signif. codes:  0 '***' 0.001 '**' 0.01 '*' 0.05 '.' 0.1 ' ' 1
```

Whereas the GenStat anova directives recognise the design as unbalanced and refuse to produce an anova table (Section 3.7), R partitions the variation among lines into a component that occurs between the blocks and a component that occurs within each blocks. However, the interpretation of this anova is not straightforward: no F test is available for the between-blocks component of the variation due to 'line', and that for the within-blocks component does not represent all the variation due to this term.

To prepare the 'padded', balanced version of the same data set for analysis by R, it is transferred from the spreadsheet 'sheet 2' in the Excel workbook 'Intro to Mixed Modelling\Chapter 3\barley progeny.xls' to the text file 'barley progeny padded.dat' in the same directory. Once again, exclamation marks (!) are removed from the ends of headings, and the asterisks (*) that represent missing values are replaced by the code 'NA'. The following commands perform the analysis:

```
detach(barleyprogeny.unbalanced)
barleyprogeny.padded <- read.table(
    "Intro to Mixed Modelling\\Chapter 3\\barley progeny padded.dat",
    header=TRUE)
attach(barleyprogeny.padded)
fline <- factor(line)
fblock <- factor(block)
barleyprogeny.model1aov <- aov(yield_g_m2 ~ fline +
    Error(fblock))
summary(barleyprogeny.model1aov)
```

The anova produced by R on the padded data is the same as that produced on the unpadded data, and hence different from that produced by GenStat on the padded data. This shows that GenStat's anova directives and R's function aov() deal with the missing values in fundamentally different ways: GenStat's directives estimate them, whereas R's function excludes the units concerned from the design.

The function lme(), *as implemented in the package nlme*, is not able to accept both of the terms 'fline' and 'fblock' in its random-effect argument, so it cannot be used to perform the mixed-model analysis on the original, unbalanced data set. In order to fit a random-effect model with both these terms, two other packages, named 'lme4' and 'Matrix', available from the R web site, must be loaded.[1] The following commands then perform the mixed-model analysis on the unbalanced data set:

---

[1] The current method of doing this is as follows. The packages must first be downloaded from the R web site, then installed, then loaded. To download the packages, go to the web site http://www.r-project.org, then follow the links to a local CRAN mirror. To obtain the packages in an appropriate version to run under Windows on a PC, follow the link to 'Windows (95 and later)'. Then follow the link 'contrib' to contributed packages, then the link to a directory whose name corresponds to the current version of R. To download the package 'lme4', follow the link

```
barleyprogeny.unbalanced <- read.table(
   "Intro to Mixed Modelling\\Chapter 3\\barley progeny.dat",
header=TRUE)
attach(barleyprogeny.unbalanced)
fline <- factor(line)
fblock <- factor(block)
barleyprogeny.model1lmer <-
   lmer(yield_g_m2 ~ 1 + (1|fblock) + (1|fline),
   data = barleyprogeny.unbalanced)
summary(barleyprogeny.model1lmer)
```

The function `lmer()` is similar to `lme()`, but can fit a wider range of models. The syntax for model specification is similar to that for `lm()` and `lme()` presented in Chapter 1 (Section 1.11), but the fixed-effect and random-effect models are specified in a single argument of the function. The fixed-effect model comes first. In the present case, the only fixed-effect term is the intercept, which is represented by the constant '1'. The term '(1|fline)' indicates that the factor 'fline' is to be regarded as a random-effect term, *and that its effects are to be estimated within each level of the fixed-effect term '1'*. If the fixed-effect term were a factor with more than one level, this refinement would be important; in the present case, it makes no difference to the model. The factor 'fblock' is specified as a random-effect term in the same way. The next argument of the function, 'data', names the data frame holding the data to be analysed.

The output of the `summary()` function is as follows:

```
Linear mixed-effects model fit by REML
Formula: yield_g_m2 ~ 1 + (1 | fblock) + (1 | fline)
   Data: barleyprogeny.unbalanced
      AIC        BIC    logLik MLdeviance REMLdeviance
 1878.362 1887.230 -936.181   1880.365    1872.362
Random effects:
 Groups   Name        Variance   Std.Dev.
 fline    (Intercept) 30643.643  175.0533
 fblock   (Intercept)    15.033    3.8772
 Residual             13218.718  114.9727
# of obs: 142, groups: fline, 83; fblock, 2
```

named 'lme4_*.zip'. (The asterisk (*) in this link name is a *wildcard* representing a string of characters that will vary according to the version of the package.) Follow the instructions to download this .zip file to a suitable folder on your computer, preferably the folder named 'R-*' (the asterisk is again a wildcard) that holds other files relating to R. To download the package 'Matrix', repeat this process, following the link named 'Matrix_*.zip'. To install the package 'lme4' select 'Packages' in the main menu of the R GUI, then, in the 'Packages' sub-menu, select 'Install package(s) from local zip files...'. A window opens headed 'Select files'. Browse in this window until you reach the file name 'lme4_*.zip', then click on the button labelled 'Open'. To install the package 'Matrix', repeat this process, substituting the file name 'Matrix_*.zip'. To load the package 'lme4', select 'Packages' in the main menu, then, in the 'Packages' sub-menu, select 'Load package...'. A window headed 'Select one' opens. In the list of packages in this window, select 'lme4', then click on the button labelled 'OK'. Because 'lme4' requires 'Matrix', the latter is loaded automatically. Another package required by 'lme4', named 'lattice', is also loaded automatically: unlike 'lme 4' and 'Matrix', this package does not have to be installed explicitly.

```
Fixed effects:
            Estimate Std. Error t value
(Intercept)  572.633     21.846  26.212
```

The estimates of the variance components are similar to those given by GenStat (Section 3.10), but not identical (Table 3.8). The square root of each variance estimate – the standard deviation – is also presented.

The following commands fit the reduced model omitting the term 'fline':

```
barleyprogeny.model2lmer <-
    lmer(yield_g_m2 ~ 1 + (1|fblock),
    data = barleyprogeny.unbalanced)
summary(barleyprogeny.model2lmer)
```

The output of these commands is as follows:

```
Linear mixed-effects model fit by REML
Formula: yield_g_m2 ~ 1 + (1 | fblock)
    Data: barleyprogeny.unbalanced
        AIC      BIC    logLik MLdeviance REMLdeviance
  1916.042 1921.953 -956.0208   1919.904     1912.042
Random effects:
 Groups    Name            Variance Std.Dev.
 fblock    (Intercept)      210.02   14.492
 Residual                 43718.38  209.089
# of obs: 142, groups: fblock, 2

Fixed effects:
            Estimate Std. Error t value
(Intercept)  582.427     20.375  28.586
```

Values of deviance calculated on two different bases are presented in this output, namely the *m*aximum *l*ikelihood deviance (MLdeviance) and the *re*sidual *m*aximum *l*ikelihood deviance (REMLdeviance). The latter is the one we require in order to

Table 3.8   Comparison between estimates of variance components given by GenStat and R, in the mixed-model analysis of yields of breeding lines of barley in a randomised block design

Term		Variance component component	GenStat		R estimate
GenStat	R		Estimate	$SE_{estimate}$	
block	fblock	σ_A^2	15	325	15.033
line	fline	σ_ϕ^2	30645	6242	30643.643
Residual	Residual	σ_E^2	13222	2431	13218.718

compare the full and reduced models. Substituting the values from the two models into Equation 3.20, we obtain

$$\text{deviance}_{\text{reduced model}} - \text{deviance}_{\text{full model}} = 1912.042 - 1872.362 = 39.680.$$

Note that neither of the individual deviances agrees with that obtained by GenStat, but that the difference between them does. This reflects the fact that a deviance includes arbitrary constants dependent on the fixed-effect model, and emphasises the importance of ensuring that the fixed-effect models are the same, and specified in the same way, when deviances are compared.

The following commands fit the alternative reduced model, omitting the term 'fblock':

```
barleyprogeny.model3lmer <-
    lmer(yield_g_m2 ~ 1 + (1|fline),
    data = barleyprogeny.unbalanced)
summary(barleyprogeny.model3lmer)
```

The output of these commands is as follows:

```
Linear mixed-effects model fit by REML
Formula: yield_g_m2 ~ 1 + (1 | fline)
   Data: barleyprogeny.unbalanced
     AIC      BIC    logLik MLdeviance REMLdeviance
 1876.365 1882.276 -936.1823   1880.351     1872.365
Random effects:
 Groups   Name        Variance Std.Dev.
 fline    (Intercept) 30666    175.12
 Residual             13223    114.99
# of obs: 142, groups: fline, 83

Fixed effects:
            Estimate Std. Error t value
(Intercept)   572.45      21.67  26.417
```

From the values of 'REMLdeviance' from this alternative reduced model and the full model, we obtain

$$\text{deviance}_{\text{reduced model}} - \text{deviance}_{\text{full model}} = 1872.365 - 1872.362 = 0.003,$$

which agrees with the value given by GenStat.

3.16 Estimation of variance components in the regression analysis on grouped data

We now return to the model of the relationship between latitude and house prices in England, introduced in Chapter 1 (Model 1.2), to interpret the variance components for the random-effect terms. The output from fitting this model (Section 1.8) gives the following estimates:

- variance component for the effect of towns: $\hat{\sigma}_T^2 = 0.01963$, $SE_{\hat{\sigma}_T^2} = 0.01081$

- residual variance: $\hat{\sigma}^2 = 0.0171$, $SE_{\hat{\sigma}^2} = 0.00332$.

We can take the square root of each of these variance-component estimates in order to convert them to standard deviations (SDs) in the units of the original model (Equation 1.2), log (house price in pounds), and then take the antilogarithm to obtain the multiple of house price represented by the SD:

- SD for the effect of towns: $\hat{\sigma}_T = \sqrt{0.01963} = 0.14011$

$$10^{\hat{\sigma}_T} = 10^{0.14011} = 1.381$$

- residual SD: $\hat{\sigma} = \sqrt{0.0171} = 0.13076$

$$10^{\hat{\sigma}} = 10^{0.13076} = 1.351.$$

The two components are similar in magnitude: that is, after adjustment for the effect of latitude, the amount of variation among mean house prices in different towns is about the same as that among individual houses in the same town. The mean house price typically varies from town to town by a factor of 1.381, after adjustment for the effect of latitude. Strictly speaking, a town typically lies this far *above or below* the relationship with latitude, so that the typical *difference* between two towns at the same latitude is a factor of $10^{\sqrt{2 \times 0.01963}} = 1.5781$. Within each town, houses typically vary in price by a factor of 1.351. The mixed-model analysis gave the estimated effect of latitude as -0.08147 (Chapter 1, Section 1.8): that is, it is estimated that $\log_{10}$(house price, pounds) decreases by 0.08147 for each degree of latitude northward. This is equivalent to a reduction in house prices by a factor of $10^{-0.08147} = 0.8290$ for each degree north, or an increase by a factor of $10^{0.08147} = 1/0.8290 = 1.2063$ for each degree south. So:

- the typical variation between towns is equivalent to the effect of $1.381/1.2063 = 1.145$ degrees of latitude, and

- the typical variation between houses in the same town is equivalent to the effect of $1.351/1.2063 = 1.120$ degrees of latitude.

When the analysis is restricted to an equal number of houses from each town, the mean squares in the anova (Chapter 1, Section 1.9) are related to the variance components as shown in Table 3.9. Because three houses are considered from each town, the variance component σ_T^2 has the coefficient 3 in the 'town' stratum. These

Table 3.9   Expected mean squares of the terms in the anova of the effect of latitude on house prices in England.

Source of variation	Expected MS
town	$3\sigma_T^2 + \sigma^2$
Residual	σ^2

expected mean squares are the same if

$$H_0 : \sigma_T^2 = 0$$

is true, so it is appropriate to perform an F test of the significance of 'town' against the residual variance. This gives

$$F = \frac{0.07700}{0.01943} = 3.96, \quad P(F_{9,22} > 3.96) = 0.0040.$$

That is, the variation among towns is highly significant, even after adjusting for the effect of latitude.

In R, the output of the commands

```
summary(houseprice.model4)
anova(houseprice.model4)
```

given in Chapter 1, Section 1.11, provides the information required to obtain these variance-component estimates. The output from the function `summary()` includes the following information:

```
Random effects:
  Formula: ~1 | town
            (Intercept)   Residual
StdDev:     0.1401131  0.1307764
```

The two SDs presented here are $\hat{\sigma}_T$ and $\hat{\sigma}$ respectively.

3.17 Estimation of variance components for block effects in the split plot experimental design

Variance components can also be estimated for the block effects in the split plot experiment considered in Chapter 2 (Model 2.2). The expected mean squares in the anova (Section 2.2) are as shown in Table 3.10. These expected mean squares show that, as in the case of the hierarchical model of the three-stage assay process (Model 3.1, Sections 3.2 to 3.4), it is appropriate to perform an F test of the significance of each block-effect term against the block-effect term below. The values obtained are as shown in Table 3.11.

Table 3.10   Expected mean squares of the terms in the anova of the perception of saltiness of commercial brands of ravioli by trained assessors.

Source of variation	Expected MS
day	$36\sigma_\Delta^2 + 9\sigma_\Pi^2 + \sigma^2$
day.presentation	$9\sigma_\Pi^2 + \sigma^2$
day.presentation.serving	σ^2

Table 3.11 *F* tests for the significance of the block-effect terms in the anova of the perception of saltiness of commercial brands of ravioli by trained assessors.

Source of variation	DF	MS	F	P
day	2	371.7	2.67	0.15
day.presentation	6	139.3	1.39	0.23
day.presentation.serving	64	100.3		

These tests show that neither

$$H_0 : \sigma_\Delta^2 = 0$$

nor

$$H_0 : \sigma_\Pi^2 = 0$$

can be rejected. Nevertheless, we can, if we wish, proceed to estimate these variance components. This is done by rearranging the formulae for the expected mean squares as follows:

$$\hat{\sigma}_\Delta^2 = \frac{MS_{day} - MS_{Residual,day.presentation\ stratum}}{36} = \frac{371.7 - 139.3}{36} = 6.4556$$

$$\hat{\sigma}_\Pi^2 = \frac{MS_{Residual,day.presentation\ stratum} - MS_{Residual,day.presentation.serving\ stratum}}{9}$$

$$= \frac{139.3 - 100.3}{9} = 4.3333$$

$$\hat{\sigma}^2 = MS_{Residual,day.presentation.serving\ stratum} = 100.3.$$

Allowing for rounding error, these values agree with those given by the mixed-model analysis (Section 2.4). They show that even if there is variation among the true means of days and presentations, it is slight compared with the variation among servings within each presentation.

In R, the following command presents the estimates of the variance components in this model:

```
summary(ravioli.model2)
```

The first part of the output of this command is as follows:

```
Linear mixed-effects model fit by REML
 Data: NULL
        AIC       BIC     logLik
    660.2387 751.3053 -290.1193

Random effects:
 Formula: ~1 | day
         (Intercept)
 StdDev:    2.540482
```

```
Formula: ~1 | presentation %in% day
        (Intercept)
StdDev:    2.083383

Formula: ~1 | serving %in% presentation %in% day
        (Intercept) Residual
StdDev:    9.425732 3.380323
```

The output also includes extensive information on the fixed effects, which does not concern us here. The estimates of variance components are obtained from these SDs as follows:

$$\hat{\sigma}_{\Delta}^2 = 2.540482^2 = 6.4540$$

$$\hat{\sigma}_{\Pi}^2 = 2.083383^2 = 4.3405$$

$$\hat{\sigma}^2 = 9.425732^2 + 3.380323^2 = 100.3.$$

Allowing for rounding error, these values agree with those produced by GenStat.

3.18 Summary

For each random-effect term in a mixed model, a variance component – the part of the total variance that is due to that term – can be estimated.

The estimation of variance components is illustrated in an experiment to study a three-stage assay process. The experiment has a hierarchical design: deliveries of a chemical paste were sampled, a sample of casks was taken from each delivery, and repeated tests were performed on each cask. The 'strength' of the paste was measured.

Estimates of the variance components can be obtained from the mean squares in the anova, together with the number of replications at each level in the hierarchy. They can also be obtained by REML.

The expected mean squares justify the F tests that are performed.

Estimates of variance components can be used to design a future sampling strategy. They provide the basis for an estimate of the optimum number of casks to be sampled from each delivery, and the optimum number of tests to be performed on each cask.

Genetic (heritable) and environmental (non-heritable) variance components are routinely estimated in plant and animal breeding, and are compared in a ratio called heritability.

The use of variance components in this way is illustrated by a field experiment to evaluate doubled-haploid breeding lines derived from a cross between two inbred varieties of barley. The experiment has a randomised block design, but is unbalanced as not all lines are represented in both blocks. The effects of breeding lines, as well as those of blocks and the residual effects, are regarded as random.

An anova can be performed on the experiment if it is 'padded' with missing values to make it balanced, but it can be analysed in its original form using REML.

Three tests for the significance of a variance component are compared:

- The F test, which determines the significance of a term in a balanced experimental design.

- Comparison of the estimate of the variance component with its standard error, to give a z statistic.

- The change in deviance due to dropping the term in question from the random-effect model, keeping the fixed-effect model unchanged. This is interpreted as a χ^2 statistic with 1 degree of freedom, but the P value associated with this statistic is halved, for reasons explained in Section 3.12.

The F statistic is to be preferred in the circumstances in which it is valid. The comparison of the estimate with its standard error should not be used as a formal test, but is useful as a preliminary screening device when many possible models are under consideration, as it does not require more than one model to be fitted. The deviance-based test is a good substitute for the F test.

The calculation of heritability, and its use to predict the genetic advance under selection in a breeding programme, are illustrated.

If the heritability is less than 1, the predicted genetic advance is less than the value given by a naïve prediction from the mean of the selected breeding lines. This 'shrinkage' of predicted values is considered more fully in Chapter 5.

3.19 Exercises

1. In research on artificial insemination of cows, a series of semen samples from each of six bulls was tested for the ability to produce conceptions (Snedecor and Cochran, 1989, Chapter 13, Section 13.7, pp 245–247). The results are presented in Table 3.12.

   (a) Arrange these data for analysis by GenStat or R.

   (b) Analyse the data by analysis of variance and by mixed modelling, making the assumption that the bulls studied have been chosen at random from a population of similar animals. Perform the best available test to determine whether the variation among bulls is significant.

   The data are percentages based on slightly different numbers of tests: the assumption, made in the analyses specified in Part (b), that the variance among samples within each bull is constant, is therefore not quite correct.

Table 3.12 Conception rates obtained from semen samples from six bulls

Bull	Percentage of conceptions to services								
1	46	31	37	62	30				
2	70	59							
3	52	44	57	40	67	64	70		
4	47	21	70	46	14				
5	42	64	50	69	77	81	87		
6	35	68	59	38	57	76	57	29	60

(c) Obtain diagnostic plots of the residuals, and investigate whether there is evidence of a serious breach of the assumptions on which your analyses are based.

(d) Estimate the following variance components:
   (i) among bulls
   (ii) among samples within each bull.

Suppose that the results from this experiment are to be used to design an assay procedure to estimate the fertility of similar bulls in the future. Suppose also that the cost of including an additional bull in the assay is three times the cost of obtaining an additional sample from a bull already included.

(e) How many samples should be tested from each bull included in the assay?

Note that the sums of squares and mean squares in the anova of these data do not agree with those presented by Snedecor and Cochran.

2. An experiment using 36 samples of Portland cement is described by Davies and Goldsmith (1984, Chapter 6, Section 6.75, pp 154–158). The samples were 'gauged' (i.e. mixed with water and worked) by three gaugers, each one gauging 12 samples. After the samples had set, their compressive strength was tested by three breakers, each breaker testing four samples from each of the three gaugers. The results are presented in Table 3.13. (Data reproduced by kind permission of Pearson Education Ltd.)

(a) Arrange these data for analysis by GenStat or R.
'Gauger' and 'Breaker' are both to be regarded as random-effect terms.

(b) Justify these decisions.

(c) Analyse these data by analysis of variance and by mixed modelling.

(d) Perform F tests for the significance of the following model terms:
   (i) Gauger
   (ii) Breaker
   (iii) Gauger $\times$ Breaker interaction.

(e) Obtain estimates of the variance components for the same model terms, and also for the residual term, from the analysis of variance. Confirm your answers from the mixed-modelling results.

Table 3.13 Compressive strength of samples of Portland cement (psi) mixed by three gaugers, measured by three breakers

Gauger	Breaker					
	1		2		3	
1	5280	5520	4340	4400	4160	5180
	4760	5800	5020	6200	5320	4600
2	4420	5280	5340	4880	4180	4800
	5580	4900	4960	6200	4600	4480
3	5360	6160	5720	4760	4460	4930
	5680	5500	5620	5560	4680	5600

(f) According to the evidence from this experiment, which of these sources of variation needs to be taken into account when obtaining an estimate of the mean strength of samples of Portland cement?

The breakers in this experiment were human assistants who operated the testing machine, and the variation in the results that they produce is due to personal factors in their preliminary adjustment of the machine.

(g) Suppose that the cost of employing an additional breaker is 10 times the cost of getting the current breaker to test an additional sample. Determine the number of samples that each breaker should test in order to obtain the most accurate estimate possible of the strength of the cement, at a fixed cost.

3. Two inbred cultivars of wheat were hybridised, and seed of 48 F_2-derived F_3 families (i.e., families in the third progeny generation, each derived by inbreeding from a single plant in the second progeny generation) was obtained. The seed was sown in field plots in competition with ryegrass, in two replications in a randomised block design, with 'family' as the treatment factor. The mean grain yield per plant was determined in each plot. The results obtained are presented in Table 3.14. (Data reproduced by kind permission of S. Mokhtari.)

In the analysis of these data, 'family' will be regarded as a random-effect term.

(a) Justify this decision.

(b) Analyse these data by analysis of variance and by mixed modelling. Determine whether the variation among families is significant according to the F test.

(c) Obtain estimates of the variance components for the following model terms:

   (i) Residual

   (ii) family

   (iii) block.

(d) Compare the estimate of the variance component for the term 'family' with its standard error. Does their relative magnitude suggest that this term might reasonably be dropped from the model?

(e) Test the significance of the term 'family' by comparing the deviances obtained with and without the inclusion of this term in the model.

(f) Obtain diagnostic plots of the residuals from your analysis.

(g) Compare the results of the significance tests conducted in Parts (b) and (e) with each other, and with the informal evaluation conducted in Part (d). Consider how fully the assumptions of each test are likely to be met. Considered together, do these tests indicate that the term 'family' should be retained in the model?

(h) Estimate the heritability of yield in this population of families. (Note that the estimate obtained using the methods described in this chapter is slightly biased downwards, as some of the residual variance is due to genetic differences among plants of the same family.)

(i) Estimate the genetic advance that is expected if the highest-yielding 10 % of the families are selected for further evaluation. Find the expected mean yield of the selected families, and compare it with the observed mean yield of the same families. Account for the difference between these values.

Table 3.14 Yield per plant (g) of F_3 families of wheat grown in competition with ryegrass in a randomised block design.

	A	B	C
1	block	family	yield
2	1	11	3.883
3	1	9	3.717
4	1	7	3.850
5	1	41	1.817
6	1	2	7.483
7	1	25	3.483
8	1	10	2.500
9	1	38	3.683
10	1	32	1.167
11	1	42	
12	1	23	0.717
13	1	8	2.883
14	1	17	5.883
15	1	43	2.900
16	1	5	2.650
17	1	16	2.517
18	1	21	1.933
19	1	18	2.417
20	1	15	3.683
21	1	44	0.940
22	1	36	4.617
23	1	20	2.817
24	1	1	4.260
25	1	22	2.667
26	1	40	6.917
27	1	35	7.883
28	1	6	3.050
29	1	14	2.533
30	1	29	5.400
31	1	45	3.150
32	1	3	2.150
33	1	28	1.367
34	1	19	0.733
35	1	26	1.700

(*Continued overleaf*)

Table 3.14   (*continued*)

	A	B	C
36	1	30	1.800
37	1	46	1.720
38	1	12	4.083
39	1	27	3.667
40	1	31	4.317
41	1	39	4.033
42	1	47	1.950
43	1	4	5.000
44	1	24	2.600
45	1	34	4.250
46	1	13	4.300
47	1	33	4.250
48	1	37	3.783
49	1	48	2.617
50	2	9	8.550
51	2	11	9.633
52	2	5	5.733
53	2	25	8.117
54	2	39	4.633
55	2	21	1.450
56	2	41	3.240
57	2	42	
58	2	14	2.850
59	2	37	3.850
60	2	19	5.020
61	2	36	8.740
62	2	43	1.300
63	2	16	4.600
64	2	29	9.100
65	2	6	4.600
66	2	8	4.900
67	2	44	4.400
68	2	1	4.900
69	2	35	4.283
70	2	3	5.233
71	2	13	3.717
72	2	26	4.717

Table 3.14 (*continued*)

	A	B	C
73	2	28	2.817
74	2	12	7.467
75	2	33	4.567
76	2	45	5.480
77	2	34	7.933
78	2	18	4.833
79	2	7	4.933
80	2	2	5.300
81	2	31	7.500
82	2	40	3.083
83	2	46	2.900
84	2	10	4.917
85	2	32	4.400
86	2	17	6.533
87	2	23	5.783
88	2	27	4.767
89	2	38	4.867
90	2	30	3.983
91	2	22	4.017
92	2	47	1.775
93	2	48	1.540
94	2	20	
95	2	24	8.033
96	2	4	4.020
97	2	15	3.183

(j) When these estimates of genetic advance and expected mean yield are obtained, what assumptions are being made about the future environment in which the families are evaluated?

These data have also been analysed and interpreted by Mokhtari *et al.* (2002).

4

Interval estimates for fixed-effect terms in mixed models

4.1 The concept of an interval estimate

We have seen in earlier chapters that the estimates of fixed effects produced by mixed-model analysis are similar to those produced by the more familiar analysis methods. Regression analysis gives coefficients that are interpreted as slopes and intercepts; anova gives treatment means. Mixed-model analysis, when the same terms are placed in the fixed-effect model, gives similar estimates, which can be interpreted in the same way (Chapter 1 – slopes and intercepts; Chapter 2 – treatment means). However, the apparent *precision* of these estimates, indicated by their SEs, is affected by the decision to treat other terms in the model as random. In this chapter we will explore the precision with which fixed effects are estimated by mixed-model analysis. The single value so far given as the estimate of each model parameter (slope, intercept or mean) is referred to as a *point estimate*, and the precision of each point estimate is indicated by enclosing it in an *interval estimate*. This indicates the range within which the true value of the parameter can reasonably be supposed to lie.

Before determining the interval estimates of the various model parameters introduced so far, we need to set out the general principles to be followed when obtaining such an estimate. Consider a model parameter β. The point estimate of this parameter is referred to as $\hat{\beta}$, and the basis for calculation of the interval estimate is the *standard error* of the point estimate, $SE_{\hat{\beta}}$. To obtain this, $\hat{\beta}$ is treated as an observation of a random variable $\hat{B}$, the variance of which, $\sigma_{\hat{\beta}}^2$, is a function of the variance components in the model fitted. In order to regard $\hat{\beta}$ in this way, it is necessary to envisage an infinite population of data sets, from which the actual data set studied has been sampled at random, each of which would yield a value $\hat{\beta}$. It is assumed that these

Introduction to Mixed Modelling: Beyond Regression and Analysis of Variance N. W. Galwey
© 2006 John Wiley & Sons, Ltd

values of $\hat{\beta}$ are normally distributed around the true value β: that is,

$$\hat{B} \sim N(\beta, \sigma_{\hat{\beta}}^2).$$

These ideas are illustrated in Figure 4.1.

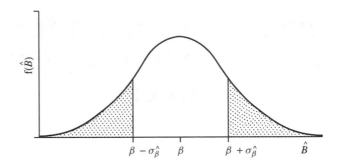

Figure 4.1   Distribution of the random variable $\hat{B}$, of which the parameter point estimate $\hat{\beta}$ is an observation.

An estimate of the variance $\sigma_{\hat{\beta}}^2$, referred to as $\hat{\sigma}_{\hat{\beta}}^2$, is obtained using methods presented in Sections 4.2 and 4.3. The SE of the point estimate of the parameter is then given by the square root of this variance estimate: that is,

$$SE_{\hat{\beta}} = \hat{\sigma}_{\hat{\beta}}. \qquad (4.1)$$

This value is added to and subtracted from the point estimate to obtain the boundaries of the interval estimate, as illustrated in Figure 4.2. This *number line* represents the range of possible values for the true parameter value β. The stippled areas indicate values outside the range in which the true value β can reasonably be supposed to lie. The boundaries of this region are symmetrical on either side of the point estimate $\hat{\beta}$. The concept of an interval estimate can be applied not only to an individual parameter, but also to the difference between two parameters.

Figure 4.2   An interval estimate of the parameter β, based on the SE of the point estimate.

4.2   SEs for regression coefficients in a mixed-model analysis

In the study of the relationship between latitude and house prices in Chapter 1, estimates of the fixed effects (constant, i.e. intercept, and effect of latitude, i.e. slope)

Table 4.1 Comparison of the parameter estimates, and their SEs, obtained from different methods of analysis of the effect of latitude on house prices in England.

Term	Method of analysis					
	Regression analysis ignoring towns (Model 1.1)		Regression analysis on town means		Mixed-model analysis (Model 1.2)	
	Estimate	$SE_{Estimate}$	Estimate	$SE_{Estimate}$	Estimate	$SE_{Estimate}$
Constant	9.68	1.00	9.44	1.87	9.497	1.9248
latitude	−0.0852	0.0188	−0.0804	0.0353	−0.08147	0.036272

obtained by three different methods were presented, together with their SEs (Table 1.6, Section 1.8); they are repeated in Table 4.1. The point estimates of the effect of latitude obtained by the three methods are very similar, as noted earlier (Section 1.8), but the SE from the regression analysis ignoring towns is considerable smaller than those from the other two analyses. This SE is misleading: it exaggerates the precision with which the effect of latitude is estimated, just as the significance test from the regression analysis ignoring towns exaggerated our confidence that the effect is real (Section 1.3). This analysis makes the assumption that every observation (every house) is independent: by ignoring the presence of groups of observations (i.e. towns), it overestimates the amount of information on which the analysis is based. The SEs from the other two analyses are both reasonable. The slight difference between them arises from the unequal numbers of houses sampled in different towns. Similar arguments apply to the estimates of the constant and their SEs.

The method of calculation of SEs of estimates in a mixed-model analysis is generally complicated, but when the number of observations per group is constant it is relatively straightforward. In order to compare the calculations for the three models in this simpler case, the estimates and their SEs from the data set 'trimmed' to three houses per town, which were presented in Chapter 1, Section 1.9, are collated in Table 4.2. The SEs from the regression analysis on town means and from the mixed-model analysis are now the same, allowing for rounding error.

Table 4.2 Comparison of the parameter estimates, and their SEs, obtained from different methods of analysis of the effect of latitude on house prices in England, using data 'trimmed' to three houses per town.

Term	Method of analysis					
	Regression analysis ignoring towns (Model 1.1)		Regression analysis on town means		Mixed-model analysis (Model 1.2)	
	Estimate	$SE_{Estimate}$	Estimate	$SE_{Estimate}$	Estimate	$SE_{Estimate}$
Constant	9.04	1.37	9.04	2.01	9.038	2.0068
latitude	−0.0726	0.0259	−0.0726	0.0378	−0.07260	0.037835

The methods for calculating these SEs are as follows. To obtain the SE of the effect of latitude from the regression analysis ignoring towns, we define

$$x_i = \text{the latitude of the } i\text{th house},$$

$$n = \text{the number of houses represented in the data set},$$

$$\bar{x} = \frac{\sum_{i=1}^{n} x_i}{n} \text{ (the mean value of latitude)}, \tag{4.2}$$

$$S_{XX} = \sum_{i=1}^{n} (x_i - \bar{x})^2 \text{ (the corrected sum of squares of latitude)} \tag{4.3}$$

and

$\hat{\sigma}^2 = \text{the estimate of } \sigma^2 \text{ from Model 1.1} = MS_{\text{Resid}}$ from the anova of this model.

We refer to the true effect of latitude as β_1, and the point estimate of this effect as $\hat{\beta}_1$. The estimate of the variance of $\hat{\beta}_1$ is then given by

$$\hat{\sigma}^2_{\hat{\beta}_1} = \frac{\hat{\sigma}^2}{S_{XX}} \tag{4.4}$$

and

$$SE_{\hat{\beta}_1} = \hat{\sigma}_{\hat{\beta}_1} = \frac{\hat{\sigma}}{\sqrt{S_{XX}}}. \tag{4.5}$$

In the present case,

$$n = 33,$$

$$\bar{x} = \frac{53.7947 + 53.7947 + \cdots + 51.7871}{33} = \frac{1749.8337}{33} = 53.0253,$$

$$S_{XX} = (53.7947 - 53.0253)^2 + (53.7947 - 53.0253)^2 + \cdots$$
$$+ (51.7871 - 53.0253)^2 = 53.7878,$$

$$\hat{\sigma}^2 = 0.03615,$$

$$\hat{\sigma}^2_{\hat{\beta}_1} = \frac{0.03615}{53.7878} = 0.000672$$

and

$$SE_{\hat{\beta}_1} = \sqrt{\frac{0.03615}{53.7878}} = 0.02592$$

as stated above.

To obtain the SE of the effect of latitude from the regression analysis on town means, the same method of calculation is used, but town means replace the individual house values, and the value of MS_{Resid} obtained from the town means is used as $\hat{\sigma}^2$. Thus we obtain

$$n_{\text{town means}} = 11,$$

$$\bar{x}_{\text{town means}} = \frac{53.7947 + 53.2591 + \cdots + 51.7871}{11} = \frac{583.2779}{11} = 53.0253,$$

$$S_{XX,\text{town means}} = (53.7947 - 53.0253)^2 + (53.2591 - 53.0253)^2 + \cdots$$
$$+ (51.7871 - 53.0253)^2$$
$$= 17.92927$$

and

$$\hat{\sigma}^2_{\text{town means}} = 0.02567.$$

We modify Equation 4.5 to give

$$SE_{\hat{\beta}_1,\text{town means}} = \hat{\sigma}_{\hat{\beta}_1,\text{town means}} = \frac{\hat{\sigma}_{\text{town means}}}{\sqrt{S_{XX,\text{town means}}}}. \qquad (4.6)$$

Substituting the numerical values into Equation 4.6, we obtain

$$SE_{\hat{\beta}_1,\text{town means}} = \sqrt{\frac{0.02567}{17.92927}} = 0.03784$$

as stated above.

The SE of the effect of latitude from the mixed-model analysis is numerically identical to that from the regression analysis on town means, but can be viewed somewhat differently. It is not based on a single variance estimate: it is calculated using both the among-towns and the within-towns variance components. To see how these are combined, we need to consider the relationship between variance components and mean squares in this analysis, discussed in Chapter 3, Section 3.16. In Table 3.9 we noted that

$$\text{expected MS}_{\text{town}} = r\sigma_T^2 + \sigma^2$$

where
r = number of houses sampled in each town = 3.

Now the value of MS_{town} is closely related to the variance among the town means. Specifically,

$$\sigma^2_{\text{town means}} = \frac{\text{Expected MS}_{\text{town}}}{r} = \sigma_T^2 + \frac{\sigma^2}{r}.$$

Replacing the expected mean squares by the observed mean squares in this formula, and the true variance components by their estimates, we obtain

$$\hat{\sigma}^2_{\text{town means}} = \frac{\text{MS}_{\text{town}}}{r} = \hat{\sigma}_T^2 + \frac{\hat{\sigma}^2}{r}.$$

We can therefore rewrite Equation 4.6 as

$$SE_{\hat{\beta}_1,\text{town means}} = \sqrt{\frac{\hat{\sigma}_T^2 + \dfrac{\hat{\sigma}^2}{r}}{S_{XX,\text{town means}}}}. \qquad (4.7)$$

This version of the formula makes explicit the dependence of the SE on both σ_T^2 and σ^2. Both variance components tend to reduce the precision of the estimate $\hat{\beta}_1$. The reduction in precision due to σ^2 can be overcome by increasing the value of r – that is, the variation among houses within each town will have less effect if more houses are sampled from each town. However, this strategy does not reduce the impact of σ_T^2, which can only be overcome by sampling more towns.

The estimate of the constant, $\hat{\beta}_0$, is of less interest than that of the slope, $\hat{\beta}_1$, but the SEs of the different estimates are presented here for completeness. The SE of $\hat{\beta}_0$ from the regression analysis ignoring towns is given by

$$\text{SE}_{\hat{\beta}_0} = \sqrt{\left(\frac{1}{n} + \frac{\bar{x}^2}{S_{XX}} \right)} \hat{\sigma} . \tag{4.8}$$

The numerical value is given by

$$\text{SE}_{\hat{\beta}_0} = \sqrt{\left(\frac{1}{33} + \frac{53.0253^2}{53.7878} \right)} \times 0.03615 = 1.375.$$

The corresponding value from the regression analysis on the town means is given by

$$\text{SE}_{\hat{\beta}_0} = \sqrt{\left(\frac{1}{n_{\text{town means}}} + \frac{\bar{x}^2_{\text{town means}}}{S_{XX,\text{town means}}} \right)} \hat{\sigma} \tag{4.9}$$

$$\text{SE}_{\hat{\beta}_0} = \sqrt{\left(\frac{1}{11} + \frac{53.0253^2}{17.92927} \right)} \times 0.02567 = 2.007.$$

The SE of the constant from the mixed-model analysis is given by

$$\text{SE}_{\hat{\beta}_0} = \sqrt{\left(\frac{1}{n_{\text{town means}}} + \frac{\bar{x}^2_{\text{town means}}}{S_{XX,\text{town means}}} \right) \left(\hat{\sigma}_T^2 + \frac{\hat{\sigma}^2}{r} \right)} , \tag{4.10}$$

which gives the same numerical value as Equation 4.9. As in the case of $\hat{\beta}_1$, the precision of the estimate is reduced by both σ_T^2 and σ^2. As before, the reduction in precision due to σ^2 can be overcome by increasing the value of r, but that due to σ_T^2 can only be overcome by sampling more towns.

4.3   SEs for differences between treatment means in the split plot design

We will next consider the precision of the treatment effects in the split plot experiment considered in Chapter 2, and how this is related to the variance components identified in that experiment. The tables of means and the accompanying SEs from this experiment (Section 2.4) are repeated here, for ease of reference.

Tables of means

Variate: saltiness

Grand mean	29.28

brand	A	B	C	D
	21.78	19.22	43.23	32.91

assessor	ALV	ANA	FAB	GUI	HER	MJS	MOI
	35.27	4.83	47.52	37.86	29.14	36.01	21.53

assessor	NOR	PER
	33.41	18.00

brand	assessor	ALV	ANA	FAB	GUI	HER	MJS
A		30.44	4.46	40.84	23.01	28.21	24.50
B		32.67	0.00	18.56	30.44	17.82	11.14
C		39.35	11.88	74.99	60.13	40.09	57.17
D		38.61	2.97	55.68	37.86	30.44	51.23

brand	assessor	MOI	NOR	PER
A		15.59	15.59	13.36
B		13.36	33.41	15.59
C		30.44	54.20	20.79
D		26.73	30.44	22.27

Standard errors of differences of means

Table	brand	assessor	brand assessor
rep.	27	12	3
s.e.d.	3.213	4.088	8.351
d.f.	6	64	66.70
Except when comparing means with the same level(s) of			
brand			8.176
d.f.			64

In a designed experiment, the natural focus of estimation is on *comparisons* between means, rather than individual means. The SEs of these differences have to take account of natural variation among main plots as well as among sub-plots in the same main plot. In the context of the experiment on the sensory evaluation of ravioli, this means that they must take account of σ_{Π}^2, the variance component representing variation among presentations, as well as σ^2, the component representing variation among servings in the same presentation. The resulting formulae are presented in Table 4.3. The numerical values in this table agree with those in the GenStat output, allowing for rounding error.

The considerations that determine which variance component(s) contribute to the precision of each comparison are similar to those that determine which stratum of the anova should be used to test the significance of each treatment term (Chapter 2, Section 2.2). The formulae in the table show that the precision of comparisons between brands is affected not only by σ^2, but also by σ_{Π}^2, because these comparisons always

Table 4.3 Standard errors of differences for the treatment terms in the split plot experiment to compare the perception of saltiness of commercial brands of ravioli by trained assessors.

Treatment term	Type of comparison	SE$_{\text{difference}}$ Formula	Numerical value
brand	—	$\sqrt{\dfrac{2}{ra}(a\hat\sigma_\Pi^2 + \hat\sigma^2)} = \sqrt{\dfrac{2}{ra}MS_{\text{Residual.day.presentation stratum}}}$	$\sqrt{\dfrac{2}{3\times9}(9\times4.3 + 100.3)}$ $= \sqrt{\dfrac{2}{3\times9}\times139.3} = 3.212$
assessor	—	$\sqrt{\dfrac{2}{rb}\hat\sigma^2} = \sqrt{\dfrac{2}{rb}MS_{\text{Residual.day.presentation.serving stratum}}}$	$\sqrt{\dfrac{2}{3\times4}\times100.3} = 4.089$
brand.assessor	between means with different levels of brand	$\sqrt{\dfrac{2}{r}(\hat\sigma_\Pi^2 + \hat\sigma^2)}$	$\sqrt{\dfrac{2}{3}(4.3 + 100.3)} = 8.351$
	between means with the same level of brand	$\sqrt{\dfrac{2}{r}\hat\sigma^2} = \sqrt{\dfrac{2}{r}MS_{\text{Residual.day.presentation.serving stratum}}}$	$\sqrt{\dfrac{2}{3}\times4.3} = 8.177$

where
b = number of brands
a = number of assessors
r = number of replications (days).

require the comparison of different presentations. However, the precision of comparisons between assessors is affected only by σ^2, because these comparisons are based entirely on comparison of servings in the same presentation. The precision of comparisons between brand.assessor means depends on the particular pair of brand.assessor combinations being compared. If they involve different brands, the comparison between them requires comparison of different presentations, and its precision depends on both σ^2 and σ_Π^2. However, if they involve the same brand, the comparison between them is based entirely on comparison of servings in the same presentation, and its precision depends on σ^2 only. For example, suppose that the treatment 'Brand D, Assessor HER' is to be compared with 'Brand D, Assessor GUI'. Both treatments involve the same brand, so $SE_{\text{Difference}} = 8.176$. On each day, this pair of treatments occurred in a single presentation. The values from the same presentation are compared, and the natural variation among presentations does not reduce the precision of the comparison. However, suppose that 'Brand B, Assessor ANA' is to be compared with 'Brand A, Assessor GUI'. These two treatments involve different brands, so $SE_{\text{Difference}} = 8.351$. On each day, these two treatments occurred in different presentations, and the natural variation among presentations reduces the precision of the comparison. The value of $SE_{\text{Difference}}$ is accordingly larger.

4.4   A significance test for the difference between treatment means

The point estimate of the difference between any pair of means, $\hat{\delta}$, can be compared with its standard error, $SE_{\hat{\delta}}$, to determine whether it is significant: that is, to assess the strength of the evidence against the null hypothesis

$$H_0 : \delta = 0$$

where δ is the difference between the true means. If H_0 is true, then the ratio between the point estimate and its SE is a random variable, referred to as T, that follows a t distribution. This statement can be expressed in symbolic shorthand as

$$T = \frac{\hat{\delta}}{SE_{\hat{\delta}}} \sim t.$$

The t distribution is compared with the standard normal distribution (introduced in Chapter 3, Section 3.11) in Figure 4.3. The two distributions are similar in shape, but the precise shape of the t distribution depends on the degrees of freedom (DF) of $SE_{\hat{\delta}}$, i.e. the precision with which the variance of $\hat{\delta}$, $\sigma_{\hat{\delta}}^2$, is estimated. If the DF were infinite, $\sigma_{\hat{\delta}}^2$ would be known exactly, and T would follow the standard normal distribution exactly. In the case of comparisons between brands, $SE_{\hat{\delta}}$ has 6 degrees of freedom, which is the DF for the residual MS against which the term 'brand' was tested in the anova (Chapter 2, Section 2.2). This rather small value gives a rather

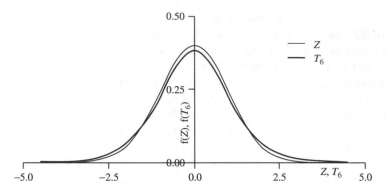

Figure 4.3 Comparison of the t distribution (DF = 6) with the standard normal distribution.

imprecise estimate of $\sigma_{\hat{\delta}}^2$, and a t distribution that is rather broader and flatter than the standard normal distribution, as shown in the figure.

In order to test H_0 using the t distribution, the observed value of the random variable T, given by the values of $\hat{\delta}$ and $SE_{\hat{\delta}}$ obtained from the data, and referred to as t, is first calculated. A parameter α is then specified, which defines a *critical region* of the t distribution, comprising an area of $\alpha/2$ in each of its tails. For a T variable with ν degrees of freedom, the *critical values*, at the boundaries of the critical region, are $\pm t_{\nu,\alpha/2}$. These ideas are illustrated in Figure 4.4 for $\alpha = 0.05$ and $\alpha = 0.01$, in a t distribution with 6 degrees of freedom. The hatched areas comprise the 5 % critical region of the distribution, an area totalling 0.05 in the two tails of the distribution. The cross-hatched areas comprise the 1 % critical region, specified in the same way.

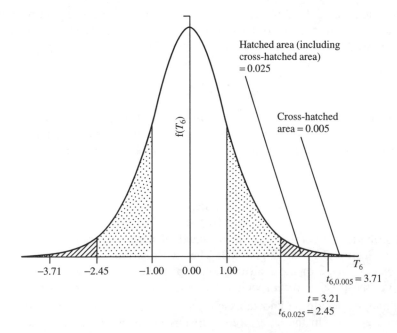

Figure 4.4 The t distribution (DF = 6), showing critical values for significance tests and the corresponding critical regions.

The evidence against H_0 is assessed by determining whether the observed value of t lies within the critical region for each value of α considered. For example, consider the comparison between the mean values of saltiness for Brands C and D. This gives

$$t = \frac{\text{Mean}_{\text{Brand C}} - \text{Mean}_{\text{Brand D}}}{\text{SE}_{\text{difference between brands}}} = \frac{43.23 - 32.91}{3.213} = \frac{10.32}{3.213} = 3.21.$$

This value exceeds $t_{6,0.025} = 2.45$, but not $t_{6,0.005} = 3.71$; that is, it lies in the critical region for $\alpha = 0.05$, but not in that for $\alpha = 0.01$, and the difference between these brands is significant at the 5 % level, but not at the 1 % level. The decision to subtract the mean saltiness of Brand D from that of Brand C is arbitrary. If the value for Brand C is subtracted from that for Brand D, the value $t = -3.21$ is obtained. This also lies in the critical region for $\alpha = 0.05$ but not in that for $\alpha = 0.01$, and the conclusion reached is the same. For a fuller introduction to the t distribution and its use for the comparison of means, see, for example, Moore and McCabe (1998, Chapter 7, Sections 7.1 and 7.2, pp 504–566).

Significance tests for comparisons between assessors, or between brand.assessor combinations, are made in the same way, using the appropriate value of DF. The DF for comparisons between assessors is determined in the same was as that for comparisons between brands: that is, it is given by the DF for the residual term against which 'assessor' is tested, namely.

$$\text{DF}_{\text{Residual,day.presentation.serving stratum}} = 64.$$

However, the DF for a comparison between brand.assessor combinations depends on the particular comparison considered. For comparisons that involve the same brand, the DF is that for the term against which 'brand.assessor' is tested: again,

$$\text{DF}_{\text{Residual,day.presentation.serving stratum}} = 64.$$

However, for comparisons that involve different brands, the precision of which depends on two variance components, the determination of the value of DF is less straightforward. It is obtained by the method of Satterthwaite (1946), which was applied to the split plot situation by Taylor (1950). (For an alternative reference on this topic, see Steel and Torrie, 1981, Chapter 16, Section 16.2, pp 381–382. Steel and Torrie's approach leads directly to a critical value for t, and does not obtain the degrees of freedom explicitly.) We will not examine this method in detail, but will note that the value obtained is always:

- greater than the minimum of the residual DF in any of the strata where effects contributing to the table are estimated: in the present case,

$$\min(\text{DF}_{\text{Residual,day.presentation stratum}}, \text{DF}_{\text{Residual,day.presentation.serving stratum}}) = \min(6, 64)$$
$$= 6;$$

and

- less than the sum of the residual DF in those strata: in the present case,

$$\text{DF}_{\text{Residual,day.presentation stratum}} + \text{DF}_{\text{Residual,day.presentation.serving stratum}} = 6 + 64 = 70.$$

The value given in the GenStat output, 66.70, meets these criteria. As noted earlier (Chapter 1, Section 1.2), the value of DF for a model term represents the number of independent pieces of information to which that term is equivalent. Normally, therefore, values of DF are integers. However, in this case, a non-integer value of DF is specified. As an example of a case in which this value is required, consider the comparison between 'Brand D, Assessor HER' and 'Brand A, Assessor GUI'. This gives the value

$$t = \frac{40.09 - 23.01}{8.351} = 2.0453.$$

The critical values of a t distribution with non-integer DF cannot be looked up in standard statistical tables. However, some items of software, including GenStat, are able to cope with this situation. The significance of this t statistic is given by the following GenStat statement:

```
PRINT CUT(2.0453; 66.70)
```

The function name CUT indicates that the Cumulative probability (the P value) for the Upper tail of the t distribution is to be obtained. The output of this statement is as follows:

```
CUT((2.045;66.7))
           0.02239
```

That is,

$$P(t_{66.70} > 2.0453) = 0.02239.$$

However, the corresponding value of t in the other tail of the distribution would be equally significant, and we must take this into account. The P value we require is therefore

$$P(t_{66.70} > 2.0453 \text{ or } t_{66.70} < -2.0453) = P(|t_{66.70}| > 2.0453)$$

$$= 2 \times 0.02239 = 0.04478.$$

The probability of obtaining so extreme a t value by chance is 0.04478, indicating that the difference between these means is significant at the 5 % level, but not at the 1 % level.

If no software that can deal correctly with non-integer values of DF is available, the P value required can be estimated by interpolation between the values given by the integers on either side of the actual value of DF, in this case 66 and 67. Thus

$$P(|t_{66}| > 2.0453) = 2 \times 0.02239 = 0.04482$$

and

$$P(|t_{67}| > 2.0453) = 2 \times 0.02239 = 0.04476.$$

The P value given by the actual DF is intermediate between those given by the integer DF values, as it should be.

4.5   The least significant difference between treatment means

As an alternative to the SE of the difference between two means, the *least significant difference* (LSD) can be calculated. This is done as follows. Consider the case

$$\frac{\hat{\delta}}{\mathrm{SE}_{\hat{\delta}}} = t_{v,\alpha/2}. \tag{4.11}$$

In this case, $\hat{\delta}$ is *just significant*: that is, this value of $\hat{\delta}$ is the least significant difference, $\mathrm{LSD}_{\hat{\delta}}$, the smallest value that will cause the observed value of t to lie in the critical region of the t distribution, providing significant evidence against H_0. Substituting $\mathrm{LSD}_{\hat{\delta}}$ for $\hat{\delta}$ in Equation 4.11, and rearranging, we obtain

$$\mathrm{LSD}_{\hat{\delta}} = t_{v,\alpha/2} \mathrm{SE}_{\hat{\delta}}. \tag{4.12}$$

For example, if we substitute into Equation 4.12 the 5 % critical value of t, namely:

$$t_{6,0.05/2} = 2.45,$$

and the SE for comparisons between brands, namely:

$$\mathrm{SE}_{\hat{\delta}} = 3.213,$$

we obtain the least significant difference between brands,

$$\mathrm{LSD}_{\hat{\delta}} = 2.45 \times 3.213 = 7.87.$$

The concept of the LSD is illustrated graphically in Figure 4.5. This number line represents the range of possible values for the true difference δ between the treatment means, with zero at the centre. The stippled areas indicate values outside the range $\pm\mathrm{SE}_{\hat{\delta}}$. Such values, if observed, provide at least tentative evidence against the null hypothesis

$$H_0 : \delta = 0.$$

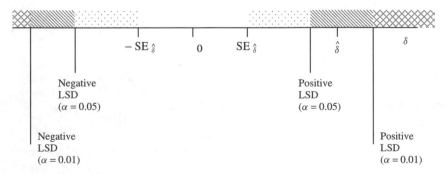

Figure 4.5   Comparison of LSDs based on two critical regions ($\alpha = 0.05$ and $\alpha = 0.01$) and the SE of the difference.

The hatched areas represent values outside the range $\pm$LSD ($\alpha = 0.05$). Such values provide significant evidence against H_0, at least at the 5 % level. This range is wider than the range $\pm$SE$_{\hat{\delta}}$ – that is, treatment means must differ by more than SE($\hat{\delta}$) in order to be significantly different. The cross-hatched areas represent values outside the range $\pm$LSD ($\alpha = 0.01$). Such values provide significant evidence against H_0, at least at the 1 % level. An observed value, $\hat{\delta}$, significant at the 5 % level but not at the 1 % level, is represented on the number line.

The following GenStat statement specifies that LSDs, rather than SEs, are to accompany the table of means from an anova:

```
ANOVA [FPROB = yes; PSE = lsd; LSDLEVEL = 5] saltiness
```

The option setting 'PSE = lsd' specifies that SEs are to be replaced by LSDs, and the option setting 'LSDLEVEL = 5' specifies that the LSDs are to be determined at the 5 % level of significance: that is, that the parameter α is to be set to 0.05. The output produced by this statement is the same as that presented in Section 4.3, except for the table of LDSs, which is as follows:

Least significant differences of means (5 % level)			
Table	brand	assessor	brand assessor
rep.	27	12	3
l.s.d.	7.861	8.167	16.670
d.f.	6	64	66.70
Except when comparing means with the same level(s) of			
brand			16.334
d.f.			64

If the option setting 'LSDLEVEL = 5' is changed to 'LSDLEVEL = 1', the table of LSDs is as follows:

Least significant differences of means (1 % level)			
Table	brand	assessor	brand assessor
rep.	27	12	3
l.s.d.	11.910	10.853	22.144
d.f.	6	64	66.70
Except when comparing means with the same level(s) of			
brand			21.706
d.f.			64

Note that each LSD at the 1 % level is larger than the corresponding LSD at the 5 % level, as indicated in Figure 4.5. Substituting the SE and LSDs for comparisons between brands into this figure, we obtain Figure 4.6. As was indicated by the corresponding significance test (Section 4.4), the observed difference between Brands C and D, $\hat{\delta} = 10.312$, is significant at the 5 % level, but not at the 1 % level.

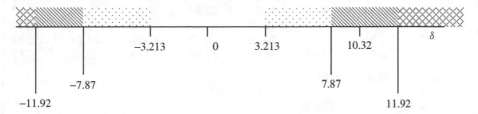

Figure 4.6   Numerical values of LSDs based on two critical regions ($\alpha = 0.05$ and $\alpha = 0.01$), and of the SE of the difference, in the split plot experiment to compare the perception of saltiness of commercial brands of ravioli by trained assessors.

In the case of a balanced experimental design such as this split plot design, for which GenStat's analysis of variance directives (BLOCKSTRUCTURE, TREATMENTSTRUC-TURE and ANOVA) can be used, these directives give a clear, simple and complete statement of the point and interval estimates of treatment means. However, we will also examine how the same information is displayed in the output from the mixed-modelling directives (VCOMPONENTS and REML), which can be applied to data sets that are not balanced. In the present case, the appropriate statements are:

```
VCOMPONENTS [FIXED = brand * assessor] \
    RANDOM = day / presentation / serving
REML [PRINT = model, components, Wald, means] saltiness
```

The first part of the output from the REML statement was presented in Chapter 2, Section 2.4. The tables of means are as follows:

Table of predicted means for Constant

29.28 Standard error: 1.855

Table of predicted means for brand

brand	A	B	C	D
	21.78	19.22	43.23	32.91

Standard error of differences: 3.213

Table of predicted means for assessor

assessor	ALV	ANA	FAB	GUI	HER	MJS	MOI	NOR
	35.27	4.83	47.52	37.86	29.14	36.01	21.53	33.41

assessor	PER
	18.00

Standard error of differences: 4.088

Table of predicted means for brand.assessor

assessor brand	ALV	ANA	FAB	GUI	HER	MJS	MOI
A	30.44	4.46	40.84	23.01	28.21	24.50	15.59
B	32.67	0.00	18.56	30.44	17.82	11.14	13.36
C	39.35	11.88	74.99	60.13	40.09	57.17	30.44
D	38.61	2.97	55.68	37.86	30.44	51.23	26.73

assessor brand	NOR	PER
A	15.59	13.36
B	33.41	15.59
C	54.20	20.79
D	30.44	22.27

Standard errors of differences

Average:	8.311
Maximum:	8.351
Minimum:	8.176

Average variance of differences: 69.08

Standard error of differences for same level of factor:

	brand	assessor
Average:	8.176	8.351
Maximum:	8.176	8.351
Minimum:	8.176	8.351

The means for the treatment terms are the same as those given by the corresponding anova, allowing for rounding error. The SEs are the same as those given by the anova, with the addition of a SE for the constant (equivalent to the grand mean from the ANOVA statement). However, the way in which the two SEs for the brand.assessor table are presented is not quite as neat as in the output of the ANOVA statement. The output first notes that the value of $SE_{Difference}$ varies according to the means being compared, and presents the average, maximum and minimum values. The average value presented does not apply to any individual comparison. A closely related value, the average variance of the differences, is also presented, because averaging of variances is more valid than averaging of SEs. This information might be the best available from an unbalanced data set, but in the present case we can interpret the various values of $SE_{Difference}$ in relation to the two treatment factors. The next part of the output does this, noting the values of $SE_{Difference}$ that occur when the means are compared have the same level of one or other of the treatment factors. As in the output of the ANOVA statement, these values show that comparisons that involve the same brand are made with greater precision. In the analysis of a balanced experimental design, as in the present case, there is only one value for each factor (8.176 for comparisons involving the same brand and 8.351 for comparisons involving the same assessor), but the output allows for the possibility of a range of values by presenting average, maximum and minimum values of $SE_{Difference}$.

4.6 SEs for treatment means in designed experiments: a difference in approach between analysis of variance and mixed-model analysis

Investigators often wish to present SEs of means, rather than of differences between means. Although this may appear to be a simpler idea, it raises issues that do not occur in relation to SEs of differences. We will therefore consider the calculation of SEs of means in the context of a simpler experimental design, namely the randomised block design.

In an experiment to test the effect of different manufacturing specifications (a range of combinations of particle size and compacting pressure) on the tensile strength of iron powder sinters, described by Bennett and Franklin (1954, Chapter 8, Section 8.52, pp 518–519), each combination of manufacturing specifications was tested in each of three furnaces. Thus the experiment had a randomised block design, with furnaces as the blocks and manufacturing specifications as the treatments. The data are presented in the spreadsheet in Table 4.4. (Data reproduced by permission of John Wiley & Sons, Inc.)

The manufacturing specifications comprised all combinations of two levels of compacting pressure (25 and 50) and six particle-size ranges (A to F). Thus the treatment effects could be partitioned into main effects of pressure and size, and pressure.size interaction effects. However, it is simpler, and sufficient for the present purpose, to regard the 12 pressure.size combinations as a single treatment factor. Thus the model to be fitted to these data (Model 4.13) is

$$y_{ij} = \mu + \phi_i + \varepsilon_{ij} + \gamma_{k|ij} \qquad (4.13)$$

where

y_{ij} = the value of tensile strength from the jth sample in the ith furnace (the ijth furnace.sample combination),

μ = the grand mean value of tensile strength,

ϕ_i = the effect of the ith furnace,

ε_{ij} = the effect of the jth sample in the ith furnace, and

$\gamma_{k|ij}$ = the effect of the kth manufacturing specification, being the specification tested in the ijth furnace.sample combination.

It is assumed that the block effects are random: that is, that the ϕ_i are independent values of a random variable Φ, such that

$$\Phi \sim N(0, \sigma_\phi^2),$$

and that the ε_{ij} are independent values of a random variable E, such that

$$E \sim N(0, \sigma^2).$$

Table 4.4 Tensile strengths of iron powder sinters manufactured to a range of specifications, investigated in an experiment with a randomised block design.

The conventions used in this spreadsheet are the same as those used in Table 1.1, Chapter 1.

	A	B	C
1	furnace!	mfng_spec!	tnsl_strngth
2	1	25A	11.3
3	1	25B	12.2
4	1	25C	12.9
5	1	25D	12.1
6	1	25E	16.9
7	1	25F	14.3
8	1	50A	21.1
9	1	50B	21.1
10	1	50C	21.7
11	1	50D	24.4
12	1	50E	23.6
13	1	50F	23.5
14	2	25A	11.9
15	2	25B	10.4
16	2	25C	12.4
17	2	25D	13.9
18	2	25E	14.9
19	2	25F	15.0
20	2	50A	21.3
21	2	50B	21.4
22	2	50C	22.0
23	2	50D	24.1
24	2	50E	25.5
25	2	50F	22.1
26	3	25A	10.0
27	3	25B	9.9
28	3	25C	11.3
29	3	25D	13.3
30	3	25E	12.4
31	3	25F	13.8
32	3	50A	18.8
33	3	50B	19.5
34	3	50C	21.6
35	3	50D	23.8
36	3	50E	23.3
37	3	50F	20.5

The following GenStat statements read the data and analyse them according to Model 4.13:

```
IMPORT 'Intro to Mixed Modelling\\Chapter 4\iron sinters.xls'; \
    SHEET = 'sheet1'
BLOCKSTRUCTURE furnace
TREATMENTSTRUCTURE mnfctrng_spec
ANOVA [FPROB = yes; PSE = differences, means] tnsl_strngth
```

The output of the ANOVA statement is as follows:

Analysis of variance

Variate: tnsl_strngth

Source of variation	d.f.	s.s.	m.s.	v.r.	F pr.
furnace stratum	2	15.6817	7.8408	10.38	
furnace.*Units* stratum					
mfng_spec	11	886.4367	80.5852	106.72	<.001
Residual	22	16.6117	0.7551		
Total	35	918.7300			

Message: the following units have large residuals.

furnace 1 *units* 4	−1.48	s.e.	0.68
furnace 1 *units* 5	1.69	s.e.	0.68
furnace 3 *units* 5	−1.40	s.e.	0.68

Tables of means

Variate: tnsl_strngth

Grand mean 17.45

mfng_spec	25A	25B	25C	25D	25E	25F	50A
	11.07	10.83	12.20	13.10	14.73	14.37	20.40

mfng_spec	50B	50C	50D	50E	50F
	20.67	21.77	24.10	24.13	22.03

Standard errors of means

Table	mfng_spec
rep.	3
d.f.	22
e.s.e.	0.502

Standard errors of differences of means

Table	mfng_spec
rep.	3
d.f.	22
s.e.d.	0.709

The F test for the term 'mnfctrng_spec' shows that the effect of the manufacturing specification on tensile strength is highly significant. The null hypothesis that there is no real variation among furnaces,

$$H_0 : \sigma_\phi^2 = 0,$$

is not explicitly tested in the GenStat output. However, the F value for this term, 10.38, provides such a test, namely:

$$P(F_{2,22} > 10.38) = 0.0006687.$$

Thus there is highly significant evidence against this null hypothesis, which can be rejected.

By default, GenStat presents only the SEs of differences between the treatment means. However, the option setting 'PSE = differences, means' in the ANOVA statement has caused the SEs both of differences and of means to be produced.

Before comparing the SEs of differences and means, we will obtain the corresponding mixed-model analysis of these data. This is produced by the following GenStat statements:

```
VCOMPONENTS [FIXED = mfng_spec] RANDOM = furnace
REML [PRINT = model, components, means, wald; PSE = differences] \
    tnsl_strngth
```

The REML statement produces the following output:

REML variance components analysis

Response variate: tnsl_strngth
Fixed model: Constant + mfng_spec
Random model: furnace
Number of units: 36

Residual term has been added to model

Sparse algorithm with AI optimisation

Estimated variance components

Random term	component	s.e.
furnace	0.5905	0.6537

Residual variance model

Term	Factor	Model(order)	Parameter	Estimate	s.e.
Residual		Identity	Sigma2	0.755	0.2277

Wald tests for fixed effects

Sequentially adding terms to fixed model

Fixed term	Wald statistic	d.f.	Wald/d.f.	chi pr
mfng_spec	1173.97	11	106.72	<0.001

Dropping individual terms from full fixed model

Fixed term	Wald statistic	d.f.	Wald/d.f.	chi pr
mfng_spec	1173.97	11	106.72	<0.001

Message: chi-square distribution for Wald tests is an asymptotic approximation (i.e. for large samples) and underestimates the probabilities in other cases.

Table of predicted means for Constant

17.45 Standard error: 0.467

Table of predicted means for mfng_spec

mfng_spec	25A	25B	25C	25D	25E	25F	50A	50B
	11.07	10.83	12.20	13.10	14.73	14.37	20.40	20.67

mfng_spec	50C	50D	50E	50F
	21.77	24.10	24.13	22.03

Standard error of differences: 0.7095

We need to check that the estimates of variance components from this analysis are consistent with the mean squares in the preceding anova. The relationships among these values are as shown in Table 4.5. The numerical equations are correct, allowing for rounding error.

It is not possible to specify in the same REML statement that SEs both of differences and of means are to be printed. However, the following additional statement will produce the SEs of means:

```
REML [PRINT = means; PSE = estimates] tnsl_strngth
```

The output of this statement is as follows:

Table of predicted means for Constant

17.45 Standard error: 0.467

Table of predicted means for mfng_spec

mfng_spec	25A	25B	25C	25D	25E	25F	50A
	11.07	10.83	12.20	13.10	14.73	14.37	20.40

mfng_spec	50B	50C	50D	50E	50F
	20.67	21.77	24.10	24.13	22.03

Standard error: 0.6697

The two types of SEs produced by each of the two methods of analysis are collated in Table 4.6. The formulae and the numerical values for the SE of differences agree

Table 4.5 Relationships between estimated variance components in the randomised block experiment on the effects of manufacturing specifications on the tensile strength of iron sinters. g = number of treatments (manufacturing specifications) = 12.

Algebraic relationship	Numerical relationship
$g\hat{\sigma}_\phi^2 + \hat{\sigma}^2 = \mathrm{MS_{Residual, furnace\ stratum}}$	$12 \times 0.5905 + 0.755 = 7.8408 = 7.8408$
$\hat{\sigma}^2 = \mathrm{MS_{Residual, furnace.*units*stratum}}$	$0.755 = 0.7551 = 0.7551$

Table 4.6 Comparison of SEs obtained by different methods of analysis of the randomised block experiment on the effects of manufacturing specifications on the tensile strength of iron sinters. r = number of replications (furnaces) = 3.

Method of analysis	SE of differences between means	SE of means
anova	$\sqrt{\dfrac{2}{r}}\hat{\sigma} = \sqrt{\dfrac{2}{3}} \times 0.755 = 0.709$	$\sqrt{\dfrac{1}{r}}\hat{\sigma} = \sqrt{\dfrac{1}{3}} \times 0.755 = 0.502$
mixed-model analysis	$\sqrt{\dfrac{2}{r}}\hat{\sigma} = \sqrt{\dfrac{2}{3}} \times 0.755 = 0.7095$	$\sqrt{\dfrac{1}{r}(\hat{\sigma}_\phi^2 + \hat{\sigma}^2)}$ $= \sqrt{\dfrac{1}{3} \times (0.5905 + 0.755)} = 0.6697$

between the two methods of analysis, but those for the SE of means do not. The SE of means from the anova depends on the residual variance only, whereas that from the mixed-model analysis depends also on the variation among blocks (furnaces). Since each treatment mean is based on information from all the blocks, the latter is the more natural view of the matter. The more variation there is among blocks, the less precise must be the treatment mean, unless it is intended to be *conditional on* the blocks, i.e. to apply only to future observations on the same blocks. This is an unrealistic restriction, since in most cases observations will never be made again on the blocks in question. But for the purpose of a *comparison* between treatment means, variation among blocks does not matter: this is why σ_ϕ^2 does not contribute to the SE of differences in either method of analysis. And even when the SE of means is presented in preference to the SE of differences, its true function is generally to assist the interpretation of comparisons between treatment means. It is in tacit recognition of this that σ_ϕ^2 is omitted from the calculation of the SE of means in the output produced by GenStat's analysis of variance directives. Mixed-model analysis is a tool for purists.

It the block term is specified as a fixed-effect term in mixed-model analysis, the SEs of means are the same as those obtained from analysis of variance, and it appears that the discrepancy between the two methods is resolved. However, the inadequacy of this approach is revealed when we return to consideration of the split plot design. In this case, specifying the main plot term as a fixed-effect term is not an option: to do so would result in entirely the wrong analysis. For example, in the case of the experiment considered in Chapter 2, if 'presentation' were included in the model in the TREATMENTSTRUCTURE statement, all terms would be tested against

$MS_{Resid,day.presentation.serving\ stratum}$, which would overestimate the precision of comparisons between brands. The question of which variance components should be taken into account when calculating SEs for estimates from mixed modelling is considered more fully by Welham *et al.* (2004).

4.7   Use of R to obtain SEs of means in a designed experiment

To prepare the data presented in Section 4.6 for analysis by R, they are transferred from the Excel workbook 'Intro to Mixed Modelling\Chapter 4\iron sinters.xls' to the text file 'iron sinters.dat' in the same directory. Exclamation marks (!) are removed from the ends of headings in this file.

The following commands import the data, fit Model 4.13, and produce the resulting anova:

```
ironsinters <- read.table(
   "Intro to Mixed Modelling\\Chapter 4\\iron sinters.dat",
   header=TRUE)
attach(ironsinters)
furnace <- factor(furnace)
mfng_spec <- factor(mfng_spec)
ironsinters.aov <- aov(tnsl_strngth ~ mfng_spec + Error(furnace))
summary(ironsinters.aov)
```

The output of these commands is as follows:

```
Error: furnace
          Df  Sum Sq Mean Sq F value Pr(>F)
Residuals  2 15.6817  7.8408
Error: Within
          Df Sum Sq Mean Sq F value      Pr(>F)
mfng_spec 11 886.44   80.59  106.72 < 2.2e-16 ***
Residuals 22  16.61    0.76
---
Signif. codes:  0 '***' 0.001 '**' 0.01 '*' 0.05 '.' 0.1 ' ' 1
```

The values are the same as those given by GenStat (Section 4.6), allowing for rounding. The following command produces the table of treatment means from this analysis:

```
model.tables(ironsinters.aov, type = "means", se = TRUE)
```

The output of this command is as follows:

```
Tables of means
Grand mean

17.45
```

```
   mfng_spec
mfng_spec
   25A     25B     25C     25D     25E     25F     50A     50B     50C
      50D     50E
11.067 10.833 12.200 13.100 14.733 14.367 20.400 20.667 21.767
   24.100 24.133
      50F
22.033
Warning message:
SEs for type   means   are not yet implemented in:
model.tables.aovlist(ironsinters.aov, type = "means", se = TRUE)
```

The mean values are the same as those given by GenStat. However, note that R is not able to give SEs for these means, though a message indicates that this may become possible in future.

The following commands use REML to fit Model 4.13, and summarise the results:

```
ironsinters.lme <- lme(tnsl_strngth ~ mfng_spec,
    random = ~ 1|furnace)
anova(ironsinters.lme)
```

Their output is as follows:

```
              numDF denDF   F-value p-value
(Intercept)       1    22 1398.0771  <.0001
mfng_spec        11    22  106.7246  <.0001
```

The F value for 'mfng_spec' and the DF are the same as those obtained by GenStat.

4.8   Summary

Each parameter estimate (slope, intercept or mean) is a *point estimate*, and can be enclosed in an *interval estimate*.

In order to obtain an interval estimate for a parameter β, the parameter estimate $\hat{\beta}$ is regarded as an observation of a random variable, with distribution

$$\hat{B} \sim N(\beta, \sigma_{\hat{\beta}}^2).$$

An estimate of $\sigma_{\hat{\beta}}^2$, designated $\hat{\sigma}_{\hat{\beta}}^2$, can be obtained from the data, and the standard error of $\hat{\beta}$ is given by

$$SE_{\hat{\beta}} = \hat{\sigma}_{\hat{\beta}}.$$

An interval estimate for the true parameter value β is given by $\hat{\beta} \pm SE_{\hat{\beta}}$.

Table 4.7 Variance components that contribute to each type of comparison between treatments in a split plot design.

Type of comparison[1]	Variance components that contribute to $SE_{difference}$
between levels of Factor A	main plot residual, sub-plot residual
between levels of Factor B	sub-plot residual only
between A.B combinations with different levels of Factor A	main plot residual, sub-plot residual
between A.B combinations with the same level of Factor A	sub-plot residual only

[1] Factor A = the treatment factor that varies only among main plots, not within each main plot.
Factor B = the treatment factor that varies among sub-plots within each main plot.

When several observations of a response variable Y are taken at each value of an explanatory variable X, an appropriate estimate of $SE_{\hat{\beta}}$ can be obtained either by performing a regression analysis on the group means, or by a mixed-model analysis.

If the data are analysed as if the observations within each group were mutually independent, $SE_{\hat{\beta}}$ will be underestimated, i.e. the precision of $\hat{\beta}$ will be overestimated.

The value of $SE_{\hat{\beta}}$ depends on the variance-component estimates for the terms that are relevant to the estimation of β. Larger variance-component estimates give a larger value of $SE_{\hat{\beta}}$.

In a designed experiment, the emphasis is on comparison between treatments. Therefore, SEs of differences between means are often presented, rather than SEs of the means themselves.

In a split plot experiment, several types of comparison can be made. The variance components that contribute to $SE_{difference}$ for each type are shown in Table 4.7.

A confidence interval for the difference between the true means, δ, is given by

$$\hat{\delta} \pm SE_{\hat{\delta}}.$$

The significance of the difference between treatment means can be determined by a t test.

The degrees of freedom of the t statistic depend on the number of pieces of information that effectively contribute to the term in question. The calculation of the appropriate value is straightforward, except in the case of a comparison between A.B combinations with different levels of Factor A, when a non-integer value must be calculated.

The difference between two means that gives a t value that is just significant at a particular level (say, $\alpha = 0.05$, the 5 % significance level) is called the least significant difference (LSD).

The LSD provides the basis for an alternative confidence interval for δ, namely

$$\hat{\delta} \pm LSD_{\hat{\delta}}.$$

In any model term, the variance components that contribute to SE_{mean} are not necessarily the same as those that contribute to $SE_{difference}$. For example, in a randomised block design, when the treatment means are considered and compared using GenStat's mixed-modelling directives, $\hat{\sigma}^2_{block}$ contributes to SE_{mean} but not to $SE_{difference}$.

When the treatment means are presented using GenStat's anova directives, $\hat{\sigma}^2_{block}$ does not contribute to either SE_{mean} or $SE_{difference}$. This is only valid if the means are intended to be conditional upon the blocks observed, which is unrealistic. However, SE_{mean} is usually presented in order to assist the comparison of means, and it is in tacit recognition of this that $\hat{\sigma}^2_{block}$ is omitted from both SEs.

4.9 Exercises

1. Refer to your results from Exercise 1 in Chapter 1, in which effects on the speed of greyhounds were modelled. Obtain standard errors for the estimates of the constant and of the effect of age that you obtained from your mixed model.

2. Refer to your results from Exercise 1 in Chapter 2. Examine the output produced by the analysis of variance to determine the effects of variety and nitrogen level in a split plot design.

   (a) Obtain the SE for the difference between the mean yields for the following factor levels or combinations of levels:

      (i) nitrogen Level 1 vs. nitrogen Level 3

      (ii) variety 'Victory' at nitrogen Level 1 vs. variety 'Victory' at nitrogen Level 3

      (iii) variety 'Victory' at nitrogen Level 1 vs. variety 'Golden Rain' at nitrogen Level 3.

   (b) Obtain the LSD at the 5 % level between variety means. Are there any two varieties that are significantly different according to this criterion?

   (c) Obtain the SE for nitrogen-level means on the following bases:

      (i) conditional on the blocks and main plots

      (ii) taking into account the random effects of blocks and main plots.

5

Estimation of random effects in mixed models: best linear unbiased predictors

5.1 The difference between the estimates of fixed and random effects

In ordinary regression analysis and anova the random-effect terms are always *nuisance variables*: residual variation or block effects. The effects of individual levels of such terms are not of interest. However, we have seen that in a mixed model, effects that are of intrinsic interest may be regarded as random – for example, the effects of the individual breeding lines in the barley field trial discussed in Chapter 3. The decision to treat a term as random causes a fundamental change in the way in which the effect of each level of that term is estimated. We will illustrate this change and its consequences in the context of the barley breeding lines. However, the concepts introduced and the arguments presented apply equally to any situation in which replicated, quantitative evaluations are available for the comparison of members of some population – for example, new chemical entities to be evaluated as potential medicines by a pharmaceutical company, or candidates for admission to a university on the basis of their examination scores.

It is easiest to illustrate the relationship between the estimates of fixed and random effects in data that are grouped by a single factor – for example, a fully randomised design leading to a one-way anova. The data from the barley field trial are classified by two factors, line and block. Fortunately, however, the effects of blocks are negligible (Chapter 3, Section 3.10). We will therefore treat this experiment as having a single-factor design, reducing Model 3.18 from

$$y_{ij} = \mu + \delta_i + \varepsilon_{ij} + \phi_{k|ij}$$

to Model 5.1, i.e.

$$y_j = \mu + \phi_{k|j} + \varepsilon_j \tag{5.1}$$

Introduction to Mixed Modelling: Beyond Regression and Analysis of Variance   N. W. Galwey
© 2006 John Wiley & Sons, Ltd

where

y_j = the grain yield of the jth plot,

μ = the grand mean (overall mean) value of grain yield,

$\phi_{k|j}$ = the effect of the kth breeding line, being the line sown in the jth plot,

ε_j = the residual effect of the jth plot.

We can then compare an analysis in which the variation among lines is modelled as a fixed-effect term with one in which it is modelled as a random-effect term. In the analysis of designed experiments, it is not normal practice to omit features of the design from the model fitted: it is justified in the present case solely in the interests of clarity.

The variation among lines in the original, 'unpadded' data set can be modelled as a fixed-effect term using GenStat's anova directives, as follows:

```
IMPORT \
    'Intro to Mixed Modelling\\Chapter 3\\barley progeny.xls'; \
    SHEET = 'sheet1'
BLOCKSTRUCTURE
TREATMENTSTRUCTURE line
ANOVA [FPROBABILITY = yes] yield_g_m2
```

The same model can be fitted using the mixed-modelling directives, with no random-effect model specified, as follows:

```
VCOMPONENTS [FIXED = line]
REML [PRINT = model, components, deviance, means; \
    PTERMS = 'constant' + line] yield_g_m2
```

In order to treat 'line' as a random-effect term, it is moved to the random-effect model in the VCOMPONENTS statement, thus:

```
VCOMPONENTS RANDOM = line
REML [PRINT = model, components, deviance, means, effects; \
    PTERMS = 'constant'+ line; METHOD = Fisher] yield_g_m2
```

In order to obtain standard errors of the differences between means, the option setting 'METHOD = Fisher' must be used (see Chapter 2, Section 2.5 and Chapter 10, Section 10.9).

The breeding-line means obtained from these two analyses are compared in Table 5.1. They are not the same: in the case of a high-yielding line such as Line 7, the random-effect mean is lower than the fixed-effect mean, whereas for a low-yielding line such as Line 16, the opposite is the case. The fixed-effect means are the simple means of the observations for the line in question. For example, the mean for Line 7 is

$$\frac{907.38 + 820.08}{2} = 863.73,$$

whereas that for Line 3, which occurs only in Block 1, is the single plot value 873.04. Each of these means is taken as an estimate of the true mean yield of the breeding

Table 5.1 Comparison between the estimates of mean yields of breeding lines of barley obtained when breeding line is regarded as either a fixed- or a random-effect model term.

Line	Fixed-eff. mean	Rand-eff. mean	Line	Fixed-eff. mean	Rand-eff. mean	Line	Fixed-eff. mean	Rand-eff. mean
1	654.5	639.9	29	769.4	734.4	57	182.0	299.6
2	483.3	510.2	30	353.3	419.3	58	478.5	495.2
3	873.0	782.5	31	196.4	263.1	59	594.0	590.2
4	719.1	674.9	32	753.1	721.0	60	428.8	454.3
5	799.0	730.7	33	808.5	766.7	61	542.0	547.4
6	802.4	761.6	34	656.3	641.4	62	730.5	682.9
7	863.7	812.1	35	464.5	483.6	63	659.4	644.0
8	468.2	486.7	36	927.5	864.5	64	664.8	648.4
9	681.0	661.7	37	724.0	697.1	65	766.1	731.7
10	760.8	727.4	38	591.0	587.8	66	767.0	732.5
11	580.8	579.4	39	179.3	249.0	67	554.9	558.0
12	436.6	477.5	40	489.6	504.3	68	355.7	394.2
13	508.1	519.5	41	429.9	455.2	69	423.1	449.6
14	922.7	860.6	42	673.0	655.2	70	950.5	883.5
15	600.1	591.7	43	757.6	724.8	71	582.7	580.9
16	225.4	286.9	44	695.4	673.6	72	212.2	276.1
17	755.5	723.1	45	547.8	552.2	73	355.3	420.7
18	315.8	361.3	46	675.6	657.3	74	520.4	529.6
19	523.2	532.0	47	192.4	306.9	75	617.3	603.8
20	632.5	614.4	48	715.6	690.2	76	344.5	413.2
21	514.6	532.1	49	826.9	781.8	77	357.9	422.6
22	340.7	410.5	50	803.4	762.4	78	260.1	315.5
23	711.6	686.9	51	719.3	693.3	79	378.4	412.8
24	567.2	568.2	52	603.0	593.8	80	483.0	498.9
25	428.6	472.0	53	559.1	563.2	81	309.7	356.3
26	609.8	603.2	54	834.4	787.9	82	251.3	348.1
27	747.1	716.2	55	555.3	560.5	83	280.4	368.4
28	679.9	647.5	56	479.1	495.7			

line in question. But the plant breeder knows that if he or she selects the highest-yielding lines this year for further evaluation, he or she is selecting on both genetic and environmental variation. The environmental component will not contribute to the selected lines' performance next year, and consequently, their mean performance will generally be somewhat lower. We have already seen how this leads to an expected genetic advance under selection that is smaller than the difference between the mean of the selected lines and that of the full set of lines (Chapter 3, Section 3.14). Mixed-model analysis provides a way of building the pessimism of the plant breeder more fully into the formal analysis of the data, giving a similar adjustment to the mean of each individual breeding line. This adjusted mean is the random-effect mean.

5.2   The method for estimation of random effects. The best linear unbiased predictor or 'shrunk estimate'

The adjustment to obtain the random-effect mean is made as follows. Following Model 5.1, the true mean of the kth breeding line is represented by

$$\mu_k = \mu + \phi_k. \tag{5.2}$$

In the table of means presented above, this value is estimated by

$$\hat{\mu}_k = \frac{\sum_{j=1}^{r_k} y_{kj}}{r_k} \tag{5.3}$$

where
y_{kj} = the jth observation of the kth breeding line,
r_k = the number of observations of the kth breeding line.

The overall mean of the population of breeding lines, μ, is estimated by

$$\hat{\mu} = 572.5.$$

Note that this is not quite the same as the mean of all the observations ($= 581.6$) or the mean of the line means ($= 569.1$). Although we are treating this estimate as given, and using it to explain the estimation of the effects of individual lines, these two estimation steps are really interconnected in a single process. Something more will be said about the estimation of μ in a moment.

Rearranging Equation 5.2, we obtain

$$\phi_k = \mu_k - \mu. \tag{5.4}$$

Similarly, an estimate of ϕ_k is given by

$$\hat{\phi}_k = \hat{\mu}_k - \hat{\mu}. \tag{5.5}$$

This ordinary estimate of the difference between a treatment mean and the overall mean is called the best linear unbiased estimate (BLUE). To allow for the expectation that high-yielding lines in the present trial will perform less well in a future trial – and that low-yielding lines will perform better – the BLUE can be replaced by a 'shrunk estimate' called the best linear unbiased predictor (BLUP). The formula for the required shrinkage is

$$\text{BLUP}_k = \text{BLUE}_k \cdot \text{shrinkage factor}_k = (\hat{\mu}_k - \hat{\mu}) \cdot \left(\frac{\hat{\sigma}_G^2}{\hat{\sigma}_G^2 + \dfrac{\hat{\sigma}_E^2}{r_k}} \right). \tag{5.6}$$

This relationship, combined with the constraint

$$\sum_{k=1}^{p} \text{BLUP}_k = 0, \tag{5.7}$$

where
p = number of breeding lines,

determines the value of $\hat{\mu}$ as well as those of the BLUPs. For Line 7, substituting the values given by Model 5.1, we obtain

$$\text{BLUP}_7 = (863.7 - 572.5) \cdot \left(\frac{30667}{30667 + \dfrac{13226}{2}} \right) = 239.54.$$

A new estimate of the mean for the kth breeding line is then given by

$$\hat{\mu}'_k = \hat{\mu}' + \text{BLUP}_k. \tag{5.8}$$

For Line 7,

$$\hat{\mu}'_7 = 572.5 + 239.54 = 812.04.$$

As Line 7 is relatively high yielding, its shrunk mean, 812.04, is lower than its unadjusted mean, 863.7.

The original estimates (BLUE_k and $\hat{\mu}_k$) are compared with the shrunk estimates (BLUP_k and $\hat{\mu}'_k$) for an arbitrary subset of the breeding lines (about a quarter of the total) in Figure 5.1. Note that BLUP_k is shrunk towards zero relative to BLUE_k, and $\hat{\mu}'_k$ is correspondingly shrunk towards the estimate of the overall mean: estimates above the overall mean are lowered, those below it, raised. (In practice, it is usually the values at one extreme that are of interest, e.g. the high-yielding breeding

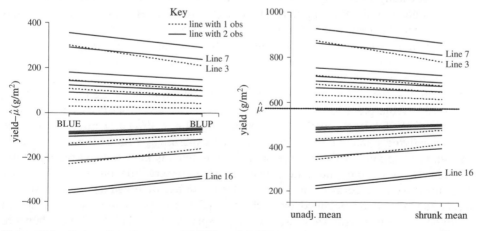

Figure 5.1 Comparison between BLUEs and BLUPs, and between unadjusted and shrunk means, for yields of breeding lines of barley.

lines.) Values far from the grand mean are shrunk more than those close to the mean. As should be expected, the amount of shrinkage specified by Equation 5.6 is large when:

- the genetic variance, $\hat{\sigma}_G^2$, is small;

- the environmental variance, $\hat{\sigma}_E^2$, is large;

- the number of replications of the breeding line under consideration, r_k, is small.

Provided that the number of replications is constant over breeding lines, the shrinkage of BLUPs does not change the ranking of the means. However, if the number of replications is unequal, *crossovers* may occur, as in the present case, where Line 3 is estimated to be higher yielding than Line 7 on the basis of their BLUEs, but lower yielding on the basis of their BLUPs.

5.3   The relationship between the shrinkage of BLUPs and regression towards the mean

The relationship between the shrunk and unadjusted means can also be illustrated by a scatter diagram, as shown in Figure 5.2. Each point represents a breeding line. The shrinkage towards the overall mean is indicated by the fact that the points representing breeding lines that have an estimated yield above $\hat{\mu}$ lie below the line

$$\text{shrunk mean} = \text{unadjusted mean},$$

whereas those representing breeding lines with an estimated yield below $\hat{\mu}$ lie above this line. That is, the points lie approximately along a line that is flatter than this line. Moreover, points based on a single observation lie on a flatter line than those based on two observations. The crossover of Lines 3 and 7 is indicated by the fact that the point representing Line 3 lies below, but to the right of, that representing Line 7.

   The flattening of the line of points on this scatter diagram is reminiscent of the commonly observed phenomenon of *regression towards the mean*, and this is no accident. An exploration of the connection between the two phenomena will help to clarify the distinction between the BLUE and the BLUP, and the sense in which each can be regarded as a 'best' statistic.

   Suppose that the values of σ_G^2 and σ_E^2 for a population of breeding lines are known with considerable precision from experiments like that just described. Then suppose that the yield of a large number of new breeding lines, drawn at random from the same underlying population, is measured in an experiment with a single replication. Though much is known about the population, not much is known about the new lines individually – only the information from a single observation on each. If another experiment were performed on the same large sample of breeding lines, a second observation would be obtained on each. The relationship between the present and future observations (designated Y_{obs} and Y_{new} respectively) would then be as shown in Figure 5.3. Each point represents the value of the present observation on an individual breeding line, and a possible value for the future observation. The figure shows that

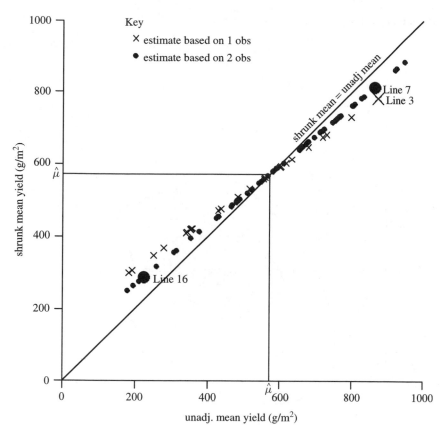

Figure 5.2   An alternative representation of the comparison between unadjusted and shrunk mean yields of breeding lines of barley.

the performance of each breeding line in the future experiment can be predicted from the available information, but not very accurately. High-yielding lines in the present experiment will generally give a high yield in the future experiment, and low-yielding lines a low future yield, but this relationship, which is due to the genetic value of each line, is blurred by the environmental component of each observation, present and future. The distribution of points in the figure – the probability distribution – is summarised by the ellipses. These are contour lines, each indicating a path along which the density of points (the probability density) is constant – a high density on the inner ellipse, a low density on the outer one. As all these density contours are the same shape, and concentric, a single contour is sufficient to indicate the general shape of the distribution, and this convention will be used in subsequent figures.

The criterion for the conversion of a BLUE to a BLUP, given in Equation 5.6, can also be represented graphically on this probability distribution, as follows. Consider Figure 5.4. The variance of the observed values (Y_{obs}) in the present experiment is $\sigma_G^2 + \sigma_E^2$, and the variance of observations in a future experiment (Y_{new}) will be the same. The genetic component of each observation contributes to new observations on the same breeding line, but the environmental component does not: hence the covariance between Y_{obs} and Y_{new} is given by

$$\text{cov}(Y_{obs}, Y_{new}) = \sigma_G^2. \tag{5.9}$$

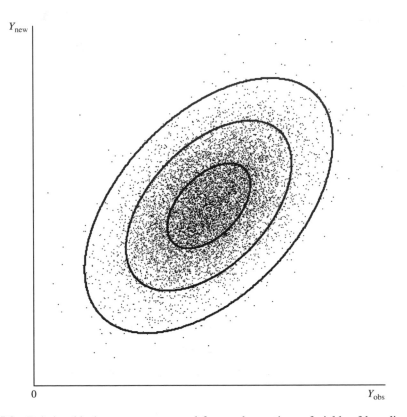

Figure 5.3   Relationship between present and future observations of yields of breeding lines of barley, when a single observation has been made on each line.

Each point represents the observations on a single line. Each ellipse is a contour, indicating a path along which the density of points (the probability density) is constant.

These variance and covariance values specify the bivariate-normal distribution of Y_{obs} and Y_{new}, and the ellipse in the figure represents the 1-standard-deviation probability density contour of this distribution. The simplest prediction of the future yield of any breeding line is its yield in the present experiment. This prediction is given by the equation

$$Y_{new} = Y_{obs},$$

which is represented by the line OA on the figure. For a breeding line that gives a yield y_{obs}, the point G is the corresponding position on OA, and the vertical coordinate of this point gives the prediction of the line's future yield. This is the unadjusted mean (based on a single observation in the present case), and the corresponding BLUE is given by its difference from the overall mean μ, i.e. the distance EG. But the figure

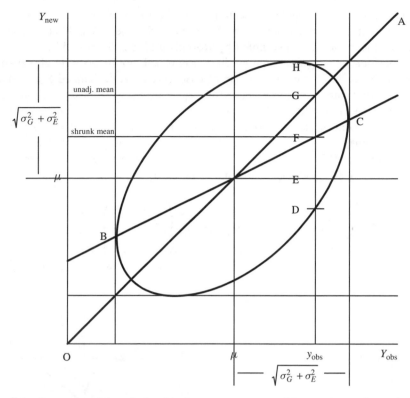

Figure 5.4 Summary of the relationship between present and future observations of yields of breeding lines of barley, when a single observation has been made on each line, showing the criterion for conversion of a BLUE to a BLUP.

For an explanation of the annotations on this figure, see the text.

shows that this will not be the mean future yield of all the lines that give a yield of y_{obs} in the present experiment. The distances DF and FH are equal: hence the distribution of Y_{new}, *among those lines for which*

$$Y_{obs} = y_{obs},$$

is symmetrical about the point F, and the vertical coordinate of this point is the mean value of this distribution. This is the shrunk mean, and the corresponding BLUP is given by its difference from μ, i.e. the distance EF. The point F lies on the line BC, which connects the leftmost and rightmost points on the ellipse, and which is the line of best fit obtained when Y_{obs} is treated as the explanatory variable, and Y_{new} as the response variable, in a regression analysis. For values of Y_{obs} above μ, BC lies below OA, and the expected value of Y_{new} is less than Y_{obs}. Conversely, for values of Y_{obs} below μ, BC lies above OA, and the expected value of Y_{new} is greater than Y_{obs}. That is, the BLUP is always shrunk towards μ, relative to the BLUE. This is the phenomenon of regression towards the mean noted by Francis Galton, which gave regression analysis its name and which was mistakenly interpreted as indicating that, over time, a biological population would converge to mediocrity unless steps were taken to prevent this. The connection between the shrinkage of BLUPs and regression towards the mean has also been noted by Robinson (1991, Section 5.2).

Now consider the situation when prediction is based, not on a single observation of each breeding line, Y_{obs}, but on the mean of r observations, designated $\overline{Y}_{obs}$. This case is illustrated in Figure 5.5. The variance of $\overline{Y}_{obs}$ is $\sigma_G^2 + \sigma_E^2/r$, less than that of Y_{obs}, due to the greater reliability of the mean of several observations and the consequent smaller contribution of environmental variation. However, the variance of individual new observations, Y_{new}, is unchanged. Consequently the ellipse is steeper, and so is the regression line BC: it is closer to the line OA. There is less regression towards the mean. Specifically, the regression line is given by

$$(Y_{new} - \mu) = (\overline{Y}_{obs} - \mu) \cdot \left(\frac{\sigma_G^2}{\sigma_G^2 + \dfrac{\sigma_E^2}{r}} \right). \tag{5.10}$$

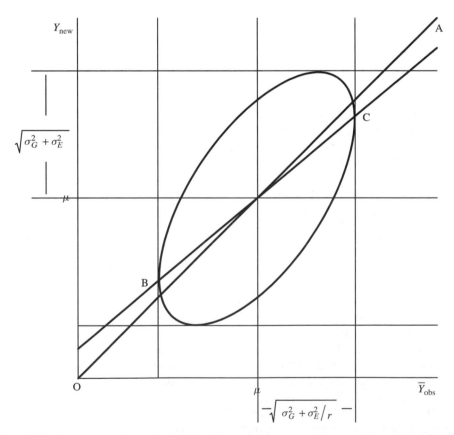

Figure 5.5 Summary of the relationship between present and future observations of yields of breeding lines of barley when r observations have been made on each line, showing the reduced discrepancy between BLUE and BLUP.

Comparison of this equation with Equation 5.6 shows that the amount of regression towards the mean is precisely equivalent to the shrinkage of the BLUP relative to the BLUE.

Finally, consider the situation when the prediction is based on a large – effectively infinite – number of observations, illustrated in Figure 5.6. The effect of environmental variation on $\overline{Y}_{obs}$ is eliminated and the variance of $\overline{Y}_{obs}$ is consequently reduced to σ_G^2, making the ellipse so steep that the regression line BC is superimposed on the line OA. $\overline{Y}_{obs}$ now needs no adjustment to give the expected value of Y_{new}: there is no regression towards the mean, and no shrinkage of the BLUE is required to obtain the BLUP. If the plant breeder could attain this situation (which would require unlimited experimental resources), he or she would have no need of pessimism in his or her predictions: they would be based on full knowledge of the genetic potential of each line, and would require no adjustment to indicate its future *mean* performance (though future individual observations would still vary as much as ever).

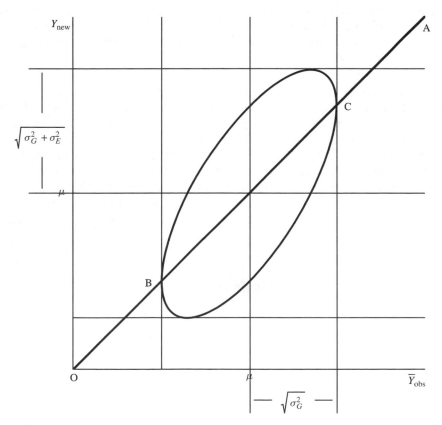

Figure 5.6 Summary of the relationship between present and future observations of yields of breeding lines of barley when an infinite number of observations have been made on each line, showing the equivalence of BLUE and BLUP in this case.

Except in this limiting case, the BLUE and the BLUP are different, so they cannot both be 'best' for the same purpose. For what purpose should each be preferred? The answer to this question lies in a comparison of the following two equations:

$$E(Y_{\text{new}} | \mu_k) = \mu_k = E(\hat{\mu} + \text{BLUE}_k) \tag{5.11}$$

$$E(Y_{\text{new}} | \bar{y}_k, \hat{\mu}, \hat{\sigma}_G^2, \hat{\sigma}_E^2) = \mu + (\bar{y}_k - \mu) \cdot \left(\frac{\sigma_G^2}{\sigma_G^2 + \dfrac{\sigma_E^2}{r_k}} \right) = E(\hat{\mu} + \text{BLUP}_k). \tag{5.12}$$

Translated into words, these equations state the following:

- Equation 5.11. The expected value of a new observation on breeding line k, given that the true mean value for this breeding line is the unknown value μ_k, is also the expected value of the unadjusted mean of the sample of observations on this breeding line.

- Equation 5.12. The expected value of a new observation on breeding line k, given the unadjusted mean for this line, *together with the specified information about the population of lines to which it belongs* ($\hat{\mu}, \hat{\sigma}_G^2$ and σ_E^2), is also the expected value of the shrunk mean.

Taken together, these statements tell us – perhaps surprisingly – that our best prediction about the future performance of Line k as a barley variety (or new chemical entity k as a medicine, or examination candidate k as a university student) depends on our reason for taking an interest in it. If we have chosen Line k because of its relative performance in the experiment under consideration – for example, because it is high yielding relative to the other lines studied – then we should be guided by the BLUP. However, if we have chosen it for reasons that have nothing to do with its performance in the present experiment, e.g. its brewing quality, or its resistance to some disease – in short, if we have chosen it 'because it is Line k' – then we should be guided by the BLUE.

5.4　Use of R for the estimation of random effects

The following R commands import the original, 'unpadded' data on the barley breeding lines, fit the model omitting the block term and treating 'line' as a random-effect term, and present the results:

```
barleyprogeny.unbalanced <- read.table(
    "Intro to Mixed Modelling\\Chapter 3\\barley progeny.dat",
    header=TRUE)
attach(barleyprogeny.unbalanced)
fline <- factor(line)
fblock <- factor(block)
barleyprogeny.model1lme <- lme(yield_gm2 ~ 1,
    data = barleyprogeny.unbalanced, random = ~ 1|fline)
summary(barleyprogeny.model1lme)
coef(barleyprogeny.model1lme)
```

The function `coef()` displays the coefficients of random-effect terms in the model specified as its argument – in the present case, the BLUPs for the breeding lines. The coefficients for the fixed-effect terms – in this case, the intercept only – are included in the output of the function `summary()`.

The output of these commands is as follows:

```
Linear mixed-effects model fit by REML
Fixed: yield_g_m2 ~ 1
 Data: barleyprogeny.unbalanced
        AIC        BIC      logLik
 1878.365  1887.232  -936.1823

Random effects:
 Groups    Name          Variance  Std.Dev.
 fline     (Intercept)   30666     175.12
 Residual                13223     114.99
# of obs: 142, groups: fline, 83

Fixed effects:
             Estimate  Std. Error  DF  t value  Pr(>|t|)
(Intercept)   572.45       21.67  141   26.417  < 2.2e-16  ***
---
Signif. codes:  0 '***' 0.001 '**' 0.01 '*' 0.05 '.' 0.1 ' ' 1
> coef(barleyprogeny.model11me)
$fline
     (Intercept)
1      67.496691
2     -62.501341
3     209.999035
4     102.396322
5     158.293836
6     189.247733
7     239.428906
8     -85.926076
9      89.296708
10    155.108083
.
.
.
82  -224.604129
83  -204.341281
```

The shrunk mean for each line can be obtained by substituting the estimate of the intercept and the appropriate BLUP into Equation 5.2. For example, for Line 7,

$$\text{shrunk mean} = 572.45 + 239.428906 = 811.88.$$

This is almost, but not exactly, the same as the value given by GenStat (812.1, Table 5.1).

5.5 Summary

In ordinary regression analysis and analysis of variance random-effect terms are always regarded as nuisance variables – residual variation or block effects – but this is not always appropriate. Effects that are of intrinsic interest may be regarded as random.

The decision to treat a term as random causes a fundamental change in the way in which the effect of each of its levels is estimated.

This is illustrated in the field experiment to evaluate breeding lines of barley (Chapter 3, Sections 3.7 to 3.15), but the same concepts and arguments apply equally to any situation in which replicated, quantitative evaluations are available for the comparison of members of some population, e.g.:

- new chemical entities to be evaluated as potential medicines;
- candidates for admission to a university on the basis of their examination scores.

We saw earlier (Chapter 3, Section 3.14 and Summary) that the predicted genetic advance due to selection among the barley breeding lines is less than the value given by a naive prediction from the mean of the selected lines. Mixed-model analysis gives a similar adjustment to the mean of each individual breeding line.

When a term is regarded as random, the ordinary difference between the mean for each level and the grand mean (the best linear unbiased estimator, BLUE) is replaced by a 'shrunk estimate' (the best linear unbiased predictor, BLUP).

The shrinkage of the BLUP, relative to the BLUE, is large when:

- the variance component for the term in question is small;
- the residual variance is large;
- the number of replications of the factor level under consideration is small.

Crossovers may occur as a result of the shrinkage, so that a level of a random-effect term that is ranked higher than another on the basis of the BLUEs is ranked lower on the basis of the BLUPs.

The shrinkage of the BLUP is equivalent to the phenomenon of regression towards the mean. This is illustrated by considering the present observations of each level of a random-effect term as predictions of future observations. The line of best fit relating present observations to future observations (the regression line) is flatter than a line of unit slope passing through the origin.

The degree of flattening of the line of best fit (the amount of regression towards the mean) is equivalent to the shrinkage of the BLUP relative to the BLUE.

The discrepancy between the BLUE and the BLUP tells us that our best prediction about the future performance of any level of a random-effect term depends on our reason for taking an interest in it, as follows:

- if we have chosen the level in question because of its relative performance in the experiment under consideration, we should be guided by the BLUP;
- however, if we have chosen it for reasons unrelated to its performance in the present experiment, we should be guided by the BLUE.

5.6 Exercises

1. Return to the data set concerning the yield of F_3 wheat families in the presence of ryegrass, introduced in Exercise 3 in Chapter 3.

   (a) Using mixed modelling, obtain an estimate of the mean yield of each of the F_3 families, regarding 'family' as a fixed-effect term.

   (b) Using mixed modelling and regarding 'family' as a random-effect term, obtain the following:

   (i) an estimate of the overall mean of the population of F_3 families

   (ii) the BLUP for the effect of each family.

   (c) From the output of the analysis performed in Part (b), obtain the following:

   (i) an estimate of the variance component for 'family'

   (ii) an estimate of the residual variance component.

   Obtain also the number of observations of each family.

   (d) From the information obtained in Part (a)–(c), compare the relationship between the BLUPs and the estimates of family means obtained regarding 'family' as a fixed-effect term with that given in Equation 5.6.

   (e) Obtain the shrunk mean for each family, and plot the shrunk means against the means obtained regarding 'family' as a fixed-effect term (the unadjusted means).

   The point representing one of the families deviates from the general relationship between these two types of mean.

   (f) What is the distinguishing feature of this family?

2. In many types of plant, exposure to low temperature at an early stage of development causes flowering to occur more rapidly: this phenomenon is called *vernalisation*. An inbred line of chickpea with a strong vernalisation response and a line with little or no vernalisation response were crossed, and the F_1 hybrid progeny were self-fertilised to produce the F_2 generation. Each F_2 plant was self-fertilised to produce an F_3 family. The seed of each F_3 family was divided into two batches. Germinating seeds of one were vernalised by exposure to low temperature ($4°C$) for four weeks. The other batch provided a control. All F_3 seeds were then sown, and allowed to grow. The plants were arranged in groups of four: within each group the plants were of the same family and had received the same low-temperature treatment. Generally there were 12 plants (i.e. three groups of four) in each family exposed to each low-temperature treatment, but in some families fewer or more plants were available. The number of days from sowing to flowering was recorded for each plant. The first and last few rows of the spreadsheet holding the data are presented in Table 5.2: the full data set is held in the file 'chickpea vernalisation.xls' (www.wiley.com/go/mixed_modelling). (Data reproduced by kind permission of S. Abbo, Field Crops and Genetics, The Hebrew University of Jerusalem.)

   (a) Divide the data between two spreadsheets, one holding only the results from the vernalised plants, the other, only those from the control plants.

Table 5.2   Time from sowing to flowering of F_3 chickpea plants with and without exposure to a vernalising stimulus.

	A	B	C	D	E
1	plant_ group	plant	family	low_ T	days_ to_ flower
2	60	1	88	vernalised	63
3	60	2	88	vernalised	64
4	60	3	88	vernalised	61
5	60	4	88	vernalised	70
6	1	1	14	vernalised	75
7	1	2	14	vernalised	
8	1	3	14	vernalised	
9	1	4	14	vernalised	
10	103	1	29	control	78
11	103	2	29	control	82
12	103	3	29	control	82
13	103	4	29	control	88
.					.
.					.
.					.
1092	41	3	44	vernalised	70
1093	41	4	44	vernalised	64

(b) Analyse the results from the vernalised plants by mixed modelling, regarding 'family', 'group' and 'plant' as random-effect terms.

(c) Obtain an estimate of the component of variance for each of the following terms:

(i) family

(ii) group within family

(iii) plant within group.

Which term in your mixed model represents residual variation?

(d) Estimate the heritability of time to flowering in vernalised plants from this population of families. (Note that the estimate obtained using the methods described in Chapter 3 is slightly biased downwards, as some of the residual variance is due to genetic differences among plants of the same family.)

(e) Obtain the unadjusted mean and the shrunk mean, and the BLUP, for the number of days from sowing to flowering in each family.

(f) Extend Equation 5.6 to the present situation, in which two components of variance contribute to the shrinkage of the BLUPs. Use the values obtained above to check your equation.

(g) Repeat the steps indicated in Parts (b)–(d) of this exercise for the control plants. Comment on the difference between the estimates of variance components and heritability obtained from the vernalised plants and the control plants.

(h) Plot the shrunk mean for each family obtained from the control plants against the corresponding value obtained from the vernalised plants. Comment on the relationship between the two sets of means.

6

More advanced mixed models for more elaborate data sets

6.1 Features of the models introduced so far: a review

The mixed models introduced in the examples presented so far are reviewed in Table 6.1. These examples illustrate several features of the range of mixed models that can be specified. Each model term may be a variate, like 'latitude', in which each observation can have any numerical value, or a factor, like 'town', in which each observation must come from a specified set of levels. Each part of the model (the fixed-effect model and the random-effect model) may have more than one term, as in the random-effect model 'block + line'. Factors may be *crossed*, as in the model 'brand * assessor', which is equivalent to

$$\text{brand} + \text{assessor} + \text{brand.assessor} = \text{main effects of brand}$$
$$+ \text{ main effects of assessor} + \text{brand} \times \text{assessor interaction effects.}$$

Alternatively, they may be *nested*, as in the model 'day / presentation / serving', which is equivalent to

$$\text{day} + \text{day.presentation} + \text{day.presentation.serving} = \text{main effects of day}$$
$$+ \text{ effects of presentation within each day}$$
$$+ \text{ effects of serving within each day.presentation combination.}$$

Crossing and nesting of factors is specified using the notation of Wilkinson and Rogers (1973).

Introduction to Mixed Modelling: Beyond Regression and Analysis of Variance   N. W. Galwey
© 2006 John Wiley & Sons, Ltd

Table 6.1 Mixed models introduced in earlier chapters.

Example	Response variable	Fixed-effect model	Random-effect model	Design illustrated	Where discussed
effects on house prices in England	logprice	latitude	town	regression analysis on grouped data	Chapter 1; Chapter 3, Section 3.16; Chapter 4, Section 4.2
sensory evaluation of ravioli	saltiness	brand * assessor	day/ presentation / serving	split plot design	Chapter 2; Chapter 3, Section 3.17; Chapter 4, Sections 4.3–4.5
effects on strength of a chemical paste	strength	–	delivery / cask / test	hierarchical design	Chapter 3, Sections 3.2–3.6
genetic variation in a barley field trial	yield_g_m2	–	block + line	randomised block design	Chapter 3, Sections 3.7–3.15; Chapter 5
tensile strength of iron powder sinters	tnsl_strngth	furnace	mfng_spec	randomised block design	Chapter 4, Section 4.6

6.2 Further combinations of model features

More elaborate combinations of these features are permitted. More than two factors may be crossed, thus:

$$a * b * c = a + b + c + a.b + a.c + b.c + a.b.c. \tag{6.1}$$

The three-way interaction term 'a.b.c' represents variation among the individual combinations of levels of 'a', 'b' and 'c' that is not adequately represented by the main effects or the two-way interaction terms. If factors 'a', 'b' and 'c' each have two levels, it is 'the difference of a difference of a difference' (Mead, 1988, Chapter 13, Section 13.1, p 348).

Factors and variates may be used in the same part of the mixed model (the random-effect model or the fixed-effect model), and factors may be crossed or nested with variates. For example, if 'f' is a factor and 'v' is a variate, the model

$$f * v = f + v + f.v \tag{6.2}$$

or

$$f/v = f + f.v \tag{6.3}$$

is permitted. These models represent a set of sloping lines relating the response variate to the explanatory variate 'v'. The term 'f' represents variation in the intercept of the line among the levels of this factor. The term 'v' represents the mean slope of the line, averaged over the levels of 'f'. In the crossed model 'f * v', the term 'f.v' represents

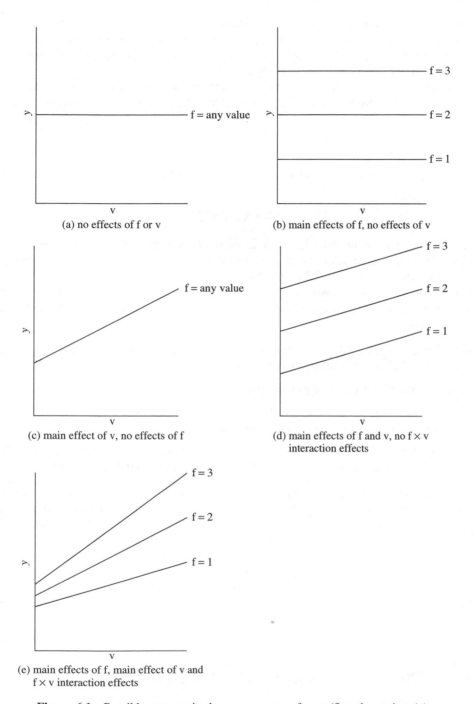

Figure 6.1 Possible patterns in the response to a factor (f) and a variate (v).

variation among the slopes at different levels of 'f' around the average slope. In the nested model 'f/v', in which no main effect of 'v' is estimated, the term 'f.v' represents all the variation among the slopes at different levels of 'f'. The meaning of each term in the model 'f ∗ v' is illustrated in Figure 6.1.

Variates may be crossed with other variates. In this case, the interaction term is equivalent to the product of the two variates: that is, if v1 and v2 are variates, and

$$v12 = v1 \times v2,$$

then the model

$$v1 * v2 = v1 + v2 + v1.v2 \tag{6.4}$$

is equivalent to

$$v1 + v2 + v12.$$

A fuller account of models that contain products and powers (squares, cubes, etc.) of explanatory variates, known as *polynomial models*, is given by Draper and Smith (1998, Chapter 12, Section 12.1, pp 251–253).

In all the examples considered so far, the significance of each fixed-effect model term is tested by comparing the mean square for that term with one other mean square. For example:

- In the regression analysis on grouped data, the effect of latitude is tested by comparing $MS_{latitude}$ with MS_{town} (Chapter 1, Section 1.5).

- In the analysis of the split plot design:

  - the effects of the brand of ravioli are tested by comparing MS_{brand} with $MS_{Residual}$ in the 'day.presentation' stratum (the main plot stratum);

  - the effects of assessor are tested by comparing $MS_{assessor}$ with $MS_{Residual}$ in the 'day.presentation.serving' stratum (the sub-plot stratum) (Chapter 2, Section 2.2).

This confinement of each fixed-effect term to a single stratum of an anova is a consequence of the *balance* of these data sets: it is not a universal feature of mixed models. If there were some towns (presumably large ones) within which houses had been sampled at more than one latitude, the effect of latitude would have had to be estimated both among and within towns, the information being pooled over these strata. If some assessors had been omitted from certain presentations of ravioli, the effects of assessor would have had to be estimated both between and within presentations. In a mixed-model analysis, these adjustments are made automatically, but because they are necessary, it is only in special cases that an equivalent analysis can be specified as a regression analysis or an anova.

6.3   The choice of model terms to be regarded as random

When constructing a mixed model, it is of course necessary to decide which terms should be allocated to the random-effects part. The first steps in this decision-making process are as follows:

- It is generally sensible to treat all variates as fixed-effect terms. To treat a variate as a random-effect term is to assume that the associated slope (the corresponding parameter estimate in the regression model) is a single observation from an infinite population of possible slopes: this assumption is occasionally helpful, but not often.

- Factors should be classified as:

  - Terms representing nuisance effects (block effects). All block terms should normally be regarded as random.

  - Terms representing effects of intrinsic interest (treatment effects). These may be fixed or random.

This leaves the question of whether a factor representing effects of intrinsic interest should be regarded as fixed or random.

If a factor is regarded as a random-effect term, it is assumed that the levels of this factor that have been studied are a random sample from an infinite population of possible levels. Hence we need to decide whether the levels of a factor of interest have been chosen in this way, and, if so, whether we prefer to *regard* them as such a sample, or as levels of individual interest a priori. The assumption that the factor levels studied comprise a random sample is stringent, but it greatly extends the range of inferences that we can make from the available data (see Chapter 1, Section 1.6). Quite often, the assumption turns out to be more reasonable than it might at first appear: see for example the justification for regarding block effects as random given in the account of the split plot design (Chapter 2, Section 2.6). Even if the assumption that factor levels are randomly sampled from an infinite population is not strictly valid, it may still be a reasonable approximation. Factor levels often comprise a representative sample from a finite set, e.g. 'English towns', 'barley breeding lines derived from the cross Chebec × Harrington'. Provided that the set of available levels (towns or breeding lines as the case may be) is several times larger than the set actually studied, such a factor can reasonably be regarded as a random-effect term. It may be appropriate to think of the levels studied as being representative not only of those actually available for study, but also of those levels that *might have* existed but that happen not to. This is more persuasive in the case of crop breeding lines than of towns. Alternatively, such a set of factor levels may be regarded as fixed, being chosen for their intrinsic interest. If one wants to know about house prices in Bradford 'because it is Bradford' (e.g. 'Because my employer has asked me to move there' or 'Because my aunt lives there'), not because it is representative of something else (e.g. 'Because it is in the north' nor 'Because houses there are inexpensive'), then the ordinary sample mean, obtained by treating town effects as fixed, gives the best available prediction (cf. the explanation of the different purposes for which the BLUE and the BLUP are 'best', given in Chapter 5, Section 5.3).

One useful test to determine whether a term should be regarded as random is to ask whether, if one of the levels chosen were replaced by a new level, the study would be essentially unchanged. The number of distinct breeding lines that could be produced from the cross Chebec × Harrington is effectively infinite, and it would clearly make no essential different to the study of this population of lines (Chapter 3, Sections 3.7–3.15) if one of those sampled were removed, and replaced by a line not currently included. However, in the study of commercial brands of ravioli (Chapter 2), the brands included were deliberately selected, and were of individual interest to manufacturers, retailers and consumers. To replace one of them with a newly chosen brand would effectively make it a different study. When we took the decision to treat 'town' as a random-effect term in the study of the effect of latitude on house prices (Chapter 1), we effectively asserted that if, for example, Bradford were replaced by Swindon, the study would be essentially unchanged.

Although additional assumptions are made when a factor is regarded as a random-effect term, this should not necessarily be thought of as a radical step: on the contrary, the inferences made about other terms are more cautious than when the factor is regarded as fixed. One consequence of this is that all block terms should normally be regarded as random, whether or not it is easy to envisage the infinite population from which the blocks studied were sampled: blocks are normally intended to be a representative (if not entirely random) sample of some range of realistic circumstances. The problems that can arise if block effects are regarded as fixed are considered in more detail in Section 6.5.

When the levels of a factor are deliberately chosen values of a continuous variable (e.g. different doses of a drug in a pharmacological study, or different levels of application of a chemical fertiliser in an agricultural experiment), the factor should be regarded as a fixed-effect term. The levels of such a factor do not constitute a random sample: there is no underlying distribution that indicates the frequency with which each value is expected to occur.

Decisions on whether a term is fixed or random need usually only be made for main effects. For interaction terms and some nested terms, the choice is determined by the following rules:

- *Rule 1.* When two terms are crossed, if both terms are fixed, the interaction must be fixed: otherwise, it must be random. That is, in the model

$$A * B = A + B + A.B$$

  – if A and B are both fixed, then A.B must be fixed;

  – if either A or B is random, or both are random, then A.B must be random.

- *Rule 2.* When one term is nested within another, the nested term may be either fixed or random if the term within which it is nested is fixed; otherwise, it must be random. That is, in the model

$$A/B = A + A.B$$

  – if A is fixed, then A.B may be fixed or random;

  – if A is random, then A.B must be random.

6.4 Disagreement concerning the appropriate significance test when fixed- and random-effect terms interact

There is an inconsistency in the statistical literature concerning the method for construction of the F test when a fixed-effect term and a random-effect term interact. We will here review the two methods that have been proposed, and consider which of them corresponds to the results obtained from mixed modelling.

Consider an experiment in which all combinations of two factors, A with n_A levels and B with n_B levels, are studied in r replications. A response variable Y is measured on each experimental unit. Thus the model to be fitted is

$$A * B \tag{6.5}$$

in the notation of Wilkinson and Rogers (1973), or in algebraic notation,

$$y_{ijk} = \mu + \alpha_i + \beta_j + (\alpha\beta)_{ij} + \varepsilon_{ijk} \tag{6.6}$$

where
y_{ijk} = the observation of the response variable on the kth replication of the ith level of A and the jth level of B,
μ = the grand mean (overall mean) value of the response variable,
α_i = the effect of the ith level of A,
β_j = the effect of the jth level of B,
$(\alpha\beta)_{ij}$ = the interaction between the ith level of A and the jth level of B, and
ε_{ijk} = the residual effect, i.e. the deviation of y_{ijk} from the value predicted on the basis of μ, α_i, β_j and $(\alpha\beta)_{ij}$.

It is assumed that the ε_{ijk} are independent values of a random variable E, such that

$$E \sim N(0, \sigma^2).$$

It is decided to regard A as a fixed-effect term and B as a random-effect term. The values β_j are then interpreted as independent values of a random variable B, such that
$$B \sim N(0, \sigma_B^2).$$

It follows from Rule 1 in Section 6.3 that the interaction term A.B is also a random-effect term: that is, the values $(\alpha\beta)_{ij}$ are interpreted as independent values of a random variable (AB), such that
$$(AB) \sim N(0, \sigma_{AB}^2).$$

The variance of the fixed effects of A is designated by κ_A.

The issue on which authorities differ is the expected mean squares (EMSs) that are implied by this model. The alternative views are illustrated by the accounts of Snedecor and Cochran (1989, Chapter 16, Section 16.14, pp 319–324) and Ridgman (1975, Chapter 8, pp 136–137). These are presented in Table 6.2, together with the F tests that they imply. Snedecor and Cochran considered that the interaction component of variance, σ_{AB}^2, does not contribute to the EMS for the main effect of B. It follows that the main effect of B should be tested against the residual term, whereas the interpretation of Ridgman indicates that this term should be tested against the A.B interaction. The other F tests are the same on either interpretation.

Table 6.2   Expected mean squares obtained when a fixed- and a random-effect factor interact, according to two authorities.

A is the fixed-effect factor and B the random-effect factor. The notations of the earlier authors have been adapted to conform to that of the present account. The pairs of mean squares to be compared by F tests are indicated by angled lines. The lines representing tests of the main effects of B are solid, for greater prominence: those representing other tests are dotted.

Source of variation	DF	Expected mean square	
		Snedecor and Cochran, 1989	Ridgman, 1975
A	$n_A - 1$	$n_B r \kappa_A^2 + r\sigma_{AB}^2 + \sigma_E^2$	$n_B r \kappa_A^2 + r\sigma_{AB}^2 + \sigma_E^2$
B	$n_B - 1$	$n_A r \sigma_B^2 + \sigma_E^2$	$n_A r \sigma_B^2 + r\sigma_{AB}^2 + \sigma_E^2$
A.B	$(n_A - 1)(n_B - 1)$	$r\sigma_{AB}^2 + \sigma_E^2$	$r\sigma_{AB}^2 + \sigma_E^2$
Residual	$n_A n_B (r - 1)$	σ_E^2	σ_E^2

We will now determine which of these interpretations is implemented in the software systems GenStat and R. The spreadsheet in Table 6.3 contains a small data set suitable for fitting Model 6.5.

The following GenStat statements read the data and analyse them according to Model 6.5, using anova methods:

```
IMPORT \
    'Intro to Mixed Modelling\\Chapter 6\\fixed A, random B.xls'
BLOCKSTRUCTURE B + A.B
TREATMENTSTRUCTURE A
ANOVA [FPROBABILITY = yes; PRINT = aovtable] y
```

The output of the ANOVA statement is as follows:

Analysis of variance

Variate: y

Source of variation	d.f.	s.s.	m.s.	v.r.	F pr.
B stratum	2	3978.08	1989.04	13.66	
B.A stratum					
A	3	5179.46	1726.49	11.85	0.006
Residual	6	873.92	145.65	5.40	
B.A.*Units* stratum	12	323.50	26.96		
Total	23	10354.96			

The fixed-effect model term A is represented explicitly in this anova table, but some effort is needed to recognise the other terms from the model. The random-effect term B is the only term in the 'B' stratum of the table, the residual variation due to E is the only term in the 'B.A.*Units*' stratum, and the term A.B is represented by the

Table 6.3 Data from an experiment to test all combinations of factors A and B in two replications.

The conventions used in this spreadsheet are the same as those used in Table 1.1, Chapter 1.

	A	B	C
1	A!	B!	y
2	1	1	30
3	1	1	19
4	1	2	51
5	1	2	56
6	1	3	65
7	1	3	53
8	2	1	49
9	2	1	41
10	2	2	72
11	2	2	79
12	2	3	74
13	2	3	67
14	3	1	72
15	3	1	65
16	3	2	91
17	3	2	95
18	3	3	70
19	3	3	66
20	4	1	22
21	4	1	14
22	4	2	61
23	4	2	54
24	4	3	42
25	4	3	43

'Residual' term within the 'B.A' stratum. Following Ridgman's interpretation of the model, the F statistic for B is therefore given by

$$\frac{MS_B}{MS_{A.B}} = \frac{1989.04}{145.65} = 13.66,$$

which agrees with the value given by GenStat.

The following statements analyse the data using the same model, but with mixed-modelling methods:

```
VCOMPONENTS [FIXED = A] RANDOM = B + A.B
REML [PRINT = Wald, components] y
```

The output of the REML statement is as follows:

Estimated variance components

Random term	component	s.e.
B	230.42	248.85
B.A	59.35	42.40

Residual variance model

Term	Factor	Model(order)	Parameter	Estimate	s.e.
Residual		Identity	Sigma2	26.96	11.01

Wald tests for fixed effects

Sequentially adding terms to fixed model

Fixed term	Wald statistic	d.f.	Wald/d.f.	chi pr
A	35.56	3	11.85	<0.001

Dropping individual terms from full fixed model

Fixed term	Wald statistic	d.f.	Wald/d.f.	chi pr
A	35.56	3	11.85	<0.001

Message: chi-square distribution for Wald tests is an asymptotic approximation (i.e. for large samples) and underestimates the probabilities in other cases.

According to Ridgman's formulae for the EMSs, σ_B^2 is estimated by

$$\frac{MS_B - MS_{A.B}}{n_A r} = \frac{1989.04 - 145.65}{4 \times 2} = 230.42,$$

and this agrees with the value given by the REML statement. Thus GenStat supports the interpretation of Ridgman, not that of Snedecor and Cochran.

To prepare these data for analysis by R, they are transferred from the Excel workbook 'Intro to Mixed Modelling\Chapter 7\osteoporosis phenotypes.xls' to the text file 'osteoporosis phenotypes.dat' in the same directory. Exclamation marks (!) are removed from the ends of headings in this file. The following commands import the data and analyse them according to Model 6.5 using anova methods:

```
AxB <- read.table(
    "Intro to Mixed Modelling\\Chapter 6\\fixed A, random B.dat",
    header=TRUE)
attach(AxB)
fA <- factor(A)
fB <- factor(B)
AxB.modelaov <-
    aov(y ~ fA + Error(fB + fA:fB))
summary(AxB.modelaov)
```

The output of the function `summary()` is as follows:

```
Error: fB
           Df Sum Sq Mean Sq F value Pr(>F)
Residuals   2 3978.1  1989.0

Error: fB:fA
           Df Sum Sq Mean Sq F value   Pr(>F)
fA          3 5179.5  1726.5  11.853 0.006215 **
Residuals   6  873.9   145.7
---
Signif. codes:  0 '***' 0.001 '**' 0.01 '*' 0.05 '.' 0.1 ' ' 1

Error: Within
           Df Sum Sq Mean Sq F value Pr(>F)
Residuals  12 323.50   26.96
```

R does not present an F test of the significance of the main effect of 'B', and hence does not here distinguish between the interpretations of Snedecor and Cochran and of Ridgman.

The following commands analyse the data using mixed-modelling methods:

```
AxB.modellmer <- lmer(y ~ fA + (1|fB) +(1|fA:fB),
    data = AxB)
summary(AxB.modellmer)
```

The output of these commands is as follows:

```
Linear mixed-effects model fit by REML
Formula: y ~ fA + (1 | fB) + (1 | fA:fB)
    Data: AxB
       AIC      BIC    logLik MLdeviance REMLdeviance
  160.5343 167.6026 -74.26713   170.8821     148.5343
Random effects:
 Groups    Name        Variance Std.Dev.
 fA:fB     (Intercept)  59.347   7.7037
 fB        (Intercept) 230.425  15.1798
 Residual               26.958   5.1921
# of obs: 24, groups: fA:fB, 12; fB, 3

Fixed effects:
             Estimate Std. Error t value
(Intercept)  45.6667    10.0540   4.5421
fA2          18.0000     6.9678   2.5833
fA3          30.8333     6.9678   4.4251
fA4          -6.3333     6.9678  -0.9089

Correlation of Fixed Effects:
      (Intr) fA2    fA3
```

```
fA2  -0.347
fA3  -0.347   0.500
fA4  -0.347   0.500   0.500
```

Again, the estimate of σ_B^2 agrees with that obtained from the EMSs according to Ridgman's formulae. Thus R also supports this interpretation, not that of Snedecor and Cochran.

Snedecor and Cochran's justification of their formula can be paraphrased as follows, in the notation used here:

If B is random and A is fixed, repeated sampling involves drawing a fresh set of n_B levels of the factor B in each experiment, retaining the same set of n_A levels of A. In the equation

$$\text{EMS}_B = n_A r \kappa_B^2 + \sigma_E^2, \tag{6.7}$$

κ_B^2 is then an unbiased estimate of σ_B^2. Hence, with B random and A fixed,

$$\text{EMS}_B = n_A r \sigma_B^2 + \sigma_E^2. \tag{6.8}$$

(In Snedecor and Cochran's presentation of this argument, A is temporarily regarded as the random-effect factor and B as the fixed-effect factor. Here, B is regarded as the random-effect factor throughout.)

It appears to the present author that this argument is flawed. Whenever a fresh set of levels of the factor B is drawn, a fresh set of effects of (AB) is also sampled, just as a fresh set of effects of E is sampled. Therefore σ_{AB}^2 should contribute to the EMS in Equation 6.7, just as σ_E^2 does, i.e.

$$\text{EMS}_B = n_A r \kappa_B^2 + r \sigma_{AB}^2 + \sigma_E^2. \tag{6.9}$$

Then taking κ_B^2 to be an unbiased estimate of σ_B^2, we obtain

$$\text{EMS}_B = n_A r \sigma_B^2 + r \sigma_{AB}^2 + \sigma_E^2, \tag{6.10}$$

the formula given by Ridgman.

Throughout this book, it is assumed that the formulae for the EMSs given by Ridgman, and implicit in the F statistics and variance-component estimates given by GenStat and R, are correct.

6.5 Arguments for regarding block effects as random

It was stated earlier (Section 6.3) that all block terms should normally be regarded as random. However, this practice is by no means universal, especially when the degrees of freedom of a block term are low and its variance is consequently poorly estimated. The case for this recommendation must therefore be argued.

The usual analysis of the randomised block design is valid in a wider range of circumstances if the block effects can be regarded as random than if they are regarded as fixed. To illustrate this, consider a simple experiment with a single fixed-effect treatment factor, A, with n_A levels, arranged in a randomised block design with n_B blocks. Suppose it is thought likely that there will be block × treatment interaction effects. It is then proper to replicate the treatments within each block, so that any such interaction can be detected, and its significance and magnitude assessed. Suppose that r replications of each treatment are set up within each block, and that a response variable Y is measured on each experimental unit. The appropriate model for this design is then

$$A * block = A + block + A.block \qquad (6.11)$$

in the notation of Wilkinson and Rogers (1973), or, in algebraic notation,

$$y_{ijk} = \mu + \alpha_i + \beta_j + (\alpha\beta)_{ij} + \varepsilon_{ijk} \qquad (6.12)$$

where
 y_{ijk} = the observation of the response variable on the kth replication of the ith treatment in the jth block,
 μ = the grand mean (overall mean) value of the response variable,
 α_i = the effect of the ith treatment,
 β_j = the effect of the jth block,
 $(\alpha\beta)_{ij}$ = the interaction between the ith treatment and the jth block,
 ε_{ijk} = the effect of the kth experimental unit that received the ith treatment in the jth block.

It is assumed that the ε_{ijk} are independent values of a random variable E, such that

$$E \sim N(0, \sigma^2).$$

Now compare the consequences of regarding 'block' as a fixed-effect term with those of regarding it as a random-effect term. The treatment factor A is a fixed-effect term, so if 'block' is regarded as a fixed-effect term the interaction term 'A.block' will also be a fixed-effect term. The variance of the effects of B is then designated by κ_B^2, and that of the A.B interaction effects by κ_{AB}^2. On the other hand, if 'block' is regarded as a random-effect term, 'A.block' will also be a random-effect term. The values β_j will then be interpreted as independent values of a random variable B, such that

$$B \sim N(0, \sigma_B^2),$$

and the values $(\alpha\beta)_{ij}$ as independent values of a random variable (AB), such that

$$(AB) \sim N(0, \sigma_{AB}^2).$$

The EMSs and F tests that follow from each decision are illustrated in Table 6.4. If 'block' is regarded as a fixed-effect term, the main effects of blocks, the main effects of treatment and the block × treatment interaction term are all tested against

Table 6.4 EMSs from a randomised block design in the presence of block × treatment interaction, with treatment replication within each block.
κ_A^2 = variance of the fixed effects of A. κ_B^2 = variance of the fixed effects of B. κ_{AB}^2 = variance of the fixed effects of (AB). The pairs of mean squares to be compared by F tests are indicated by angled lines.

Source of variation	DF	Expected mean square	
		Block effects fixed	Block effects random
block	$n_B - 1$	$n_A r \kappa_B^2 + \sigma_E^2$	$n_A r \sigma_B^2 + r \sigma_{AB}^2 + \sigma_E^2$
A	$n_A - 1$	$n_B r \kappa_A^2 + \sigma_E^2$	$n_B r \kappa_A^2 + r \sigma_{AB}^2 + \sigma_E^2$
A.block	$(n_A - 1)(n_B - 1)$	$r \kappa_{AB}^2 + \sigma_E^2$	$r \sigma_{AB}^2 + \sigma_E^2$
Residual		σ_E^2	σ_E^2

Table 6.5 EMSs from a randomised block design in the presence of block × treatment interaction, but with no treatment replication within each block.
Conventions are as in Table 6.4.

Source of variation	DF	Expected mean square	
		Block effects fixed	Block effects random
block	$n_B - 1$	$n_A \kappa_B^2 + \sigma_E^2$	$n_A \sigma_B^2 + \sigma_{AB}^2 + \sigma_E^2$
A	$n_A - 1$	$n_B \kappa_A^2 + \sigma_E^2$	$n_B \kappa_A^2 + \sigma_{AB}^2 + \sigma_E^2$
A.block	$(n_A - 1)(n_B - 1)$	$\kappa_{AB}^2 + \sigma_E^2$	$\sigma_{AB}^2 + \sigma_E^2$

the residual term, whereas if 'block' is regarded as a random-effect term, the main effects are tested against the interaction term.

Now consider the case where there is no replication of each treatment within each block (i.e. where $r = 1$), or where the analysis is performed using the mean value of each treatment in each block instead of the raw data. The EMSs and F tests are then as shown in Table 6.5. If 'block' is regarded as a fixed-effect term, no term in the anova can now be tested against any other term, whereas if 'block' is regarded as a random-effect term, the main effects can still be tested against the interaction.

If the block effects are to be regarded as fixed, then in order to perform significance tests it is now necessary to drop the interaction term 'A.block' from the model, simplifying it to

$$A + block \tag{6.13}$$

in the notation of Wilkinson and Rogers (1973), or, in algebraic notation,

$$y_{ij} = \mu + \alpha_i + \beta_j + \varepsilon_{ij}, \tag{6.14}$$

where
y_{ij} = the observation of the response variable on the ith treatment in the jth block,
μ = the grand mean (overall mean) value of the response variable,
α_i = the effect of the ith treatment,

Table 6.6 EMSs from a randomised block design in the absence of block × treatment inter-action.

Conventions are as in Table 6.4.

Source of variation	DF	Expected mean square	
		Block effects fixed	Block effects random
block	$n_B - 1$	$n_A \kappa_B^2 + \sigma_E^2$	$n_A \sigma_B^2 + \sigma_E^2$
A	$n_A - 1$	$n_B \kappa_A^2 + \sigma_E^2$	$n_B \kappa_A^2 + \sigma_E^2$
Residual	$(n_A - 1)(n_B - 1)$	σ_E^2	σ_E^2

β_j = the effect of the jth block,

ε_{ij} = the effect of the experimental unit that received the ith treatment in the jth block.

The EMSs and F tests are then as shown in Table 6.6: the standard F tests for the randomised block design are now appropriate whether block effects are regarded as fixed or random. However, if they are regarded as random, the inclusion or omission of the interaction term makes no difference to the appropriate F test. The interpretation of the bottom stratum of the anova changes from 'block.A' (Table 6.5) to 'Residual' (Table 6.6), but arithmetically the analysis is unchanged. In summary, this argument shows that the standard interpretation of the randomised block design is appropriate whether or not block × treatment interaction is present, provided that the blocks studied are representative of the population for which inferences are to be made.

Even if it is possible to interpret an anova in which block effects are regarded as fixed, the omission of the corresponding variance components from the calculation of significance tests and SEs may lead to overestimation of the significance of other effects, and of the precision of their estimates. This can be illustrated by specifying all model terms in the ravioli split plot experiment (except the residuals) as fixed-effect terms, using the following statements:

```
MODEL saltiness
FIT [FPROB = yes; PRINT = accumulated] \
   brand * assessor + day / presentation
```

This analysis cannot be specified using GenStat's anova directives (BLOCKSTRUC-TURE, TREATMENTSTRUCTURE and ANOVA) because the variation represented by the term 'brand.assessor' overlaps with (is partially aliased with) that represented by the term 'day.presentation'; therefore the analysis is specified using the regression analy-sis directives. Note that the term 'day.presentation.serving' is not included explicitly in the model: it is omitted so that it will be used as the residual term.

The anova produced by these statements is compared with the correct one in Table 6.7. The angled lines on this table indicate the mean squares that are com-pared in the F test of each treatment term. The mean squares in the two analyses are identical. However, when the block effects are regarded as fixed, 'brand' is incor-rectly tested against the same mean square as 'assessor' and 'brand.assessor'. This small mean square with its large degrees of freedom leads to an inflated value of F_{brand}, and an overestimation of the significance of this term.

Table 6.7 Comparison between the correct analysis of a split plot experiment and an analysis with block effects regarded as fixed.

	Correct analysis				Analysis with block effects regarded as fixed				
Source of variation	DF	MS	F	P	Source of variation	DF	MS	F	P
day stratum	2	371.7	2.67		day	2	371.7	3.71	0.0300
day.presentation stratum									
brand	3	3286.6	23.59	0.00101	brand	3	3286.6	32.78	5.86×10^{-13}
Residual	6	139.3	1.39		day.presentation	6	139.3	1.39	0.233
day.presentation.serving stratum									
assessor	8	1934.3	19.29	2.12×10^{-14}	assessor	8	1934.3	19.29	2.12×10^{-14}
brand.assessor	24	253.1	2.52	0.002	brand.assessor	24	253.1	2.52	0.00171
Residual	64	100.3			Residual	64	100.3		

In other circumstances, a decision to regard block effects as fixed can lead to *under*estimation of the significance of other effects and of the precision of their estimates. This is illustrated by the analysis of a balanced incomplete block design presented in Chapter 8 (Sections 8.1 and 8.2). In this design, each block contains only a subset of the treatments being compared, and the information concerning treatment effects is therefore distributed between two strata of the anova, the among-blocks and within-block strata. If the block effects are regarded as fixed, the estimates of treatment effects are based on the within-blocks information only. The significance of the treatment term is then lower, and $SE_{Difference}$ for comparisons between treatment means is larger than if the block effects are regarded as random. The case for regarding incomplete block effects as random, and hence recovering interblock information, is presented more rigorously by Robinson (1991, Section 5.1).

The argument that block effects should be regarded as random may apply also to factor levels that do not obviously constitute blocks. For example, suppose that in an educational experiment, a new method is tested in some schools, while others are used as controls, and that the response variable measured is the attainment of individual children within each school. It is important to treat 'school' as a random-effect term, otherwise variation among schools will not contribute to the SE of the estimate of the effect of educational method, and the precision of this estimate will be exaggerated. As for the issue of low degrees of freedom and poor estimation of block variance components, this is less of a concern if attention is focused on the *effects* of treatments and the *differences* between their means, rather than the means themselves. This distinction has been more fully discussed earlier (Chapter 4, Section 4.6).

6.6 Examples of the choice of fixed- and random-effect terms

Some examples will illustrate how the guidelines and rules for identifying random-effect terms, presented in Section 6.3, are applied. In the treatment of asthma, two classes of drug, corticosteroids and beta-agonists, may be prescribed singly or in combination. A clinical trial to compare the efficacy of several doses of each might have the design shown in Table 6.8. Each cell in this table represents a treatment, to which the same number of patients will be allocated.

Table 6.8 Factorial design to study the efficacy of two drug types in the treatment of asthma.

Dose of corticosteroid	Dose of beta-agonist			
	none	low	medium	high
none				
low				
high				

The model to be fitted is

$$corticosteroid * beta\text{-}agonist.$$

A measure of lung function called the forced expiratory volume (FEV) will be measured on each patient. The levels of both treatment factors, corticosteroid and beta-agonist, are values on numerical scales, so the main effects of both factors should be regarded as fixed-effect terms. Therefore the interaction term 'corticosteroid.beta-agonist' should also be regarded as fixed. It is likely that the interaction effects are not a set of unrelated values, but that they follow some fairly simple functional form. For example, there is some evidence that the effects of these two classes of drugs are *synergistic*, i.e. they work better in combination than either does on its own. In this case, there may be more response to corticosteroid at higher doses of beta-agonist, and vice versa. Such a relationship would vindicate the decision to treat the interaction effects as fixed.

In another clinical trial, the four doses of beta-agonist might all be given to each of 12 patients, during successive time periods. In such a trial the order in which the doses are given will be randomised, a different randomisation being used in each patient, and there will be a 'wash-out period' between treatment periods, to eliminate possible carry-over effects. Suppose that two replicate observations of FEV are made on each patient in each time period. The experiment will then have the design illustrated in Figure 6.2.

Patient 1	medium dose		no dose		high dose		low dose	
	Obs 1	Obs 2	Obs 1	Obs 2	Obs 1	Obs 2	Obs 1	Obs 2

Patient 2	no dose		low dose		high dose		medium dose	
	Obs 1	Obs 2	Obs 1	Obs 2	Obs 1	Obs 2	Obs 1	Obs 2

Patient 12	medium dose		low dose		no dose		high dose	
	Obs 1	Obs 2	Obs 1	Obs 2	Obs 1	Obs 2	Obs 1	Obs 2

Figure 6.2　The arrangement, in a factorial design, of an experiment to study the efficacy of a range of doses of a beta-agonist in the treatment of asthma.

The model to be fitted is

$$beta\text{-}agonist * patient.$$

The term 'beta-agonist' is a fixed-effect term as before, but the patients studied can (if they have been properly selected) be considered as a random sample from a population of asthma sufferers eligible to participate in the trial; hence it is natural to consider 'patient' as a random-effect term. It follows from Rule 1 (Section 6.3) that the interaction term 'beta-agonist.patient' is also a random-effect term. If Patient 1 were replaced by a different patient, this patient would respond to any particular dose of beta-agonist in a different way (assuming that any real beta-agonist.patient interaction effects were present); thus the interaction effects, as well as the main effects of patients, are a random sample from an underlying population. The mixed model specified is then as follows:

Fixed-effect model: beta-agonist
Random-effect model: patient + beta-agonist.patient.

This clinical trial effectively has a randomised block design, each patient comprising a block. The random-effect model can be expressed more succinctly as

patient/beta-agonist.

However, note that in the present case this does *not* mean

main effects of patient + effects of beta-agonist within each patient

but

main effects of patient + beta-agonist × patient interaction effects,

because the main-effect term 'beta-agonist' is present in the fixed-effect model.

Note that it is assumed that the interaction effects at each level of beta-agonist in the same patient are independent values of the underlying random variable. This assumption is reasonable if the different levels of beta-agonist are unrelated to each other. However, if, as in the present case, there is some structure among the levels, it may be desirable to reflect this in the model. If so, it should be done both for the main effect term 'beta-agonist' and for the beta-agonist.patient interaction term. For example, if it is expected that there is a linear trend in FEV in response to a higher dose of beta-agonist, this can be represented by dividing 'beta-agonist' into two terms, namely

lin(beta-agonist): the linear effect of the dose of beta-agonist
dev(beta-agonist): the deviation of the effects of individual doses of beta-agonist
from the linear effect.

The model to be fitted is then

(lin(beta-agonist) + dev(beta-agonist)) * patient.

It is partitioned between the fixed- and random-effect models as follows:

Fixed-effect model: lin(beta-agonist) + dev(beta-agonist)
Random-effect model: patient + lin(beta-agonist).patient+
dev(beta-agonist).patient

The significance of each fixed-effect term is tested against the corresponding random-effect term, as indicated by the angled lines in Table 6.9.

If it appears reasonable to assume that the variation due to the terms 'lin(beta-agonist).patient' and 'dev(beta-agonist).patient' is *homogeneous* – that is, that the variances due to these two terms are equal – then it may be decided to pool them, to give significance tests with more degrees of freedom in the denominator and more power, as shown in Table 6.10. Whether such pooling is permissible is a matter of

Table 6.9 Specification of the terms to be compared by significance tests in the anova of an experiment to study the effect of a beta-agonist on asthma.

Source of variation
lin(beta-agonist)
dev(beta-agonist)
patient
lin(beta-agonist).patient
dev(beta-agonist).patient

Table 6.10 Alternative specification of the terms to be compared by significance tests in the anova of an experiment on treatments for asthma, with the terms used in the denominator pooled.

Source of variation
lin(beta-agonist)
dev(beta-agonist)
patient
beta-agonist.patient

judgement. The two interaction mean squares can be compared formally using an F test, to see if one is significantly larger than the other, but it should be remembered that a failure to find a significant difference does not prove that two quantities are the same.

Next consider an ecological study in which soil samples are taken at several sites in each of several areas of woodland. In each sample the abundance of springtails (a small, very common type of insect) is measured. The model to be fitted is

area/site.

Suppose that it is decided to treat 'area' as a random-effect term: the areas sampled are to be considered as representative of a larger set of woodland areas. The effect of site within area (the term 'area.site') must then also be regarded as random. The choice of a nested model, with no main effect of site, means that Site 1 in Area 1 is presumed to have nothing in common with Site 1 in Area 2. If another area were substituted for Area 1, the effect of Site 1 within the new Area 1 would be different; thus the effect of site within area, like that of area, is sampled from an underlying population.

It may be argued that there *are* features in common between sites in different areas. For example, the abundance of springtails may be influenced by soil pH, and each area may comprise high- and low-pH sites. Such effects should be recognised by a main-effect term in the model, which would then become

area * site type.

The factor 'site type' might then be regarded as a fixed-effect term; or, in the case of soil pH, if measurements on a numerical scale were available, it might be replaced by a variate, which would be a fixed-effect term. The model would then be

area * pH.

Although the main effect of any variate should usually be regarded as a fixed-effect term, the interaction of a variate with a factor, or the effect of a variate nested within levels of a factor, may be either fixed or random. For example, consider a study in which the degree of anxiety indicated by the behaviour of mice is scored daily for several days. The mean score is likely to vary between mice, but it is also likely that each mouse will become more or less anxious during the period of the experiment. However, this trend is also likely to vary between mice: some may become more anxious, others less. The model to be fitted is then

mouse/time.

If the variation among mice at time zero (the main-effect term 'mouse') is random, then the variation among mice in the trend over time (the term 'mouse.time') is also random, following Rule 1.

It may be that there is some component of the trend over time that is expected to be common to all mice. For example, mice may generally become less anxious as they become used to the test situation. Such a common trend, around which the trend lines for individual mice vary, should be represented in the model by a main effect of time. The model to be fitted is then

mouse * time.

An example of a pattern of results that might be obtained from fitting this model is shown in Figure 6.3.

In the next chapter, we will apply the methods, principles and rules introduced in this chapter to real data sets, and interpret the results.

6.7 Summary

The mixed models introduced in earlier chapters illustrate the following features of mixed models in general:

- Each model term may be:
  - a variate, in which each observation can have any numerical value, or
  - a factor, in which each must come from a specified set of levels.
- Each part of the model (fixed and random) may have more than one term.

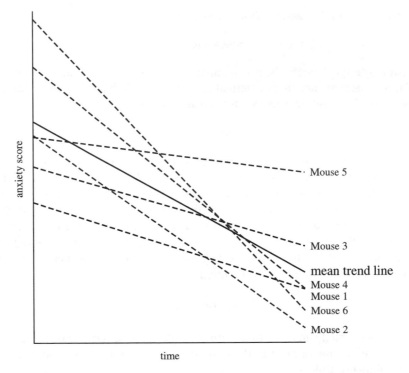

Figure 6.3 Possible pattern of results in an experiment to determine the effect of a factor (mouse) and a variate (time) on a response variate (anxiety score).

- Factors may be:

  - crossed, e.g.
$$a * b = a + b + a.b$$

    where term a.b represents the a × b interaction effects, or

  - nested, e.g.
$$a/b = a + a.b$$

where term a.b represents the effects of b within each level of a.

More elaborate combinations of these features are permitted:

- more than two factors may be crossed;
- factors and variates may be used in the same part of the mixed model;
- factors may be crossed or nested with variates.

If 'f' is a factor and 'v' is a variate, the model 'f * v' or 'f/v' represents a set of sloping lines relating the response variate to the explanatory variate 'v'.

The term 'f' represents variation in the intercepts of the lines among the levels of this factor. The term 'v' represents the mean slope of the line, averaged over the levels of 'f'. In the crossed model 'f $*$ v', the term 'f.v' represents variation among the slopes at different levels of 'f' around the average slope. In the nested model 'f/v', this term represents all the variation among the slopes at different levels of 'f'.

Variates may be crossed with other variates. In this case, the interaction term is equivalent to the product of the two variates.

In the balanced data sets considered in earlier chapters, the significance of each fixed-effect model term is tested by comparing the mean square for that term with one other mean square. However, this is not possible in all cases: information about a term may be distributed over two or more strata defined by the random-effect terms. The information from these strata must be pooled when estimating fixed effects.

In a mixed-model analysis, these adjustments are made automatically, but it is only in special cases that an equivalent analysis can be specified as a regression analysis or an anova.

Rules are presented for deciding which terms should be allocated to the random-effects part of a mixed model:

- It is generally sensible to regard all variates as fixed-effect terms.

- Factors should be classified as:

  - Terms representing nuisance effects – block effects, which should normally be regarded as random. For a justification of this see Chapter 2, Section 2.6 and Summary. If block effects are regarded as fixed, significance tests may be mis-specified.

  - Terms representing effects of interest – treatment effects, which may be fixed or random.

To decide whether a treatment factor should be regarded as random, we need to consider whether the levels of this factor that have been studied are a random sample from an infinite population of possible levels, and also whether we prefer to *regard* them as such a sample, or as levels of individual interest a priori.

Even if the assumption that factor levels are randomly sampled from an infinite population is not strictly valid, it may still be a reasonable approximation.

Although additional assumptions are made when a factor is regarded as a random-effect term, this should not necessarily be thought of as a radical step. On the contrary, the inferences made about other terms are more cautious than when the factor is regarded as fixed.

Decisions on whether a term is fixed or random need usually only be made for main effects. For interaction terms and some nested terms, the choice is determined by the following rules:

- *Rule 1.* When two terms are crossed, if both terms are fixed the interaction must be fixed; otherwise, it must be random.

- *Rule 2.* When one term is nested within another, the nested term may be either fixed or random if the term within which it is nested is fixed; otherwise, it must be random.

There is an inconsistency in the statistical literature concerning the method for construction of the F test when a fixed-effect term (A) and a random-effect term

(B) interact. Some authorities consider that B should be tested against the residual term, others that it should be tested against the A.B interaction term. It is argued that the latter view is correct.

Examples of the application of the rules for classifying terms as fixed or random are given.

6.8 Exercises

1. In the experiment to compare four commercial brands of ravioli described in Chapter 2, it may be argued that 'assessor' should be regarded as a random-effect factor.

   (a) Consider the case for this decision, and present the arguments for and against.

   (b) If 'brand' is regarded as a fixed-effect term and 'assessor' as a random-effect term, how should 'brand.assessor' be regarded?

   (c) Make these changes to the mixed model fitted to these data. Perform the new analysis, and interpret the results. Explain the effect of the changes on the SEs of differences between brands.

   (d) Perform appropriate tests to determine whether the new random-effect terms are significant.

2. Return to the data set concerning the effect of oil type on the amount of wear suffered by piston rings, introduced in Exercise 2 in Chapter 2.

   (a) For each oil type, plot the value of wear against the ring number (1 to 4). Omit the values from the 'oil ring' from this plot. Is there evidence of a trend in the amount of wear from Ring 1 to Ring 4?

   (b) Repeat the mixed-model analysis performed on these data previously, but exclude the values from the 'oil ring'.

   (c) Now specify 'ring' as a variate instead of a factor. Fit this new model to the data by mixed modelling. Do the results confirm that there is a linear trend from Ring 1 to Ring 4? Does this trend vary significantly depending on the oil type?

   (d) Note that when this change is made to the model, DF_{ring} is reduced from 3 to 1, and $DF_{oil.ring}$ from 6 to 2. What source of variation is included in these terms in the previous model, but excluded in the present model?

   Now suppose that the three types of oil tested are a representative sample of the types used in practice.

   (e) When this change is made, which parts of the expression 'oil $*$ ring' should be regarded as fixed-effect terms, and which as random-effect terms?

   (f) Fit this new model to the data by mixed modelling. Does the new model indicate that the linear trend from Ring 1 to Ring 4 varies significantly depending on the oil type?

7

Two case studies

7.1 Further development of mixed-modelling concepts through the analysis of specific data sets

In this chapter we will apply the concepts introduced earlier to two data sets more elaborate than those considered so far. These were obtained from specialised investigations, one into the causes of variation in bone mineral density among human patients, the other into the causes of variation in oil content in a grain crop. However, during the analysis of these data, new concepts and additional features of the mixed-modelling process will be introduced which are widely applicable. In summary, these are as follows:

- *In the invetsigation of the causes of variation in bone mineral density*:

  - a fixed-effects model with several variates and factors

  - a polynomial model with quadratic and cross-product terms

  - further interpretation of diagnostic plots of residuals, leading to the exclusion of outliers

  - exploration and building of the fixed-effects model

  - use of the marginality criterion when determining the terms to be retained in a polynomial model

  - predicted values from a model; their use as an aid to understanding the results of the modelling process

  - graphical representation of predicted values

  - the consequences of extrapolation of predictions outside the range of the data.

Introduction to Mixed Modelling: Beyond Regression and Analysis of Variance   N. W. Galwey
© 2006 John Wiley & Sons, Ltd

- *In the investigation of the causes of variation in oil content of a grain crop*:

  - a random-effects model with several factors

  - deviations from a linear trend regarded as random effects

  - the marginality criterion applied to the linear trend and deviations from it

  - exploration and building of the fixed-effects model

  - the effect of one term within and between the levels of another; recognition of this distinction

  - a (random factor) × (fixed variate) interaction term

  - dependence of the magnitude of the variance component for the (random factor) × (fixed variate) term on the units of measurement of the variate

  - comparison of the magnitudes of variance components

  - graphical representation of the variation accounted for by the mixed model.

7.2   A fixed-effect model with several variates and factors

In an extensive study of osteoporosis, a disease in which the strength of the bones is abnormally low and the patient is at high risk of fractures, data were obtained from about 3700 individuals. One of the main observations taken was the bone mineral density (BMD), a quantitative trait closely associated with osteoporosis, measured by means of an X-ray scan at various skeletal sites including the lumbar region of the spine. The main purpose of the study was to identify genetic factors associated with osteoporosis, and related individuals (parents and offspring, siblings, etc.) were therefore sampled. An extensive analysis of the genetic information in the data was reported by Ralston *et al.* (2005). Here we will focus on the possible relationships between BMD and other variables for which data were obtained, namely the sex, age, height and weight of the patient, and the hospital at which the scan was performed. These variables were observed in the expectation that adjustment for them would improve the precision with which the genetic influences on BMD could be estimated.

The variables to be considered here are listed in Table 7.1, and the first and last few rows of the spreadsheet holding the data are shown in Table 7.2. Each row represents an individual. (Data reproduced and used by kind permission of the FAMOS consortium.)

BMD is known to change with age, but this relationship is not expected to be linear: evidence from other studies, including studies over time on the same individual, indicates that a person's BMD reaches a peak at about age 40 and then declines. We will allow for this non-linearity by considering age^2, as well as age per se, for inclusion in the model fitted. Similarly, the relationship between BMD and height is not expected to be independent of that between BMD and weight: for example, a particular below-average value of BMD would be more extreme in a tall individual than in a short

Table 7.1   Variables recorded for each individual in a survey of osteoporosis in a human population.

Name	Type	Description
Gender	factor	1 = male; 2 = female
CentreNumber	factor	Number indicating the hospital (the centre) at which the individual was observed
Proband	factor	Whether an individual was a *proband*, i.e. was recruited on the basis that the individual suffered from osteoporosis. (Other individuals were recruited on the basis that they were related to a proband.) Valid factor levels are 'No' and 'Yes'
EverHadFractures	factor	Whether the individual had ever had fractures. Valid factor levels are 'No', 'Unknown' and 'Yes'
Age	variate	Age in years at the time of measurement
Height	variate	Height in cm
Weight	variate	Weight in kg
BMD_Outlier	factor	Whether the individual is identified as having unusually high or low values of BMD for his or her age, height and weight. Valid factor levels are 'No' and 'Yes'
CalL2-L4BMD	variate	BMD, averaged over lumbar vertebrae 2, 3 and 4, calibrated to adjust for differences between the X-ray machines with which different individuals were measured

Table 7.2   Data from a survey of osteoporosis in a human population.
Only the first and last few rows of the data are shown. The conventions used in this spreadsheet are the same as those used in Table 1.1, Chapter 1.

	A	B	C	D	E	F	G	H	I
1	Gender!	CentreNumber!	Proband!	EverHadFractures!	Age	Height	Weight	BMD_Outlier!	CalL2-L4BMD
2	2	02	Yes	No	62.00	150.00	53.10	No	0.772
3	1	02	No	Yes	61.00	177.00	76.10	No	0.970
4	2	02	No	No	40.00	157.50	56.90	No	1.238
5	1	02	No	Yes	39.00	182.00	73.00	No	0.950
.									
.									
3690	2	11	Yes	Yes	52.00	178.00	100.00	No	0.753
3691	1	11	No	Yes	36.00	185.00	98.60	No	1.099
3692	1	11	No	Yes	32.00	186.10	86.30	No	1.351

one. For this reason, height and weight are often combined into a single composite measure called *body mass index* (BMI). Thus:

$$BMI = \frac{weight\ (kg)}{(height\ (cm))^2} \times 10000.$$

However, it may not be reasonable to assume that the interaction between height and weight is of the particular form specified by BMI. In order to give a good fit to a wide range of possible patterns of interaction, the following terms will be considered for inclusion in the model fitted: height, weight, height^2, height $\times$ weight and weight^2. Because the probands have been selected on the basis that they suffer from osteoporosis, they are expected to give a distorted indication of the relationship between BMD and other variables. They will therefore be excluded from the modelling process. The following GenStat statements import the data and perform these preliminary manipulations:

```
IMPORT \
    'Intro to Mixed Modelling\\Chapter 7\\ost phenotypes.xls'
CALCULATE Agesq = Age ** 2
CALCULATE Htsq = Height ** 2
CALCULATE HtxWt = Height * Weight
CALCULATE Wtsq = Weight ** 2
RESTRICT CalL2_L4BMD; Proband .eq. 1
```

The `CALCULATE` statements obtain the additional variables required, and the `RESTRICT` statement confines attention to those individuals for which the value of the factor 'Proband' is 1, i.e. 'No'. The restriction is applied directly only to the variate 'CalL2_L4BMD', which is the measure of BMD on which we will focus here: when this variate is used in an analysis, the restriction will automatically be extended to the other variates and factors used.

It is believed that after allowing for the fact that different hospitals may have a preponderance of one or other sex, or may differ in the mean age of the patients scanned, etc., the remaining variation between hospitals can be regarded as a random variable – that is, the hospitals can be regarded as a sample from a large population of hospitals available for study. Therefore the initial model fitted (Model 7.1), comprising all the terms that are candidates for inclusion, is as follows:

Response variate:	CalL2_L4BMD
Fixed-effect model:	Gender + Age + Agesq + Height + Weight + Htsq + HtxWt + Wtsq
Random-effect model:	CentreNumber

Note that all the fixed-effect terms in this model, unlike those in the examples considered so far, vary both within and between levels of the random-effect term 'CentreNumber' (hospital). The fixed effects are therefore estimated *within* each hospital, but they nevertheless explain part of the variation *between* hospitals. The following statements fit this model and display diagnostic plots of the residuals, but do not produce any other output:

```
VCOMPONENTS \
    [FIXED = Gender + Age + Agesq + \
    Height + Weight + Htsq + HtxWt + Wtsq] \
    RANDOM = CentreNumber
REML [PRINT = *] CalL2_L4BMD
VPLOT [GRAPHICS=high] fittedvalues, normal, halfnormal, histogram
```

The diagnostic plots are presented in Figure 7.1.

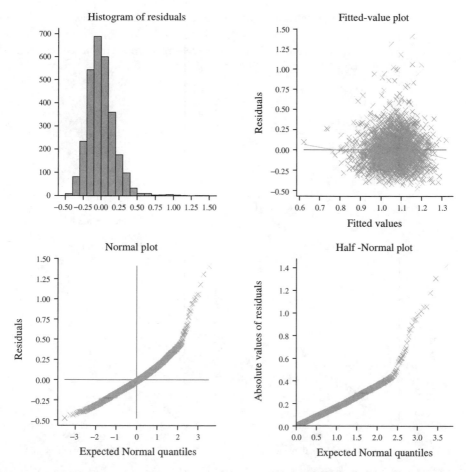

Figure 7.1   Diagnostic plots of residuals from the mixed model fitted to the osteoporosis data.

The residuals are clearly not normally and homogeneously distributed. The histogram has a long upper tail, and the fitted-value plot shows that this is due to a group of very large positive residual values associated with intermediate fitted values. This group of values also causes a kink in the distribution of the points in the normal and half-normal plots. Individuals with outlying values of BMD, assessed on the basis of all the skeletal sites, have already been identified in these data. It is believed that many of these observations represent regions of dense bone where previous fractures have healed, and are therefore unrepresentative of the overall composition of the patient's bones. The effect of excluding these values from the analysis is next explored. To do this, the RESTRICT statement is changed to

```
RESTRICT CalL2_L4BMD; Proband .eq. 1 .and. BMD_Outlier .eq. 1
```

and the VCOMPONENTS and REML statements are re-executed. The diagnostic plots are now as shown in Figure 7.2. The exclusion of the outliers has nearly, though not quite entirely, eliminated the problem.

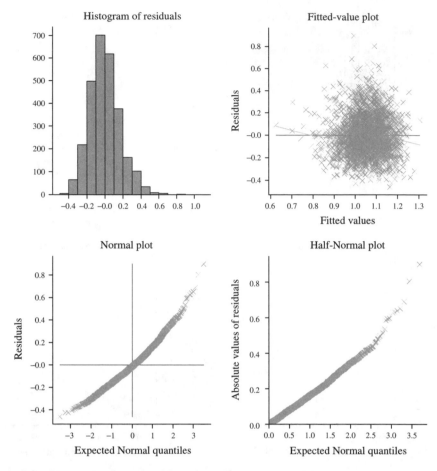

Figure 7.2 Diagnostic plots of residuals from the mixed model fitted to the osteoporosis data, following exclusion of probands and BMD outliers.

The REML statement is then run with option settings that specify the usual items of output, namely

```
REML [PRINT = model, components, Wald] CalL2_L4BMD
```

The output from this statement is as follows:

REML variance components analysis

Response variate:	CalL2_L4BMD
Fixed model:	Constant + Gender + Age + Agesq + Height + Weight + Htsq + HtxWt + Wtsq
Random model:	CentreNumber
Number of units:	2801 (117 units excluded due to zero weights or missing values)

Residual term has been added to model

Sparse algorithm with AI optimisation
All covariates centred
Analysis is subject to the restriction on CalL2_L4BMD

Estimated variance components

Random term	component	s.e.
CentreNumber	0.00065	0.00041

Residual variance model

Term	Factor	Model(order)	Parameter	Estimate	s.e.
Residual		Identity	Sigma2	0.0286	0.00077

Wald tests for fixed effects

Sequentially adding terms to fixed model

Fixed term	Wald statistic	d.f.	Wald/d.f.	chi pr
Gender	67.33	1	67.33	<0.001
Age	253.06	1	253.06	<0.001
Agesq	14.32	1	14.32	<0.001
Height	72.35	1	72.35	<0.001
Weight	109.91	1	109.91	<0.001
Htsq	33.48	1	33.48	<0.001
HtxWt	7.74	1	7.74	0.005
Wtsq	10.49	1	10.49	0.001

Dropping individual terms from full fixed model

Fixed term	Wald statistic	d.f.	Wald/d.f.	chi pr
Wtsq	10.49	1	10.49	0.001
HtxWt	17.19	1	17.19	<0.001
Htsq	45.51	1	45.51	<0.001
Weight	7.85	1	7.85	0.005
Height	47.75	1	47.75	<0.001
Agesq	0.50	1	0.50	0.479
Age	2.76	1	2.76	0.097
Gender	2.70	1	2.70	0.100

Message: chi-square distribution for Wald tests is an asymptotic approximation (i.e. for large samples) and underestimates the probabilities in other cases.

The variance component for 'CentreNumber' is somewhat larger than its SE, suggesting that there is real variation among the centres, though this component is much smaller than the residual variance among individuals. The Wald tests indicate that each successive term added to the fixed-effect model accounts for a highly significant amount of variation. However, in the fixed-effect part of a mixed model (as in an ordinary multiple regression model) the significance of a term depends on the order in which it is added to the model, except in certain special cases. Therefore, when all terms have been added, the effect of dropping each in turn is tested. In most cases, the term is still significant, though sometimes less so than when it was added in sequence. For example, 'Gender' gives a chi-square value of 67.33 when it is added as the first term in the model, but only 2.70 when it is dropped, all the other terms being retained. This is because much of the variation accounted for by 'Gender' can alternatively be accounted for by other terms: males tend to be taller and heavier than females. The term 'Agesq' is non-significant when all other terms are present in the model, and can safely be dropped. The term 'Age' is also not quite significant when all other terms are present, but it is not necessarily safe to drop it. This is because it has only been found to be non-significant in a model that includes 'Agesq', and it is a general rule of statistical modelling that when one term in the model is a power of another, then if the higher-power term ('Agesq' in the present case) is included in a model, the lower-power term ('Age' in this case) must be retained. This is an instance of the *marginality criterion*, an important concept which merits a brief digression.

The marginality criterion states that when a *higher-order term*, namely an interaction or nested term in a factorial model, or a power greater than 1 or a product term in a polynomial model, is retained in the model, then the corresponding lower-order terms must also be retained, whether or not they are significant. For example:

- if the A × B interaction is retained in the factorial model A * B, then the main-effect terms A and B must also be retained;

- if the nested term B-within-A is retained in the factorial model A/B, then the main effect term A must also be retained;

- if the term x^2 is retained in a polynomial model, then the term x must also be retained;

- if the term $x_1 x_2$ is retained in a polynomial model, then the terms x_1 and x_2 must also be retained.

The marginality criterion applies recursively. Thus:

- If the A × B × C interaction is retained in the factorial model A * B * C, then the terms A × B and C must also be retained. Because A × B is retained, A and B must be retained.

- If the term $x_1^2 x_2$ is retained in a polynomial model, the terms x_1^2 and x_2 must be retained. Because x_1^2 is retained, x_1 must be retained.

A choice of model that may appear to breach the marginality criterion, though it does not really do so, is the replacement of the model

$$A * B = A + B + A.B$$

by the model

$$A/B = A + A.B.$$

When this change is made, the main-effect term B appears to be dropped while the interaction term A.B is retained. In reality, however, B is not deleted but assimilated: the variation and the degrees of freedom that it represents are *absorbed* into A.B, and the meaning of A.B changes from the crossed term 'A × B interaction' to the nested term 'B-within-A'. For a more formal account of the marginality criterion see McCullagh and Nelder (1989, Chapter 3, Sections 3.5.1 and 3.5.2, pp 63–67 and Section 3.9, p 89).

In order to see whether 'Age' can be dropped from the model, we must first drop 'Agesq' and then refit the model. This is done by the following statements:

```
VCOMPONENTS [FIXED = Gender + Age + \
   Height + Weight + Htsq + HtxWt + Wtsq] \
   RANDOM = CentreNumber
REML [PRINT = model, components, Waldtests, deviance] CalL2_L4BMD
```

These produce the following output:

REML variance components analysis

Response variate: CalL2_L4BMD
Fixed model: Constant + Gender + Age + Height + Weight + Htsq + HtxWt + Wtsq
Random model: CentreNumber
Number of units: 2801 (117 units excluded due to zero weights or missing values)

Residual term has been added to model

Sparse algorithm with AI optimisation
All covariates centred
Analysis is subject to the restriction on CalL2_L4BMD

Estimated variance components

Random term	component	s.e.
CentreNumber	0.00066	0.00041

Residual variance model

Term	Factor	Model(order)	Parameter	Estimate	s.e.
Residual		Identity	Sigma2	0.0286	0.00077

Deviance: −2*Log-Likelihood

Deviance	d.f.
−7013.78	2791

Note: deviance omits constants which depend on fixed model fitted.

Wald tests for fixed effects

Sequentially adding terms to fixed model

Fixed term	Wald statistic	d.f.	Wald/d.f.	chi pr
Gender	67.34	1	67.34	<0.001
Age	253.11	1	253.11	<0.001
Height	77.59	1	77.59	<0.001
Weight	116.22	1	116.22	<0.001
Htsq	35.40	1	35.40	<0.001
HtxWt	7.83	1	7.83	0.005
Wtsq	10.80	1	10.80	0.001

Dropping individual terms from full fixed model

Fixed term	Wald statistic	d.f.	Wald/d.f.	chi pr
Wtsq	10.80	1	10.80	0.001
HtxWt	17.52	1	17.52	<0.001
Htsq	47.23	1	47.23	<0.001
Weight	7.94	1	7.94	0.005
Height	49.67	1	49.67	<0.001
Age	178.06	1	178.06	<0.001
Gender	2.93	1	2.93	0.087

Message: chi-square distribution for Wald tests is an asymptotic approximation (i.e. for large samples) and underestimates the probabilities in other cases.

The effect of 'Age' is now highly significant, even when all other terms are present in the model. The variance component for 'CentreNumber' is almost unchanged by dropping 'Agesq' from the model, and the diagnostic plots of residuals (not presented here) are also almost unchanged. The effect of 'Gender' is now nearly significant, but not quite; nevertheless, it may be prudent to retain this term in the model, and we will do so.

As the variance component for 'CentreNumber' is only slightly larger than its standard error, it seems advisable to perform a more rigorous test of its significance. For this purpose, this term is dropped from the random-effect model (leaving only the residual term), and the deviance obtained by fitting this reduced model is noted. This is done by the following statements:

```
VCOMPONENTS [FIXED = Gender + Age + \
   Height + Weight + Htsq + HtxWt + Wtsq]
REML [PRINT = deviance] CalL2_L4BMD
```

The output of these statements is as follows:

Deviance: $-2*$Log-Likelihood

Deviance	d.f.
−6979.32	2792

Note: deviance omits constants which depend on fixed model fitted.

The difference between the deviances from the models with and without 'CentreNumber' is obtained, namely

$$\text{deviance}_{\text{reduced model}} - \text{deviance}_{\text{full model}} = -6979.32 - (-7013.78) = 34.46,$$

and

$$P(\text{deviance}_{\text{reduced model}} - \text{deviance}_{\text{full model}} > 34.46) = \frac{1}{2} \times P(\chi_1^2 > 34.46)$$
$$= 2.18 \times 10^{-9},$$

as explained earlier (Chapter 3, Section 3.12). Thus the variance due to 'CentreNumber' is highly significant: we can be very confident that there is real variation among the centres.

Having arrived at a satisfactory model, we can obtain estimates of its parameters using the following statements:

```
VCOMPONENTS [FIXED = Gender + Age + \
   Height + Weight + Htsq + HtxWt + Wtsq] \
   RANDOM = CentreNumber
REML [PRINT = effects] CalL2_L4BMD
```

These produce the following output:

Table of effects for Constant

1.054   Standard error: 0.0113

Table of effects for Gender

Gender	1	2
	0.000000	0.016726

Standard error of differences: 0.009773

Table of effects for Age

−0.002942   Standard error: 0.0002205

Table of effects for Height

0.07436   Standard error: 0.010550

Table of effects for Weight

−0.01397   Standard error: 0.004957

Table of effects for Htsq

−0.0002451   Standard error: 0.00003566

> ## Table of effects for HtxWt
> 0.0001464   Standard error: 0.00003497
>
> ## Table of effects for Wtsq
> −0.00004883   Standard error: 0.000014857

These parameter estimates can be used to construct an arithmetical model giving predicted values of the response variable. However, in order to do this, we must take account of GenStat's default parameterisation of the model, which is centred on the mean values of the explanatory variates (cf. Chapter 1, Section 1.8, where an option was set in the VCOMPONENTS statement to prevent this adjustment). These mean values are presented in Table 7.3.

The parameter estimates, together with the mean values of the explanatory variables, tell us that the best-fitting model (among those explored) is as follows:

$$\text{CalL2\_L4BMD} = 1.054 + 0.016726 \times \text{Gender} - 0.002942 \times (\text{Age} - 48.85)$$
$$+ 0.07436 \times (\text{Height} - 168.1) - 0.01397 \times (\text{Weight} - 70.42)$$
$$- 0.0002451 \times (\text{Height}^2 - 28353)$$
$$+ 0.0001464 \times (\text{Height} \times \text{Weight} - 11923)$$
$$- 0.00004883 \times (\text{Weight}^2 - 5159) + \Gamma + E \qquad (7.1)$$

where
Gender $= 0$ for a male, 1 for a female,
$\Gamma = $ the effect of the centre,
$E = $ the residual effect.

However, the individual parameters of a model as elaborate as this are not easy to interpret. It is more informative to inspect the fitted values given by the model over a

Table 7.3   Mean values of the explanatory variables in the osteoporosis data.
Probands and BMD outliers are excluded from the calculation of these means.

Variable	Mean
Age	48.85
Height	168.1
Weight	70.42
Htsq	28353
HtxWt	11923
Wtsq	5159

range of realistic values of the explanatory variables. For this purpose, we first define
variates to hold this set of values. Appropriate variates are shown in the spreadsheets
in Tables 7.4 and 7.5. Note that it is necessary to specify every desired combination
of 'Height' and 'Weight' explicitly, in order to specify the corresponding value of
Height × Weight.

Table 7.4 Range of values of 'Height', 'Weight' and the higher-order terms derived from
them to be used for obtaining fitted values of BMD

	A	B	C	D	E
	Htlev	Wtlev	Htsqlev	HtxWtlev	Wtsqlev
1
2	133.0	35.000	17689.00	4655.000	1225.000
3	133.0	45.556	17689.00	6058.889	2075.309
4	133.0	56.111	17689.00	7462.778	3148.457
5	133.0	66.667	17689.00	8866.667	4444.444
6	133.0	77.222	17689.00	10270.556	5963.272
7	133.0	87.778	17689.00	11674.444	7704.938
8	133.0	98.333	17689.00	13078.333	9669.444
9	133.0	108.889	17689.00	14482.222	11856.790
10	133.0	119.444	17689.00	15886.111	14266.975
11	133.0	130.000	17689.00	17290.000	16900.000
12	166.5	35.000	27722.25	5827.500	1225.000
13	166.5	45.556	27722.25	7585.000	2075.309
14	166.5	56.111	27722.25	9342.500	3148.457
15	166.5	66.667	27722.25	11100.000	4444.444
16	166.5	77.222	27722.25	12857.500	5963.272
17	166.5	87.778	27722.25	14615.000	7704.938
18	166.5	98.333	27722.25	16372.500	9669.444
19	166.5	108.889	27722.25	18130.000	11856.790
20	166.5	119.444	27722.25	19887.500	14266.975
21	166.5	130.000	27722.25	21645.000	16900.000
22	200.0	35.000	40000.00	7000.000	1225.000
23	200.0	45.556	40000.00	9111.111	2075.309
24	200.0	56.111	40000.00	11222.222	3148.457
25	200.0	66.667	40000.00	13333.333	4444.444
26	200.0	77.222	40000.00	15444.444	5963.272
27	200.0	87.778	40000.00	17555.556	7704.938
28	200.0	98.333	40000.00	19666.667	9669.444
29	200.0	108.889	40000.00	21777.778	11856.790
30	200.0	119.444	40000.00	23888.889	14266.975
31	200.0	130.000	40000.00	26000.000	16900.000

Table 7.5 Range of values of 'Age' to be used for obtaining fitted values of BMD

	A
1	Agelev
2	16
3	96

These variates are imported into GenStat. The predicted values given by these values of the explanatory variables are then obtained by the following statement:

```
VPREDICT [PREDICTIONS = BMDpredict; SE = BMDpredictse] \
    CLASSIFY = Age, Height, Weight, Htsq, HtxWt, Wtsq; \
    LEVELS = Agelev, Htlev, Wtlev, Htsqlev, HtxWtlev, Wtsqlev; \
    PARALLEL = Age, Height, Height, Height, Height, Height
```

The options PREDICTIONS and SE in this statement specify the names of tables to hold the predictions and their SEs respectively. The parameter CLASSIFY indicates the model terms for which values are to be specified when making predictions, and the parameter LEVELS indicates the values of each term for which predictions are to be made. By default, the model terms specified in CLASSIFY are crossed: that is, all combinations of their specified levels are considered. However, this is not always appropriate: for example, each value of 'Height' should be combined only with the corresponding value of 'Htsq'. These two terms are to be considered *in parallel*, and this is specified by the parameter PARALLEL. In the present case, 'Height', 'Weight' and all the terms derived from them are considered in parallel with each other, whereas 'Age' is in parallel only with itself.

A representative sample of the output from this statement is as follows:

Predictions from REML analysis

Model terms included for prediction: Constant + Gender + Age + Height + Weight + Htsq + HtxWt + Wtsq
Model terms excluded for prediction: CentreNumber

Status of model variables in prediction:

Variable	Type	Status
Wtsq	variate	Classifies predictions
HtxWt	variate	Classifies predictions
Htsq	variate	Classifies predictions
Weight	variate	Classifies predictions
Height	variate	Classifies predictions
Age	variate	Classifies predictions
Gender	factor	Averaged over - equal weights
Constant	factor	Included in prediction
CentreNumber	factor	Ignored

Response variate: CalL2_L4BMD

Predictions

Wtsq_HtxWt_Htsq_Weight_Height	1225 , 4655 , 17679, 35.00 , 133.0
Age	
16.00	0.7704
96.00	0.5350

Wtsq_HtxWt_Htsq_Weight_Height	2075 , 6059 , 17689 , 45.56 , 133.0
Age	
16.00	0.7869
96.00	0.5515

.

.

.

Standard errors

Wtsq_HtxWt_Htsq_Weight_Height	1225 , 4655 , 17689, 45.56 , 133.0
Age	
16.00	0.04208
96.00	0.04044

Wtsq_HtxWt_Htsq_Weight_Height	2075 , 6059 , 17689 , 45.56 , 133.0
Age	
16.00	0.04057
96.00	0.03831

.

.

.

Approximate average standard error of difference: 0.06207 (calculated on variance scale).

The way in which each model term is used when constructing the predictions is first specified. It is noted that 'Wtsq', 'HtxWt', etc., classify the predictions. The constant is also used in forming the predictions, by default. The term 'Gender' has not been mentioned in the VPREDICT statement, so the predictions are averaged over this term. When performing the averaging, equal weight is given to the males and the females, regardless of their relative numbers in the sample. The random-effect term 'CentreNumber', also not mentioned in the VPREDICT statement, is ignored when forming the predictions: that is, the effect of 'CentreNumber' is set to zero (though this term still contributes to the SEs of the predictions). The first of the predictions tells us that when Weight = 35.00 kg, Height = 133.0 cm and Age = 16.00 years, the predicted value of CalL2_L4BMD is 0.7704. When the value of Age is changed to 96.00 years, the predicted value is 0.5350, and so on. The table of predictions is followed by a table giving the SE of each prediction, following the same format. These predictions and their SEs are displayed graphically in Figure 7.3.

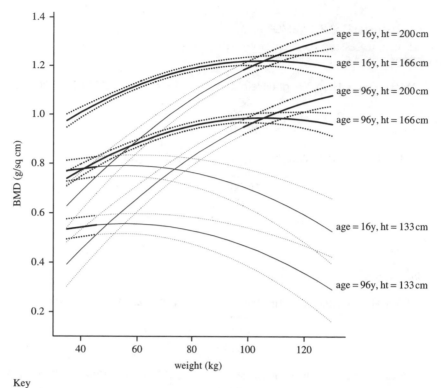

Key
solid line: prediction
dashed line: prediction $\pm$ SE$_{predicton}$
heavy line: values within the range of the explanatory variables
light line: values outside the range of the explanatory variables

Figure 7.3 Predictions of BMD on the basis of weight, age and height, from the osteoporosis data.

In every case, the curve for individuals aged 16 years predicts higher values of BMD than the corresponding curve for individuals aged 96 years, and this is consistent with information from other sources. This figure shows that in most circumstances, a greater weight is associated with a higher value of BMD, but that this trend levels off at the greatest weights. At first glance the figure appears to show that in some circumstances, BMD is actually lower in heavy individuals, which is implausible. However, closer inspection shows that the lines that show this pattern strongly represent very short individuals, only 133 cm tall. The combination of this height with a weight of 100 or 120 kg lies well outside the range of the data: the heavy lines on the figure show that the greatest weight actually observed in such a short individual was below 50 kg. It is not to be expected that a model will give good predictions far outside the range of the data. This is particularly true of polynomial models, which are *curves of convenience*: they have no functional connection to the processes underlying the relationships in the data. The range of observed values of 'CalL2_L4BMD' in the data on which the predictions are based is from 0.5326 to 1.896. Thus the pattern of predictions presented here appears to account for individuals with low BMD fairly well, but fails to predict the occurrence of some individuals with very high values of BMD.

Other sets of predictions could be obtained from this model, and other curves could be drawn from them: in particular, the effect of 'Gender' could be represented. However, it is necessary to keep any display of predictions fairly simple in order to avoid a confusing mass of detail.

7.3 Use of R to fit the fixed-effect model with several variates and factors

To prepare the osteoporosis data for analysis by R, the value 'NA' (without quotation marks) is entered in each empty cell within the range of the data in the Excel workbook 'Intro to Mixed Modelling\Chapter 7\ost phenotypes.xls'. The data are then transferred from this workbook to the text file 'ost phenotypes.dat' in the same directory. Exclamation marks (!) are removed from the ends of headings in this file.

The following commands import the data, convert Gender and CentreNumber to factors, calculate the derived variables to be used in the models (Agesq, Htsq, HtxWt and Wtsq), and fit Model 7.1 to the data from the non-proband individuals:

```
osteoporosis.phenotypes <- read.table(
    "Intro to Mixed Modelling\\Chapter 7\\ost phenotypes.dat",
    header=TRUE)
attach(osteoporosis.phenotypes)
fGender <- factor(Gender)
fCentreNumber <- factor(CentreNumber)
Agesq = Age ** 2
Htsq = Height ** 2
HtxWt = Height * Weight
Wtsq = Weight ** 2
ostpheno.model1lmer <- lmer(CalL2.L4BMD ~
    fGender + Age + Agesq + Height +
    Weight + Htsq + HtxWt + Wtsq + (1|fCentreNumber),
    data = osteoporosis.phenotypes, subset = Proband == 'No')
```

The function `lmer()` requires that the package lme4 be loaded (see Chapter 3, Section 3.15). In this function, the argument 'subset' indicates that the probands are to be omitted from the modelling process.

It might be expected that the command

```
resostpheno <- residuals(ostpheno.model1lmer)
```

would extract the residual values from this model, to be used in diagnostic plots. However, at the time of writing it produces the following message:

```
Error: 'residuals' is not implemented yet
```

The following commands overcome this problem:

```
ostpheno.dframe <- data.frame(CalL2.L4BMD, Proband, fGender,
    Age, Agesq, Height, Weight, Htsq, HtxWt, Wtsq, fCentreNumber)
ostpheno.no.nas <- na.omit(ostpheno.dframe)
valid.response <-
    ostpheno.no.nas[[1]][ostpheno.no.nas[[2]] == 'No']
fitostpheno <- fitted(ostpheno.model1lmer)
resostpheno <- valid.response - fitostpheno
```

The last of these commands calculates each residual as the difference between an observed value of the response variable and the corresponding fitted value. (No attempt will be made to explain other aspects of the syntax of these commands. Their complexity is due to the fact that the data contain missing values, and the probands are excluded from the model-fitting process.) The following commands then produce the diagnostic plots of residuals, with the exception of the half-normal plot:

```
hist(resostpheno)
par(pty = "s")
qqnorm(resostpheno)
qqline(resostpheno)
plot(fitostpheno, resostpheno)
```

The plots obtained are equivalent to those in Figure 7.1.

The following command fits the same model, this time omitting the individuals with outlying values of BMD as well as the probands:

```
ostpheno.model1lmer <- lmer(CalL2.L4BMD ~
    fGender + Age + Agesq + Height +
    Weight + Htsq + HtxWt + Wtsq + (1|fCentreNumber),
    data = osteoporosis.phenotypes,
    subset = Proband == 'No' & BMD_Outlier == 'No')
```

Once again, special measures must be taken to obtain the residuals because the function `residuals()` is not available. In this case, the required commands are

```
ostpheno.dframe <-
    data.frame(CalL2.L4BMD, Proband, BMD_Outlier, fGender,
    Age, Agesq, Height, Weight, Htsq, HtxWt, Wtsq, fCentreNumber)
ostpheno.no.nas <- na.omit(ostpheno.dframe)
valid.response <-
    ostpheno.no.nas[[1]][ostpheno.no.nas[[2]] == 'No' &
    ostpheno.no.nas[[3]] == 'No']
fitostpheno <- fitted(ostpheno.model1lmer)
resostpheno <- valid.response - fitostpheno
```

The diagnostic plots commands from the `hist()` function onwards are then re-executed, giving plots equivalent to those in Figure 7.2. As the distribution of the residuals is now satisfactory, the results of the model are displayed by the following commands:

```
summary(ostpheno.model1lmer)
anova(ostpheno.model1lmer)
```

The output of these commands is as follows:

```
Linear mixed-effects model fit by REML
Formula: CalL2.L4BMD ~ fGender + Age + Agesq + Height + Weight +
    Htsq +
HtxWt + Wtsq + (1 | fCentreNumber)
    Data: osteoporosis.phenotypes
  Subset: Proband == "No" & BMD_Outlier == "No"
        AIC       BIC    logLik MLdeviance REMLdeviance
  -1840.231 -1780.854 930.1155  -1998.444    -1860.231
Random effects:
  Groups         Name        Variance    Std.Dev.
  fCentreNumber (Intercept) 0.00065455 0.025584
  Residual                  0.02860951 0.169143
# of obs: 2801, groups: fCentreNumber, 8

Fixed effects:
                Estimate  Std. Error t value
(Intercept) -4.8221e+00  8.1374e-01 -5.9259
fGender2      1.6117e-02  9.8110e-03  1.6428
Age          -2.0758e-03  1.2476e-03 -1.6638
Agesq        -8.8728e-06  1.2576e-05 -0.7056
Height        7.3461e-02  1.0629e-02  6.9112
Weight       -1.3891e-02  4.9586e-03 -2.8014
Htsq         -2.4217e-04  3.5897e-05 -6.7463
HtxWt         1.4515e-04  3.5016e-05  4.1453
Wtsq         -4.8176e-05  1.4884e-05 -3.2367

Correlation of Fixed Effects:
           (Intr) fGndr2 Age    Agesq  Height Weight Htsq   HtxWt
fGender2 -0.063
Age       0.085 -0.040
Agesq    -0.100  0.087 -0.984
Height   -0.980  0.021 -0.112  0.121
Weight    0.387 -0.017  0.020 -0.022 -0.554
Htsq      0.927 -0.008  0.111 -0.114 -0.981  0.682
HtxWt    -0.358  0.078 -0.055  0.048  0.514 -0.923 -0.665
Wtsq      0.077 -0.147  0.077 -0.059 -0.120  0.226  0.233 -0.581
Analysis of Variance Table
         Df Sum Sq Mean Sq
fGender   1 1.9263  1.9263
Age       1 7.2393  7.2393
Agesq     1 0.4092  0.4092
Height    1 2.0700  2.0700
Weight    1 3.1450  3.1450
Htsq      1 0.9580  0.9580
HtxWt     1 0.2215  0.2215
Wtsq      1 0.2997  0.2997
```

The anova table does not include F values, but these can be obtained from the corresponding mean squares together with the estimate of the residual variance. Thus

$$F_{age} = \frac{7.2393}{0.02860951} = 253.0382$$

which is nearly (but not quite) identical to the value of (Wald statistic)/DF given by GenStat when the terms are sequentially added to the fixed-effect model. R does not give the anova table obtained by dropping each term, but this is equivalent to the t values given by the summary() function. Thus for the term Agesq, $t = -0.7056$ and $F = t^2 = 0.50$. This agrees with the Wald statistic given by GenStat. As the effect of Agesq falls far short of significance, this term is dropped from the model. The resulting model is fitted, and its results displayed, by the following commands:

```
ostpheno.model2lmer <- lmer(CalL2.L4BMD ~
   fGender + Age + Height +
   Weight + Htsq + HtxWt + Wtsq + (1|fCentreNumber),
   data = osteoporosis.phenotypes,
   subset = Proband == 'No' & BMD_Outlier == 'No')
summary(ostpheno.model2lmer)
anova(ostpheno.model2lmer)
```

The output of these commands is as follows:

```
Linear mixed-effects model fit by REML
Formula: CalL2.L4BMD ~ fGender + Age + Height + Weight + Htsq +
   HtxWt + Wtsq + (1 | fCentreNumber)
   Data: osteoporosis.phenotypes
 Subset: Proband == "No" & BMD_Outlier == "No"
      AIC        BIC    logLik MLdeviance REMLdeviance
 -1862.463 -1809.023 940.2315  -1997.940   -1880.463
Random effects:
 Groups          Name         Variance    Std.Dev.
 fCentreNumber (Intercept) 0.00065774 0.025646
 Residual                    0.02860406 0.169127
# of obs: 2801, groups: fCentreNumber, 8

Fixed effects:
                Estimate   Std. Error   t value
(Intercept) -4.8793e+00  8.0962e-01   -6.0267
fGender2     1.6725e-02  9.7732e-03    1.7113
Age         -2.9421e-03  2.2049e-04  -13.3432
Height       7.4367e-02  1.0551e-02    7.0486
Weight      -1.3969e-02  4.9569e-03   -2.8182
Htsq        -2.4507e-04  3.5658e-05   -6.8727
HtxWt        1.4635e-04  3.4971e-05    4.1848
Wtsq        -4.8795e-05  1.4857e-05   -3.2843

Correlation of Fixed Effects:
              (Intr) fGndr2 Age     Height Weight Htsq    HtxWt
```

```
fGender2 -0.055
Age       -0.076  0.257
Height    -0.980  0.011   0.040
Weight     0.387 -0.015 -0.012 -0.556
Htsq       0.926  0.002 -0.009 -0.981  0.684
HtxWt     -0.355  0.074 -0.039  0.513 -0.923 -0.665
Wtsq       0.071 -0.143  0.107 -0.114  0.225  0.228 -0.580
Analysis of Variance Table
          Df Sum Sq Mean Sq
fGender   1 1.9263  1.9263
Age       1 7.2393  7.2393
Height    1 2.2194  2.2194
Weight    1 3.3246  3.3246
Htsq      1 1.0127  1.0127
HtxWt     1 0.2241  0.2241
Wtsq      1 0.3085  0.3085
```

Once again, the F values obtained from the mean squares and the residual variance component given by R correspond to the values of (Wald statistic)/DF given by GenStat, and the estimated coefficients of the fixed effects also agree, except for the intercept (constant), which differs due to the different parameterisation used by GenStat and R (Chapter 1, Section 1.11). Equation 7.1 can be adapted to give the best-fitting model in terms of R's parameterisation, namely

$$
\begin{aligned}
\text{CalL2\_L4BMD} = {} & -4.8793 + 0.016725 \times \text{Gender} - 0.0029421 \times \text{Age} \\
& + 0.074367 \times \text{Height} - 0.013969 \times \text{Weight} \\
& - 0.00024507 \times \text{Height}^2 + 0.00014635 \times \text{Height} \times \text{Weight} \\
& - 0.000048795 \times \text{Weight}^2 + \Gamma + E.
\end{aligned} \tag{7.2}
$$

Suitable values of the explanatory variables can be substituted into this equation, in order to obtain predicted values of the response variable.

In order to compare the deviances from models with and without 'CentreNumber', the model omitting this term must be fitted. However, it cannot simply be dropped from the model, as this would give a model with no random-effect terms, which would produce an error. In order to overcome this problem, a random-effect term that accounts for no variation is added to the model, as follows:

```
constant <- factor(rep(1, each = length(CalL2.L4BMD)))
ostpheno.model3lmer <- lmer(CalL2.L4BMD ~
    fGender + Age + Height +
    Weight + Htsq + HtxWt + Wtsq + (1|constant),
    data = osteoporosis.phenotypes,
    subset = Proband == 'No' & BMD_Outlier == 'No')
summary(ostpheno.model3lmer)
```

The function `rep()` repeats the value '1' as many times as there are values in the object 'CalL2.L4BMD', and the resulting list of ones is then stored in a factor named 'constant'. This constant term is used as the random-effect term in the mixed model that follows. The relevant part of the output from the function summary() is as follows:

```
Linear mixed-effects model fit by REML
Formula: CalL2.L4BMD ~ fGender + Age + Height + Weight + Htsq +
   HtxWt + Wtsq + (1 | constant)
  Data: osteoporosis.phenotypes
 Subset: Proband == "No" & BMD_Outlier == "No"
      AIC       BIC   logLik MLdeviance REMLdeviance
 -1828.035 -1774.595 923.0173  -1964.474    -1846.035
```

Comparing these results with those obtained from the full model (including the term 'CentreNumber'), we obtain

$$\text{deviance}_{\text{reduced model}} - \text{deviance}_{\text{full model}} = -1846.035 - (-1880.463) = 34.428$$

which agrees with the value given by GenStat.

7.4   A random-effect model with several factors

During a period when the area of canola (oilseed rape) grown in Western Australia was expanding rapidly, anecdotal evidence suggested that the variety that performed best at one location or season was not necessarily the best at another. A set of field trials was therefore undertaken with the aim of identifying a pattern in these genotype × environment interactions, in order to guide the choice of variety to be sown at different locations in future years. Eleven varieties, representative of those being grown in the region, were included in the trials, which were carried out at six representative locations. At each location, plots were sown at a range of dates, ranging from early to late in relation to the beginning of the growing season. The number of sowing dates at each location varied from 3 to 6, and not all varieties were sown at every location.

The first and last few rows of the spreadsheet holding the data are shown in Table 7.6. Each row represents the mean of all the plots of a particular variety sown in a particular location on the same date. In addition to the actual sowing date, the data include a nominal value for the sowing occasion (an integer from 1 to 6). The values of rainfall represent the amount of rainfall (mm) encountered by the plant during the growing season. They vary from row to row even within a location.sowing occasion combination, because plots of different varieties generally matured and were harvested on different dates. The variate 'mrainfall' is the mean value of rainfall over the location.sowing date combination in question. The variate 'oil' indicates the percentage of oil in the grain harvested. (Data reproduced by kind permission of G. Walton, Department of Agriculture, Western Australia. An alternative analysis of these data has been presented by Si and Walton (2004).)

Table 7.6 Data from a set of field trials of canola.

Only the first and last few rows of the data are shown. The conventions used in this spreadsheet are the same as those used in Table 1.1, Chapter 1.

	A	B	C	D	E	F	G
1	location!	sowing_occasion!	sowing_date!	variety!	rainfall	mrainfall	oil
2	Mullewa	1	02-May	Karoo	79.6	80.73	43.82
3	Mullewa	1	02-May	Monty	93.6	80.73	46.99
4	Mullewa	1	02-May	Hyola 42	93.6	80.73	44.34
5	Mullewa	1	02-May	Oscar	62.4	80.73	42.89
6	Mullewa	1	02-May	Drum	51.6	80.73	41.12
7	Mullewa	1	02-May	Mustard	103.6	80.73	
8	Mullewa	2	19-May	Karoo	62.4	65.97	40.05
9	Mullewa	2	19-May	Monty	88.6	65.97	44.68
·							·
·							·
·							·
137	Beverley	6	14-Jul	Hyola 42	11.8	19.00	37.20
138	Beverley	6	14-Jul	Oscar	22.0	19.00	35.00
139	Beverley	6	14-Jul	Rainbow	22.0	19.00	37.00

Because the locations and varieties are representative of those used in the region, it is reasonable to treat them as random-effect terms. It is reasonable to expect that there will be a trend over the sowings, from early to late, and it is therefore appropriate to fit sowing occasion as a fixed effect. The initial model (Model 7.2) is then as follows:

Response variate: oil
Fixed-effect model: sowing_occasion
Random-effect model: location + variety + sowing_occasion.location
                       + sowing_occasion.variety + location.variety
                       + sowing_occasion.location.variety.

The following statements import the data and fit this model:

```
IMPORT 'Intro to Mixed Modelling\\Chapter 7\\canola oil gxe.xls'
VCOMPONENTS [FIXED = sowing_occasion] \
   RANDOM = location + variety + \
   sowing_occasion.location + sowing_occasion.variety + \
   location.variety +  sowing_occasion.location.variety
REML oil
```

The complete model fitted here is 'sowing_occasion * location * variety'; however, in order to place 'sowing_occasion' in the fixed-effect model and the remaining terms in the random-effect model, it is necessary to spell out the terms individually. The output from the REML statement is as follows:

REML variance components analysis

Response variate: oil
Fixed model: Constant + sowing_occasion
Random model: location + variety + location.sowing_occasion
 + variety.sowing_occasion + location.variety + location.variety.sowing_occasion
Number of units: 126 (12 units excluded due to zero weights or missing values)

location.variety.sowing_occasion used as residual term

Sparse algorithm with AI optimisation

Estimated variance components

Random term	component	s.e.
location	3.8538	2.6620
variety	2.1905	1.0549
location.sowing_occasion		
	1.1202	0.4467
variety.sowing_occasion	0.1183	0.0703
location.variety	0.1148	0.0713

Residual variance model

Term	Factor	Model(order)	Parameter	Estimate	s.e.
location.variety.sowing_occasion					
		Identity	Sigma2	0.293	0.0608

Wald tests for fixed effects

Sequentially adding terms to fixed model

Fixed term	Wald statistic	d.f.	Wald/d.f.	chi pr
sowing_occasion	41.32	5	8.26	<0.001

Dropping individual terms from full fixed model

Fixed term	Wald statistic	d.f.	Wald/d.f.	chi pr
sowing_occasion	41.32	5	8.26	<0.001

Message: chi-square distribution for Wald tests is an asymptotic approximation (i.e. for large samples) and underestimates the probabilities in other cases.

GenStat detects that each location.variety.sowing_occasion specifies a unique observation, and this is therefore used as the residual term. As such, it need not be specified explicitly in future models. Each of the other random-effect terms is larger than its own SE, suggesting that they should all be retained in the model, and the Wald test shows that the main effect of 'sowing_occasion' is highly significant.

Because 'sowing_occasion' is a factor (as indicated by the exclamation mark (!) in its heading in the spreadsheet), the numbers used to specify its levels have been treated as arbitrary: the amount of variation among levels has been calculated, but no linear trend from Level 1 to Level 6 has been sought. In order to seek such a trend it is necessary to create another data structure, 'vsowing_order', holding the same values but specified as a variate, not a factor, and to include this in the fixed-effect model. This is done by the following statements:

```
CALCULATE vsowing_occasion = sowing_occasion
VCOMPONENTS [FIXED = vsowing_occasion + sowing_occasion] \
    RANDOM = location + variety + sowing_occasion.location + \
    sowing_occasion.variety + location.variety
REML [PRINT = model, components, Wald, effects; \
    PTERMS = vsowing_occasion] oil
```

It might be thought that the actual sowing date, rather than the nominal sowing occasion, would provide a better estimate of any trend from early to late sowings. However, the effective difference between any two sowing dates in relation to plant development depends not only on the time elapsed, but also on the amount and distribution of rainfall in the interval. The sowing occasions were chosen to span the range from early to late as effectively as possible, and the numbers that designate them therefore constitute a reasonable explanatory variable. The output from these statements is as follows:

REML variance components analysis

Response variate: oil
Fixed model: Constant + vsowing_occasion + sowing_occasion
Random model: location + variety + location.sowing_occasion
+ variety.sowing_occasion + location.variety
Number of units: 126 (12 units excluded due to zero weights or missing values)

Residual term has been added to model

Sparse algorithm with AI optimisation
All covariates centred

Estimated variance components

Random term	component	s.e.
location	3.8538	2.6620
variety	2.1905	1.0549
location.sowing_occasion		
	1.1202	0.4467
variety.sowing_occasion	0.1183	0.0703
location.variety	0.1148	0.0713

Residual variance model

Term	Factor	Model(order)	Parameter	Estimate	s.e.
Residual		Identity	Sigma2	0.293	0.0608

Wald tests for fixed effects

Sequentially adding terms to fixed model

Fixed term	Wald statistic	d.f.	Wald/d.f.	chi pr
vsowing_occasion	39.37	1	39.37	<0.001
sowing_occasion	1.95	4	0.49	0.745

Dropping individual terms from full fixed model

Fixed term	Wald statistic	d.f.	Wald/d.f.	chi pr
sowing_occasion	1.95	4	0.49	0.745
vsowing_occasion	8.48	1	8.48	0.004

Message: chi-square distribution for Wald tests is an asymptotic approximation (i.e. for large samples) and underestimates the probabilities in other cases.

Table of effects for vsowing_occasion

−1.890   Standard error: 0.6492

The variance-component estimates for the random-effect terms are unchanged: the additional term affects only the fixed-effect part of the model. Moreover, the sum of the Wald statistics for sequentially adding 'vsowing_occasion' and 'sowing_occasion' to the model is equal to the Wald statistic for 'sowing_occasion' in the previous model, namely

$$39.37 + 1.95 = 41.32.$$

The term 'vsowing_occasion' does not account for any new variation: it accounts for the linear trend which is part of the variation due to 'sowing_occasion', and 'sowing_occasion' now accounts for the deviations from this trend. The Wald test shows that these deviations are non-significant. We are also given a Wald test for dropping 'vsowing_occasion' from the model while retaining 'sowing_occasion', but we should note that is not legitimate: the linear trend is marginal to the deviations from the trend, and must be retained in the model if the term representing the deviations is retained (see the discussion of marginality in Section 7.2). The table of effects shows that on average, the oil content of the canola grain harvested falls by 1.890 % from each sowing occasion to the next.

There is a strong relationship between the sowing date and the mean rainfall encountered by the plants at each location, as shown in Figure 7.4. An ordinary multiple regression analysis (not shown here) indicates that on average, the rainfall declines by 11.78 mm from each sowing occasion to the next. This may suggest that 'rainfall' could effectively replace 'vsowing_occasion' in the fixed-effect model. This change is made in the following statements:

```
VCOMPONENTS [FIXED = rainfall + sowing_occasion] \
    RANDOM = location + variety + \
    sowing_occasion.location + sowing_occasion.variety + \
    location.variety
REML [PRINT = model, components, Wald, effects; \
    PTERMS = rainfall] oil
```

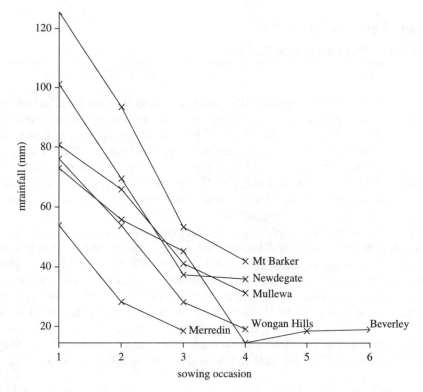

Figure 7.4 Relationship between sowing date and mean rainfall in field trials of canola at different locations.

This change to the model has little effect on the estimates of variance components (results not shown). However, the Wald tests and the estimate of the effect of rainfall from these statements are as follows:

Wald tests for fixed effects

Sequentially adding terms to fixed model

Fixed term	Wald statistic	d.f.	Wald/d.f.	chi pr
rainfallg	18.47	1	18.47	<0.001
sowing_occasion	24.49	5	4.90	<0.001

Dropping individual terms from full fixed model

Fixed term	Wald statistic	d.f.	Wald/d.f.	chi pr
sowing_occasion	24.49	5	4.90	<0.001
rainfall	1.02	1	1.02	0.313

Message: chi-square distribution for Wald tests is an asymptotic approximation (i.e. for large samples) and underestimates the probabilities in other cases.

> ## Table of effects for rainfall
>
> 0.006882   Standard error: 0.0068206

This outcome is perhaps surprising. The Wald tests indicate that the effect of rainfall is significant, as expected, but also that the effect of 'sowing_occasion' is once again highly significant. Moreover, the effect of rainfall on oil content is much weaker than might have been expected. The difference between successive sowings in the oil content of the canola grain and in the rainfall suggests that this effect should be about $(1.893\,\%)/(11.78\,\text{mm}) = 0.1707\,\%$ per mm, whereas the present analysis gives a value of only $0.006882\,\%$ per mm. The reason for this discrepancy is as follows.

Because the terms 'location', 'sowing_occasion' and 'location.sowing_occasion' are included in the model, and because the rainfall varies among varieties within each 'location.sowing_occasion' combination, the effect of rainfall is used to account only for variation within each combination. Variation in oil content among 'location.sowing_occasion' combinations is accounted for by the other terms: this gives the closest overall fit to the data. If we wish to discover the extent to which rainfall can account for the variation *among* 'location.sowing_occasion' combinations, we must specify a single rainfall value for each combination – that is, we must replace 'rainfall' by 'mrainfall' in the model. This is done in the following statements:

```
VCOMPONENTS [FIXED = mrainfall + sowing_occasion] \
   RANDOM = location + variety + \
   sowing_occasion.location + sowing_occasion.variety + \
   location.variety
REML [PRINT = model, components, Wald, effects; \
   PTERMS = mrainfall] oil
```

The Wald tests and the estimate of the effect of 'mrainfall' from these statements are now as follows:

> ## Wald tests for fixed effects
>
> Sequentially adding terms to fixed model
>
> | Fixed term | Wald statistic | d.f. | Wald/d.f. | chi pr |
> |---|---|---|---|---|
> | mrainfall | 34.83 | 1 | 34.83 | <0.001 |
> | sowing_occasion | 4.11 | 5 | 0.82 | 0.534 |
>
> Dropping individual terms from full fixed model
>
> | Fixed term | Wald statistic | d.f. | Wald/d.f. | chi pr |
> |---|---|---|---|---|
> | sowing_occasion | 4.11 | 5 | 0.82 | 0.534 |
> | mrainfall | 2.39 | 1 | 2.39 | 0.122 |

Message: chi-square distribution for Wald tests is an asymptotic approximation (i.e. for large samples) and underestimates the probabilities in other cases.

Table of effects for mrainfall

0.04173   Standard error: 0.027012

The Wald statistic for 'sowing_occasion' (sequentially added to the model) is once again reduced to non-significance, as it now accounts only for the small part of the variation among sowing occasions that is not related to rainfall. Conversely, the effect of rainfall is considerably increased, albeit not to the level that our preliminary calculation suggested.

Because the main effect of sowing_occasion is now non-significant, we may drop this term from the model. The variation that it accounted for will be absorbed by the terms 'location.sowing_occasion' and 'variety.sowing_occasion', and into the residual variation. (Note that this change does not breach the marginality criterion, for reasons explained in Section 7.2.) We should also consider the possibility that some varieties will respond more to rainfall than others, by adding the term 'variety.mrainfall' to the model. This term belongs in the random-effect model, because 'variety' is a random-effect term. Incorporating these changes, our model is specified by the following statements:

```
VCOMPONENTS [FIXED = mrainfall] RANDOM = location + \
    variety / mrainfall +  sowing_occasion.location + \
    sowing_occasion.variety + location.variety
REML [PRINT = model, components, Wald, effects; \
    PTERMS = mrainfall] oil
```

The variance components from these statements are as follows:

Estimated variance components

Random term	component	s.e.
location	1.6286	1.2957
variety	2.3654	1.1339
variety.mrainfall	0.0002	0.0001
location.sowing_occasion		
	1.3043	0.4561
variety.sowing_occasion	−0.0047	0.0337
location.variety	0.1458	0.0823

Residual variance model

Term	Factor	Model(order)	Parameter	Estimate	s.e.
Residual		Identity	Sigma2	0.286	0.0574

The estimated variance components due to 'location', 'variety', 'location.sowing_ occasion' and 'location.variety' are considerably larger than their respective SEs, indicating that these terms should be retained in the model. However, the estimated variance component due to 'variety.sowing_occasion' is now negative, and smaller than its SE, suggesting that the true value of this component is zero: this term can clearly be dropped from the model. The estimated variance component due to 'variety.mrainfall' appears very small, but it cannot be compared directly with the other components, as its magnitude depends on the units in which 'mrainfall' is measured. It requires further investigation, and the following statements obtain the deviance from models with and without this term:

```
VCOMPONENTS [FIXED = mrainfall] \
   RANDOM = location / sowing_occasion + variety / mrainfall + \
   location.variety
REML [PRINT = deviance] oil
VCOMPONENTS [FIXED = mrainfall] \
   RANDOM = location / sowing_occasion + variety + \
   location.variety
REML [PRINT = deviance] oil
```

Note that the terms 'location' and 'location.sowing_occasion' are now expressed more succinctly as 'location/sowing_occasion'. The output of these statements is as follows:

Deviance: −2*Log-Likelihood

Deviance	d.f.
138.29	118

Note: deviance omits constants which depend on fixed model fitted.

Deviance: −2*Log-Likelihood

Deviance	d.f.
148.03	119

Note: deviance omits constants which depend on fixed model fitted.

The difference between the deviances from the models with and without 'variety. mrainfall' is obtained, namely

$$\text{deviance}_{\text{reduced model}} - \text{deviance}_{\text{full model}} = 148.03 - 138.29 = 9.74$$

and

$$P(\text{deviance}_{\text{reduced model}} - \text{deviance}_{\text{full model}} > 9.74) = \frac{1}{2} \times P(\chi_1^2 > 9.74) = 0.000901.$$

The variance component 'variety.mrainfall' is significant, and should be retained in the model.

The model-building process is now complete, and the full results from fitting our final model are obtained by the following statements:

```
VCOMPONENTS [FIXED = mrainfall] \
    RANDOM = location / sowing_occasion + variety / mrainfall + \
    location.variety
REML [PRINT = model, components, Wald, effects; \
    PTERMS = mrainfall + variety.mrainfall; METHOD = Fisher] oil
REML [PRINT = means; PTERMS = 'constant' + location + variety; \
    METHOD = Fisher] oil
```

Note that we fit the model twice, the first time to obtain the effects of 'mrainfall' and 'variety.mrainfall' (together with the rest of the usual output), the second time to obtain the estimated mean oil content at each location and in each canola variety. In order to obtain SEs for these effects and means, it is necessary to fit the model by Fisher scoring, not by the more rapid average information (AI) algorithm. The output from these statements is as follows:

REML variance components analysis

Response variate: oil
Fixed model: Constant + mrainfall
Random model: location + location.sowing_occasion + variety
+ variety.mrainfall + location.variety
Number of units: 126 (12 units excluded due to zero weights or missing values)

Residual term has been added to model

Non-sparse algorithm with Fisher scoring
All covariates centred

Estimated variance components

Random term	component	s.e.
location	1.6279	1.2936
location.sowing_occasion		
	1.3065	0.4560
variety	2.3610	1.1370
variety.mrainfall	0.0002	0.0001
location.variety	0.1451	0.0825

Residual variance model

Term	Factor	Model(order)	Parameter	Estimate	s.e.
Residual		Identity	Sigma2	0.283	0.0490

Wald tests for fixed effects

Sequentially adding terms to fixed model

Fixed term	Wald statistic	d.f.	Wald/d.f.	chi pr
mrainfall	29.47	1	29.47	<0.001

Dropping individual terms from full fixed model

Fixed term	Wald statistic	d.f.	Wald/d.f.	chi pr

Message: chi-square distribution for Wald tests is an asymptotic approximation (i.e. for large samples) and underestimates the probabilities in other cases.

Table of effects for mrainfall

0.05612   Standard error: 0.010338

Table of effects for variety.mrainfall

variety	Drum	Dunkeld	Grouse	Hyola 42	Karoo
	−0.017340	0.000112	0.001748	0.008079	0.010514

variety	Monty	Mustard	Narendra	Oscar	Pinnacle
	0.012201	−0.024426	0.007548	0.001360	0.004589

variety	Rainbow
	−0.004385

Standard errors of differences

Average:	0.009594
Maximum:	0.01282
Minimum:	0.005642

Average variance of differences: 0.00009505

Table of predicted means for Constant

40.61   Standard error: 0.746

Table of predicted means for location

location	
Beverley	39.25
Merredin	40.80
Mt Barker	42.70
Mullewa	41.22
Newdegate	39.26
Wongan Hills	40.44

Standard errors of differences

Average:	0.4176
Maximum:	0.5766
Minimum:	0.3202

Average variance of differences: 0.1789

Table of predicted means for variety

variety	Drum	Dunkeld	Grouse	Hyola 42	Karoo
	37.91	42.66	42.36	41.44	38.90

variety	Monty	Mustard	Narendra	Oscar	Pinnacle
	42.31	39.92	40.13	40.08	40.10

variety	Rainbow
	40.95

Standard errors of differences

Average:	0.4938
Maximum:	0.6935
Minimum:	0.1524

Average variance of differences: 0.2690

In order to compare the amounts of variation accounted for by the various terms, we must convert the variance component for 'variety.mrainfall' to the same scale as the other variance components. For this purpose we must obtain the following values:

$$n = \text{number of observations in the data set,}$$

$$\text{mean(mrainfall)} = \frac{\sum_{i=1}^{n} \text{mrainfall}_i}{n}$$

where

$$\text{mrainfall}_i = \text{the } i\text{th value of mrainfall}$$

and

$$\text{var(mrainfall)} = \frac{\sum_{i=1}^{n} (\text{mrainfall}_i - \text{mean(mrainfall)})^2}{n}.$$

Note that var(mrainfall) is calculated on the basis of the individual values of 'mrainfall', using the number of values as the divisor: no attempt is made to determine the degrees of freedom of 'mrainfall', i.e. the number of independent pieces of information to which the variation in 'mrainfall' is equivalent.

We can then define the transformed variable

$$\text{mrainfall}' = \frac{\text{mrainfall} - \text{mean(mrainfall)}}{\sqrt{\text{var(mrainfall)}}},$$

which has a variance of 1. If we fit our mixed model using 'mrainfall'' in place of 'mrainfall', we will obtain the transformed variance component

$$\sigma'^2_{\text{variety.mrainfall}} = \text{var}(\text{mrainfall}) \cdot \sigma^2_{\text{variety.mrainfall}},$$

which is directly comparable with the other variance components. Applying these formulae to the data we obtain

$$n = 138$$

$$\text{mean}(\text{mrainfall}) = \frac{80.733 + 80.733 + \cdots + 19.000}{138} = 52.429$$

$$\text{var}(\text{mrainfall})$$
$$= \frac{(80.733 - 52.429)^2 + (80.733 - 52.429)^2 + \cdots + (19.000 - 52.429)^2}{138}$$
$$= 811.142$$

and

$$\sigma'^2_{\text{variety.mrainfall}} = 811.142 \times 0.000179847 = 0.1459$$

(giving $\sigma^2_{\text{variety.mrainfall}}$ with more precision than that with which it is presented in the GenStat output). The amount of variation accounted for by each term in the model can then be displayed in a fairly intuitive way, as shown in Table 7.7.

Table 7.7 Amount of variation in the oil content of canola accounted for by each term in the final model fitted

Source of variation in oil content (%)	Amount of variation accounted for
Total range	$47.95 - 34.95 = 13.00$
$2 \times SD_{\text{Total of random effects}}$	$2 \cdot \sqrt{\sigma^2_{\text{location}} + \sigma^2_{\text{location.sowing occasion}} + \sigma^2_{\text{variety}} + \sigma'^2_{\text{variety.mrainfall}} + \sigma^2_{\text{location.variety}} + \sigma^2_{\text{Residual}}} = 2 \times \sqrt{1.6279 + 1.3065 + 2.3610 + 0.1459 + 0.1451 + 0.2827} = 4.845$
$2 \times SD_{\text{location}}$	$2 \cdot \sqrt{\sigma^2_{\text{location}}} = 2 \times \sqrt{1.6279} = 2.552$
$2 \times SD_{\text{location.sowing occasion}}$	$2 \cdot \sqrt{\sigma^2_{\text{location.sowing occasion}}} = 2 \times \sqrt{1.3065} = 2.286$
$2 \times SD_{\text{variety}}$	$2 \cdot \sqrt{\sigma^2_{\text{variety}}} = 2 \times \sqrt{2.3610} = 3.073$
$2 \times SD_{\text{variety.mrainfall}}$	$2 \cdot \sqrt{\sigma'^2_{\text{variety.mrainfall}}} = 2 \times \sqrt{0.1459} = 0.764$
$2 \times SD_{\text{location.variety}}$	$2 \cdot \sqrt{\sigma^2_{\text{location.variety}}} = 2 \times \sqrt{0.1451} = 0.762$
$2 \times SD_{\text{Residual}}$	$2 \cdot \sqrt{\sigma^2_{\text{Residual}}} = 2 \times \sqrt{0.2827} = 1.063$
Range due to effect of rainfall	Rainfall effect $\times$ rainfall range $= 0.05612 \times (125.6 - 14.45) = 6.238$

When interpreting this table, it must be remembered that:

- $2 \times$ SD does not cover the full range of values of a random variable, and
- although variance components can be summed, the corresponding SDs cannot.

With these caveats, the table can be used to gain an impression of the relative magnitude of the different sources of variation. Of the total range in oil content of the harvested grain – 13 percentage points, or 32 % of its mean value of 40.62 % – only 1.254 percentage points are due to residual variation: the terms in the model account for most of the variation observed. About half the variation is accounted for by the effect of rainfall: on average, each additional millimetre of rainfall results in an additional 0.05671 percentage points of oil in the harvested grain. Somewhat less variation is due to the remaining effects of location and sowing occasion, and somewhat less again to the varieties. Much less variation is due to the variety.mrainfall interaction. These values give an indication of the extent to which the decisions of farmers and researchers, in choosing varieties or developing new ones, can be expected to influence the outcome of a growing season.

A line of best fit describing the relationship between oil content and rainfall can be constructed for each variety, as follows:

fitted oil content = variety mean oil content

+ (rainfall effect + variety.rainfall effect) $\times$ (rainfall − mean(mrainfall)).

For example, at the lowest value of rainfall encountered by the variety 'Drum', 11 mm, the fitted value of oil content of this variety is

$$37.93 \% + (0.05612 \%/mm - 0.017340 \%/mm) \times (11\,mm - 52.43\,mm) = 36.32 \%$$

and at the highest value, 86.4 mm, the fitted value is

$$37.93 \% + (0.05612 \%/mm - 0.017340 \%/mm) \times (86.4\,mm - 52.43\,mm)$$
$$= 39.24 \%.$$

The fitted line for each variety, together with an indication of the range of variation due to terms not represented in the fitted lines, and the observed values for the contrasting varieties 'Drum' and 'Monty', are presented graphically in Figure 7.5.

This graph will give researchers and farmers a clear idea of the relative importance of rainfall, the choice of variety and other factors in determining the oil content of a harvested canola crop. It may indicate that particular varieties should be chosen for particular environments: Mustard (not strictly a canola variety) is among the low-oil varieties, but this disadvantage is less pronounced in low-rainfall environments, and may be compensated by other advantages, making Mustard a more suitable choice

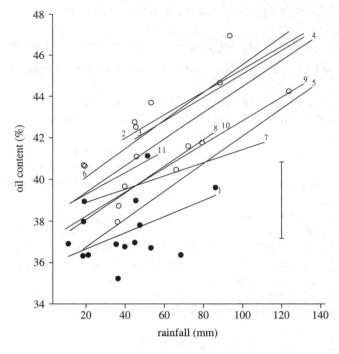

Figure 7.5 Predictions of oil content of canola on the basis of rainfall and variety, from the set of field trials.

for such environments. However, the graph shows that in general the variety.rainfall effects are small relative to the main effects of variety, so that such *crossovers* are fairly rare. The vertical bar, and the scatter of the observed values around their respective fitted lines, indicate the magnitude of the variation not accounted for by the lines. The variance-component estimates show that most of this is not residual variation, but consistent effects of location and location.sowing date combination, which may be amenable to further investigation.

7.5   Use of R to fit the random-effect model with several factors

To prepare the canola oil data for analysis by R, the value 'NA' (without quotation marks) is entered in each empty cell within the range of the data in the Excel workbook 'Intro to Mixed Modelling\Chapter 7\canola oil gxe.xls'. The data are then transferred from this workbook to the text file 'canola oil gxe.dat' in the same directory. Exclamation marks (!) are removed from the ends of headings in this file, and double quotation marks (") are placed around each item of data that contains a space (Hyola 42, Wongan Hills and Mt Barker).

The following statements import the data, convert 'location', 'sowing_occasion', 'sowing_date' and 'variety' to factors, and fit Model 7.2:

```
canola.oil <- read.table(
   "Intro to Mixed Modelling\\Chapter 7\\canola oil gxe.dat",
   header=TRUE)
attach(canola.oil)
flocation <- factor(location)
fsowing_occasion <- factor(sowing_occasion)
fsowing_date <- factor(sowing_date)
fvariety <- factor(variety)
canola.model1lmer <- lmer(oil ~ fsowing_occasion +
   (1|flocation) + (1|fvariety) +
   (1|fsowing_occasion:flocation) + (1|fsowing_occasion:fvariety) +
   (1|flocation:fvariety),
   data = canola.oil)
summary(canola.model1lmer)
anova(canola.model1lmer)
```

Note that the three-way interaction term 'sowing_occasion:location:variety' is not included explicitly in the model, as it was in the first GenStat model, but is recognised from the outset as the residual term. This is necessary in order to obtain a correct analysis from the function lmer().

The output of the summary() and anova() functions is as follows:

```
Linear mixed-effects model fit by REML
Formula: oil ~ fsowing_occasion + (1 | flocation) + (1 |
   fvariety) + (1 |
fsowing_occasion:flocation) + (1 | fsowing_occasion:fvariety) +
   (1 |
flocation:fvariety)
   Data: canola.oil
     AIC       BIC     logLik MLdeviance REMLdeviance
 376.1034 407.3025 -177.0517   362.5562     354.1034
Random effects:
 Groups                      Name        Variance Std.Dev.
 fsowing_occasion:fvariety  (Intercept) 0.11829  0.34393
 flocation:fvariety         (Intercept) 0.11481  0.33883
 fsowing_occasion:flocation (Intercept) 1.12021  1.05840
```

```
fvariety                        (Intercept) 2.19054   1.48005
flocation                       (Intercept) 3.85380   1.96311
Residual                                    0.29342   0.54168
# of obs: 126, groups: fsowing_occasion:fvariety, 51;
    flocation:fvariety, 31; fsowing_occasion:flocation, 25;
    fvariety, 11; flocation, 6


Fixed effects:
                    Estimate Std. Error t value
(Intercept)          42.57696    1.03038   41.321
fsowing_occasion2 -1.89020       0.64919   -2.912
fsowing_occasion3 -2.46637       0.64919   -3.799
fsowing_occasion4 -3.65347       0.68839   -5.307
fsowing_occasion5 -5.07736       1.29310   -3.927
fsowing_occasion6 -5.20236       1.29310   -4.023


Correlation of Fixed Effects:
              (Intr) fswn_2 fswn_3 fswn_4 fswn_5
fswng_ccsn2 -0.315
fswng_ccsn3 -0.315   0.500
fswng_ccsn4 -0.297   0.472   0.472
fswng_ccsn5 -0.158   0.251   0.251   0.250
fswng_ccsn6 -0.158   0.251   0.251   0.250   0.269
Analysis of Variance Table
                    Df   Sum Sq Mean Sq
fsowing_occasion    5 12.1246   2.4249
```

The variance-component estimates agree with those given by GenStat, and comparison of the mean square for 'fsowing_occasion' with the estimate of the residual variance gives

$$F_{\text{fsowing\_occasion}} = \frac{2.4249}{0.29342} = 8.2643$$

which agrees with the value of (Wald statistic)/DF given by GenStat.

In an attempt to add 'sowing_occasion' to the model, to represent the linear trend over sowing occasions, the following command is run:

```
canola.model21mer <- lmer(oil ~
   sowing_occasion + fsowing_occasion +
   (1|flocation) + (1|fvariety) +
   (1|fsowing_occasion:flocation) +
   (1|fsowing_occasion:fvariety) + (1|flocation:fvariety),
   data = canola.oil)
```

However, this produces the following message:

```
Error in lmer(oil ~ sowing_occasion + fsowing_occasion + (1 |
    flocation) + :
Leading minor of order 7 in downdated X:'X is not
    positive definite
```

Because of the close relationship between 'sowing_occasion' and 'fsowing_occasion' (technically known as *partial aliasing*), these terms cannot both be included in the fixed-effect model. However, if 'fsowing_occasion' is moved to the random-effect model, the resulting model can be successfully fitted, i.e.

```
canola.model3lmer <- lmer(oil ~ sowing_occasion +
   (1|fsowing_occasion) + (1|flocation) + (1|fvariety) +
   (1|fsowing_occasion:flocation) + (1|fsowing_occasion:fvariety) +
   (1|flocation:fvariety),
   data = canola.oil)
summary(canola.model3lmer)
anova(canola.model3lmer)
```

The output of these commands is as follows:

```
Linear mixed-effects model fit by REML
Formula: oil ~ sowing_occasion + (1 | fsowing_occasion) + (1 |
   flocation) +
(1 | fvariety) + (1 | fsowing_occasion:flocation) + (1 |
fsowing_occasion:fvariety) + (1 | flocation:fvariety)
   Data: canola.oil
      AIC      BIC    logLik MLdeviance REMLdeviance
 380.4288 403.1191 -182.2144   364.4071    364.4288
Random effects:
 Groups                        Name        Variance    Std.Dev.
 fsowing_occasion:fvariety  (Intercept) 1.1436e-01 3.3817e-01
 flocation:fvariety         (Intercept) 1.1430e-01 3.3808e-01
 fsowing_occasion:flocation (Intercept) 9.9284e-01 9.9641e-01
 fvariety                   (Intercept) 2.1929e+00 1.4808e+00
 flocation                  (Intercept) 3.8423e+00 1.9602e+00
 fsowing_occasion           (Intercept) 1.4760e-10 1.2149e-05
 Residual                               2.9520e-01 5.4332e-01
# of obs: 126, groups: fsowing_occasion:fvariety, 51;
   flocation:fvariety, 31;
fsowing_occasion:flocation, 25; fvariety, 11; flocation, 6;
   fsowing_occasion, 6

Fixed effects:
                Estimate Std. Error t value
(Intercept)      43.36076    1.04196  41.615
sowing_occasion  -1.11023    0.16762  -6.623

Correlation of Fixed Effects:
            (Intr)
sowing_ccsn -0.418

Analysis of Variance Table
                Df Sum Sq Mean Sq
sowing_occasion  1  12.95   12.95
```

Comparison of the mean square for 'sowing_occasion' with the estimate of the residual variance gives

$$F_{\text{sowing\_occasion}} = \frac{12.95}{0.29520} = 43.87$$

The output does not specify the degrees of freedom of the denominator of this F statistic, so we will convert it to the corresponding Wald statistic by rearranging Equation 2.3 (Chapter 2):

$$\text{Wald statistic} = F \text{ statistic} \times \text{DF}_{\text{Wald statistic}} = 43.87 \times 1 = 43.87 \tag{7.3}$$

where

$$\text{DF}_{\text{Wald statistic}} = \text{DF}_{\text{sowing\_occasion}} = 1.$$

Interpreting the Wald statistic as a χ^2 statistic, its significance is given by

$$P(\chi_1^2 > 43.87) < 0.001.$$

The linear trend over sowing occasions is highly significant, and should be retained in the model, unless 'mrainfall' is substituted for it.

This substitution is made in the following commands:

```
canola.model41mer <- lmer(oil ~ mrainfall + fsowing_occasion +
   (1|flocation) + (1|fvariety) +
   (1|fsowing_occasion:flocation) + (1|fsowing_occasion:fvariety) +
   (1|flocation:fvariety),
   data = canola.oil)
summary(canola.model41mer)
anova(canola.model41mer)
```

It is now possible to return 'fsowing_occasion' to the fixed-effect part of the model, to obtain the same model as was fitted using GenStat at the corresponding stage of the modelling process. (It is debatable whether the effects of 'fsowing_occasion', *after allowing for the linear trend over sowing occasion or rainfall*, belong in the fixed- or the random-effect part of the model.) The output of these commands is as follows:

```
Linear mixed-effects model fit by REML
Formula: oil ~ mrainfall + fsowing_occasion + (1 | flocation) +
(1 | fvariety) + (1 | fsowing_occasion:flocation) + (1 |
   fsowing_occasion:fvariety) + (1 |
flocation:fvariety)
   Data: canola.oil
       AIC       BIC    logLik MLdeviance REMLdeviance
 381.9939 416.0293 -178.9970    361.303      357.9939
Random effects:
 Groups                     Name        Variance Std.Dev.
 fsowing_occasion:fvariety  (Intercept) 0.12077  0.34752
 flocation:fvariety         (Intercept) 0.11474  0.33873
```

```
fsowing_occasion:flocation (Intercept) 1.30155  1.14086
fvariety                   (Intercept) 2.19657  1.48208
flocation                  (Intercept) 2.14492  1.46455
Residual                               0.29218  0.54054
# of obs: 126, groups: fsowing_occasion:fvariety, 51;
   flocation:fvariety, 31; fsowing_occasion:flocation, 25;
   fvariety, 11; flocation, 6


Fixed effects:
                    Estimate Std. Error t value
(Intercept)        39.024629   2.472518 15.7834
mrainfall           0.041717   0.027013  1.5443
fsowing_occasion2  -0.889750   0.949647 -0.9369
fsowing_occasion3  -0.465341   1.468722 -0.3168
fsowing_occasion4  -1.075048   1.810014 -0.5939
fsowing_occasion5  -2.550808   2.179384 -1.1704
fsowing_occasion6  -2.694581   2.169983 -1.2418


Correlation of Fixed Effects:
            (Intr) mrnfll fswn_2 fswn_3 fswn_4 fswn_5
mrainfall   -0.932
fswng_ccsn2 -0.738  0.682
fswng_ccsn3 -0.887  0.881  0.774
fswng_ccsn4 -0.905  0.914  0.763  0.896
fswng_ccsn5 -0.766  0.774  0.645  0.758  0.772
fswng_ccsn6 -0.764  0.772  0.644  0.756  0.770  0.703
Analysis of Variance Table
                 Df  Sum Sq Mean Sq
mrainfall         1 10.1767 10.1767
fsowing_occasion  5  1.2012  0.2402
```

The variance components agree with those given by GenStat, and comparison of the mean squares in the anova table with the estimate of the residual variance gives F statistics that agree with the values of (Wald statistic)/DF given by GenStat.

In order to drop the non-significant main effect of 'fsowing_occasion' from the model, and to attempt to add the term 'fvariety:mrainfall', the following command is run next:

```
canola.model5lmer <- lmer(oil ~ mrainfall +
   (1|flocation) + (1|fvariety) + (1|fvariety:mrainfall) +
   (1|fsowing_occasion:flocation) + (1|fsowing_occasion:fvariety) +
   (1|flocation:fvariety),
   data = canola.oil)
```

However, this produces the following message:

```
Error in lmer(oil ~ mrainfall + (1 | flocation) + (1 | fvariety)
   + (1 |  :
       Ztl[[1]] must have 76 columns
```

The function `lmer()` is intended for fitting terms in which *factors* are nested or crossed: it cannot, at the time of writing, be used to nest the variate 'mrainfall' within the factor 'fvariety'. The term 'fvariety:mrainfall' is therefore omitted from the model, thus:

```
canola.model61mer <- lmer(oil ~ mrainfall +
    (1|flocation) + (1|fvariety) +
    (1|fsowing_occasion:flocation) + (1|fsowing_occasion:fvariety) +
    (1|flocation:fvariety),
    data = canola.oil)
summary(canola.model61mer)
```

The output of these commands is as follows:

```
Linear mixed-effects model fit by REML
Formula: oil ~ mrainfall + (1 | flocation) + (1 | fvariety) +
(1 | fsowing_occasion:flocation) +    (1 |
    fsowing_occasion:fvariety) + (1 |
flocation:fvariety)
    Data: canola.oil
     AIC      BIC    logLik MLdeviance REMLdeviance
 385.6088 405.4628 -185.8044   365.4062      371.6088
Random effects:
 Groups                      Name        Variance Std.Dev.
 fsowing_occasion:fvariety  (Intercept) 0.11518  0.33937
 flocation:fvariety         (Intercept) 0.11403  0.33769
 fsowing_occasion:flocation (Intercept) 1.29126  1.13634
 fvariety                   (Intercept) 2.19682  1.48217
 flocation                  (Intercept) 1.76561  1.32876
 Residual                                0.29523  0.54335
# of obs: 126, groups: fsowing_occasion:fvariety, 51;
    flocation:fvariety, 31;
fsowing_occasion:flocation, 25; fvariety, 11; flocation, 6

Fixed effects:
             Estimate Std. Error t value
(Intercept) 37.4890691  0.9003290  41.639
mrainfall    0.0577746  0.0096219   6.004

Correlation of Fixed Effects:
          (Intr)
mrainfall -0.553
```

The estimated variance components for 'fsowing_occasion:fvariety' and 'flocation: fvariety' are small. Dropping these terms from the model we obtain

```
canola.model7lmer <- lmer(oil ~ mrainfall +
   (1|flocation) + (1|fvariety) +
   (1|fsowing_occasion:flocation),
   data = canola.oil)
summary(canola.model7lmer)
anova(canola.model7lmer)
```

Note that the model terms 'flocation' and 'fsowing_occasion:flocation' must be represented explicitly: they cannot be represented by the composite term 'flocation/fsowing_occasion' as they were in GenStat. The output of these commands is as follows:

```
Linear mixed-effects model fit by REML
Formula: oil ~ mrainfall + (1 | flocation) + (1 | fvariety) + (1 |
fsowing_occasion:flocation)
   Data: canola.oil
     AIC      BIC    logLik MLdeviance REMLdeviance
 388.6436 402.825 -189.3218   372.3615     378.6436
Random effects:
 Groups                     Name          Variance Std.Dev.
 fsowing_occasion:flocation (Intercept) 1.22425  1.10646
 fvariety                   (Intercept) 2.28165  1.51051
 flocation                  (Intercept) 1.75217  1.32370
 Residual                                0.46225  0.67989
# of obs: 126, groups: fsowing_occasion:flocation, 25; fvariety,
   11; flocation, 6

Fixed effects:
            Estimate Std. Error t value
(Intercept) 37.4524847  0.8885251  42.151
mrainfall    0.0587187  0.0092853   6.324

Correlation of Fixed Effects:
          (Intr)
mrainfall -0.541
> anova(canola.model7lmer)
Analysis of Variance Table
          Df Sum Sq Mean Sq
mrainfall  1 18.486  18.486
```

Comparison of the mean square for 'mrainfall' with the residual variance gives

$$F_{\text{mrainfall}} = \frac{18.426}{0.46225} = 39.86,$$

Wald statistic$_{\text{mrainfall}}$ = $F_{\text{mrainfall}} \times DF_{\text{mrainfall}}$ = $39.86 \times 1 = 39.86$,

$$P(\chi_1^2 > 39.86) < 0.001,$$

Table 7.8 Amount of variation in the oil content of canola accounted for by each term in the final model fitted using the software R

Source of variation in oil content (%)	Amount of variation
Total range	$47.95 - 34.95 = 13.00$
$2 \times \text{SD}_{\text{Total of random effects}}$	$2 \cdot \sqrt{\sigma^2_{\text{location}} + \sigma^2_{\text{sowing occasion.location}} + \sigma^2_{\text{variety}} + \sigma^2_{\text{Residual}}} =$ $2 \times \sqrt{1.75217 + 1.22425 + 2.28165 + 0.46225} = 4.7834$
$2 \times \text{SD}_{\text{location}}$	$2 \cdot \sqrt{\sigma^2_{\text{location}}} = 2 \times \sqrt{1.75217} = 2.6474$
$2 \times \text{SD}_{\text{sowing occasion.location}}$	$2 \cdot \sqrt{\sigma^2_{\text{sowing occasion.location}}} = 2 \times \sqrt{1.22425} = 2.2129$
$2 \times \text{SD}_{\text{variety}}$	$2 \cdot \sqrt{\sigma^2_{\text{variety}}} = 2 \times \sqrt{2.28165} = 3.0210$
$2 \times \text{SD}_{\text{Residual}}$	$2 \cdot \sqrt{\sigma^2_{\text{Residual}}} = 2 \times \sqrt{0.46225} = 1.3598$
Range due to effect effect of rainfall	Rainfall effect $\times$ rainfall range $= 0.0587187 \times$ $(125.6 - 14.45) = 6.5266$

confirming that this term is highly significant. The estimated effect of rainfall, 0.0587187 oil percentage points per mm, is quite similar to that given by the final model fitted in GenStat, 0.05612, although the terms 'fvariety:mrainfall' and 'flocation:fvariety', retained in the GenStat model, have been dropped from the model fitted here. The variance-component estimates are used to display the amount of variation accounted for by each term in Table 7.8, equivalent to Table 7.7 constructed from the GenStat output.

The following command obtains the estimates of the random effects:

```
coef(canola.model7lmer)
```

The output from this command is as follows:

```
$"fsowing_occasion:flocation"
                 (Intercept)  mrainfall
1:Beverley          38.61728 0.05871868
1:Merredin          37.89113 0.05871868
1:Mt Barker         36.71599 0.05871868
1:Mullewa           38.87169 0.05871868
1:Newdegate         38.12855 0.05871868
1:Wongan Hills      35.47648 0.05871868
2:Beverley          37.61858 0.05871868
2:Merredin          37.27487 0.05871868
2:Mt Barker         36.98796 0.05871868
2:Mullewa           37.56505 0.05871868
2:Newdegate         37.00939 0.05871868
2:Wongan Hills      36.51009 0.05871868
```

```
3:Beverley            37.15666 0.05871868
3:Merredin            37.30432 0.05871868
3:Mt Barker           38.86144 0.05871868
3:Mullewa             37.53932 0.05871868
3:Newdegate           36.97246 0.05871868
3:Wongan Hills        39.08136 0.05871868
4:Beverley            37.50711 0.05871868
4:Mt Barker           38.60685 0.05871868
4:Mullewa             36.21643 0.05871868
4:Newdegate           36.84296 0.05871868
4:Wongan Hills        38.62794 0.05871868
5:Beverley            36.53329 0.05871868
6:Beverley            36.39493 0.05871868

$fvariety
            (Intercept)   mrainfall
Drum          34.88742 0.05871868
Dunkeld       39.44312 0.05871868
Grouse        39.16710 0.05871868
Hyola 42      38.31666 0.05871868
Karoo         35.80168 0.05871868
Monty         39.13012 0.05871868
Mustard       36.38658 0.05871868
Narendra      36.97191 0.05871868
Oscar         36.95291 0.05871868
Pinnacle      36.91611 0.05871868
Rainbow       38.00373 0.05871868

$flocation
              (Intercept)   mrainfall
Beverley         36.18291 0.05871868
Merredin         37.61402 0.05871868
Mt Barker        39.40223 0.05871868
Mullewa          37.99999 0.05871868
Newdegate        36.22652 0.05871868
Wongan Hills     37.28922 0.05871868
```

The coefficients in this output can be used to construct lines of best fit. For example, the lines describing the relationship between oil content and rainfall for each variety can be constructed as follows:

$$\text{fitted oil content}_{\text{Variety } i} = \text{Intercept}_{\text{Variety } i} + \text{mrainfall effect}_{\text{Variety } i} \times \text{rainfall}.$$

Thus at the lowest value of rainfall encountered by the variety 'Drum', 11 mm, the fitted value of oil content of this variety is

$$34.88742\,\% + (0.05871868\,\%/\text{mm}) \times (11\,\text{mm}) = 35.53\,\%$$

and at the highest value, 86.4 mm, the fitted value is

$$34.88742\,\% + (0.05871868\,\%/\text{mm}) \times (86.4\,\text{mm}) = 39.96\,\%.$$

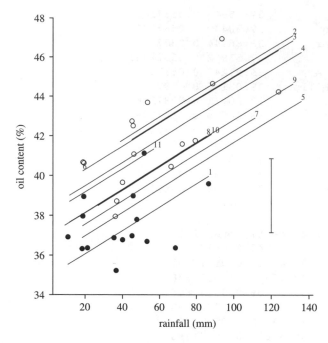

Figure 7.6 Predictions of oil content of canola on the basis of rainfall and variety, from the set of field trials, from the final model fitted using the software R.
For key see Figure 7.5.

The fitted line for each variety is presented in Figure 7.6, which is equivalent to Figure 7.5 constructed from the GenStat output. The two figures are fairly similar, despite the omission of the terms 'fvariety:mrainfall' and 'flocation:fvariety' from the final model fitted in R. The main difference is that the lines for all varieties are parallel in Figure 7.6, due to the omission of the 'flocation:fvariety' term.

7.6 Summary

The concepts introduced in earlier chapters are applied to two more elaborate data sets, one concerning the causes of variation in bone mineral density among human patients, the other concerning the causes of variation in oil content in a grain crop.

Some widely applicable new concepts and additional features of the mixed-modelling process are also introduced. These are summarised in Section 7.1.

7.7 Exercises

1. Return to the data presented in Exercise 2 in Chapter 1.
   (a) Fit the following model to these data:

Response variate: available chlorine
Fixed-effect model: linear, quadratic and cubic effects of time
Random-effect model: deviation of mean available chlorine at each
                               time from the value predicted by the
                               fixed-effect model.

(b) Investigate whether any terms can reasonably be dropped from this model.

(c) Obtain diagnostic plots of the residuals, and consider whether the assumptions
on which the analysis is based are reasonable.

(d) Plot the estimate of each random effect against the corresponding time, and
consider whether there is any evidence of a trend over time that is not accounted
for by the fixed-effect model.

Note that, as the level of available chlorine declines, the rate of decline becomes
lower, and this exponential decay process is more naturally modelled by a non-
linear function than by the polynomial function used for convenience here. The
model fitted by Draper and Smith (1998, Chapter 24, Section 24.3, pp 518–519) is

$$Y = \alpha + (0.49 - \alpha)e^{-\beta(X-8)} + \varepsilon$$

where
Y = available chlorine
X = time,

α and β are parameters to be estimated.

2. An experiment was performed to determine the effect of the enzyme lactase (which
hydrolyses the sugar lactose) on the composition of the milk in a sheep's udder. In
each side of the udder in each of eight lactating sheep, the levels of fat and of lactose
in the milk (%) were measured during a pre-experimental period. This was followed
by a treatment period, at the beginning of which lactase was injected into one side
of the udder, chosen at random. It was assumed that this injection did not affect
the composition of the milk in the other side, which therefore served as a control.
A day later, the levels of fat and lactose were measured again. This was followed
a week later by a second treatment period, at the beginning of which lactase was
injected into the other side of the udder. Again, a day later the fat and lactose
levels were measured. The results are presented in Table 7.9. (Data reproduced by
kind permission of Roberta Bencini, Faculty of Natural and Agricultural Sciences,
The University of Western Australia.)

The final levels of fat and lactose are the response variables in this experiment.

(a) Identify the block and treatment terms in this experiment. ('Period' might be
placed in either category: classify it as a treatment term.) Specify the block
structure and treatment structure models.

(b) Determine the effects of the block and treatment terms on each response variable, by analysis of variance and by mixed modelling. For each significant
relationship observed, state the nature of the effect, e.g. does lactase raise or
lower the level of lactose?

In each mixed-model analysis, the mean initial value of the response variable for
each sheep can be added to the model. The difference between this mean and the
corresponding individual initial value on each side of the udder in each sheep can
also be added.

Table 7.9 Fat and lactose content of sheep's milk, with and without injection of lactase into the udder

Sheep	Side	Period	Treatment	Fat (%)		post-treatment	Lactose (%)		post-treatment
				initial indiv. value	sheep mean		initial indiv. value	sheep mean	
Surprise	Right	1	lactase	6.06	5.990	4.81	3.17	3.600	3.57
Surprise	Left	1	control	5.92	5.990	5.70	4.03	3.600	4.06
159	Right	1	lactase	5.92	6.095	3.58	3.93	3.930	2.00
159	Left	1	control	6.27	6.095	5.06	3.93	3.930	2.98
338	Right	1	control	4.35	5.130	3.46	3.06	3.705	3.19
338	Left	1	lactase	5.91	5.130	5.82	4.35	3.705	3.22
356	Right	1	lactase	6.39	6.410	4.99	4.50	4.540	3.54
356	Left	1	control	6.43	6.410	4.86	4.58	4.540	4.01
369	Right	1	lactase	6.27	6.485	4.73	4.28	4.325	2.16
369	Left	1	control	6.70	6.485	5.21	4.37	4.325	4.03
389	Right	1	lactase	5.80	6.045	5.04	4.44	4.540	3.53
389	Left	1	control	6.29	6.045	5.11	4.64	4.540	4.56
477	Right	1	control	6.15	6.085	5.52	3.89	4.085	4.44
477	Left	1	lactase	6.02	6.085	5.43	4.28	4.085	3.62
486	Right	1	control	4.48	4.480	3.50	3.66	3.660	3.26
486	Left	1	lactase	4.48	4.480	2.95	3.66	3.660	0.37
Surprise	Right	2	control	6.06	5.990	5.79	3.17	3.600	3.93
Surprise	Left	2	lactase	5.92	5.990	7.00	4.03	3.600	0.66
159	Right	2	control	5.92	6.095	5.46	3.93	3.930	3.73
159	Left	2	lactase	6.27	6.095	5.48	3.93	3.930	4.60
338	Right	2	lactase	4.35	5.130	3.88	3.06	3.705	2.35
338	Left	2	control	5.91	5.130	4.37	4.35	3.705	3.73
356	Right	2	control	6.39	6.410	4.76	4.50	4.540	3.97
356	Left	2	lactase	6.43	6.410	6.12	4.58	4.540	3.71
369	Right	2	control	6.27	6.485	4.75	4.28	4.325	3.75
369	Left	2	lactase	6.70	6.485	5.52	4.37	4.325	3.79
389	Right	2	control	5.80	6.045	3.41	4.44	4.540	4.13
389	Left	2	lactase	6.29	6.045	4.11	4.64	4.540	4.39
477	Right	2	lactase	6.15	6.085	3.67	3.89	4.085	0.59
477	Left	2	control	6.02	6.085	3.66	4.28	4.085	3.73
486	Right	2	lactase	4.48	4.480	4.02	3.66	3.660	0.68
486	Left	2	control	4.48	4.480	4.01	3.66	3.660	3.76

(c) Should these terms be placed in the fixed-effect or the random-effect model?

(d) Make this change to each mixed model, and repeat the analysis. Is the final value of each response variate related to its initial value? Does the adjustment for the initial value give any improvement in the precision with which the effects of the treatments are estimated?

Table 7.10 Factors that were varied in an experiment to investigate influences on predation of seeds

	Factor						
species		cage_type		distance (m)	residue		
level	label	level	level no.		level	level no.	
wild oat (20 seeds/cage)	wo	no exclusion	1	0	no residue	1
wild radish (20 seeds/cage)	wr	exclusion by 2 cm mesh	2	25	standing wheat stubble	2
annual ryegrass (90 seeds/cage)	arg	exclusion by 1 cm mesh	3	50	standing wheat stubble + cut wheat straw (i.e. cages obscured by straw)	3
		exclusion by 'tac' gel around sides, but top remained open	4	100		

3. An experiment was performed to investigate the factors that influence predation of seeds lying on the ground in an area on which a crop has been grown. Seeds of three species were placed on the ground in cages of four designs that excluded different types of predators, in areas on which three types of crop residue were present, and at four distances from an area of bushland. The levels of these treatment factors are given in Table 7.10.

  This experimental area was divided into three blocks, and each block was further divided into three strips. Each strip had a boundary with the bushland at one end. Each strip received a crop residue treatment, each treatment being applied to one randomly chosen strip in each block. In each strip, one cage of each design was placed at each distance from the edge of the bushland. At each distance, the four cages in each strip were positioned at random. Each cage contained seeds of all three species. After 15 days, the number of seeds of each species that remained in each cage was recorded, and the percentage predation was calculated. The layout of the experiment is illustrated in Figure 7.7.

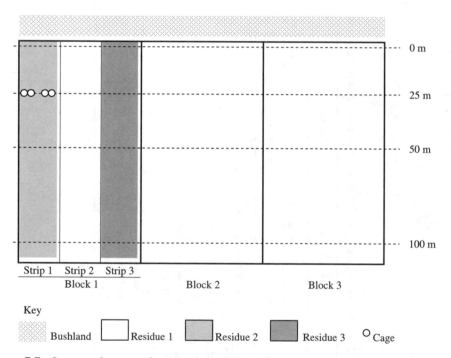

Figure 7.7   Layout of an experiment to investigate influences on predation of seeds. Crop residues are shown only in Block 1. Cages are shown only at 25 m from the bush in Strip 1 of Block 1.

  The first and last few rows of the spreadsheet holding the data are presented in Table 7.11: the full data set is held in the file 'seed predation.xls' (www.wiley.com/go/mixed_modelling). (Data reproduced by kind permission of Robert Gallagher.)

  Distance, residue, cage_type and species are to be classified as treatment factors.

(a) Identify any treatment factors for which randomisation cannot be performed.

Table 7.11 Percentage of predation of seeds of three species at different levels of three experimental variables

	A	B	C	D	E	F
1	distance	block	residue	cage_type	species	%pred
2	0	1	2	1	wo	100.0
3	0	1	2	2	wo	95.0
4	0	1	2	3	wo	30.0
5	0	1	2	4	wo	10.0
6	0	1	1	1	wo	95.0
7	0	1	1	2	wo	100.0
8	0	1	1	3	wo	100.0
9	0	1	1	4	wo	10.0
10	0	1	3	1	wo	100.0
11	0	1	3	2	wo	30.0
12	0	1	3	3	wo	100.0
13	0	1	3	4	wo	10.0
14	0	2	3	1	wo	10.0
15	0	2	3	2	wo	0.0
16	0	2	3	3	wo	5.0
17	0	2	3	4	wo	0.0
.						.
.						.
.						.
430	100	3	3	1	arg	100.0
431	100	3	3	2	arg	12.2
432	100	3	3	3	arg	23.3
433	100	3	3	4	arg	11.1

(b) Determine the block and treatment structures of this experiment.

(c) Analyse the data by analysis of variance.

(d) Perform an equivalent analysis on the data by mixed modelling, and obtain diagnostic plots of the residuals. Note any indications that the assumptions underlying the analysis may not be fulfilled, and consider why this may be the case.

(e) Create a variate holding the same values as the factor 'distance', and add this to the mixed model. Consider carefully whether the factor 'distance' should now be regarded as a fixed-effect term or a random-effect term. Consider also the consequences of your decision with regard to the interaction of 'distance' with other terms. Fit your new model to the data.

(f) Assuming that any departures from the assumptions underlying analysis of variance and mixed modelling are not so serious as to render these analyses invalid, interpret the results of the analyses performed in Part (e). In particular, consider the following points:

(i) Is there a linear trend in the level of predation over distance? If so, does the level increase or decrease with increasing distance from the bushland?

(ii) Does this trend vary among the species? If so, what is the nature of this variation?

(iii) Does this trend vary according to the crop residue? If so, what is the nature of this variation?

(iv) Does this trend vary according to the cage type? If so, what is the nature of this variation?

(v) Is there evidence of variation among the levels of distance that cannot be accounted for by the linear trend?

(vi) Is there evidence of three-way or four-way interactions among the treatment factors?

(g) Obtain predicted values of predation at representative distances from the bushland for each species, crop residue and cage type. Make a graphical display of the relationship between these variables.

A refinement of this analysis, taking account of the problems identified in the diagnostic plots, is introduced in Exercise 2, Chapter 9. These data have also been analysed and interpreted by Spafford-Jacob *et al.* (2005).

4. An experiment was conducted to determine the effect of anoxia (lack of oxygen) on the porosity of roots in nine genotypes of wheat. At an early stage of development, control plants of each genotype were grown in a well-aerated solution, while treated plants of each genotype were grown in stagnant agar which deprived the roots of oxygen. Following this treatment, the plants were grown in pots on a glasshouse bench, in a randomised complete block design with three replications. At the end of the experiment, seminal roots (i.e. those that develop directly from the seed) and nodal roots (i.e. those that originate from a node of the plant at a later stage) were harvested from each plant, and their porosity (% air space per root volume) was measured. On three additional plants of each genotype, the porosity of the seminal root was measured at an early stage of development, prior to the stagnant agar/aerated solution treatment. One of these three plants was assigned to each of the three blocks. The first and last few rows of the spreadsheet holding the data are presented in Table 7.12: the full data set is held in the file 'root porosity.xls' (www.wiley.com/go/mixed_modelling). (Data reproduced by kind permission of Michael McDonald.)

Genotype, treatment and root type are to be classified as treatment factors.

(a) Determine the block and treatment structures of this experiment.

(b) Analyse the response variable 'porosity_final' by analysis of variance and by mixed modelling.

The term 'porosity_initial' can be added to the mixed model.

(c) Should this term be placed in the fixed-effect or the random-effect model?

(d) Make this change to the mixed model, and repeat the analysis. Is the final value of porosity related to its initial value? Does the adjustment for the initial value

Table 7.12 Porosity of roots of a range of wheat genotypes with and without exposure to anoxia

	A	B	C	D	E	F	G
1	block	plant	genotype	treatment	root_type	porosity_initial	porosity_final
2	1	1	1	anoxic	seminal	3.572	4.630
3	1	1	1	anoxic	nodal	3.572	17.145
4	1	2	7	anoxic	seminal	2.722	2.589
5	1	2	7	anoxic	nodal	2.722	19.791
6	1	3	9	anoxic	seminal	3.477	7.048
7	1	3	9	anoxic	nodal	3.477	17.819
8	1	4	2	anoxic	seminal	3.341	5.110
9	1	4	2	anoxic	nodal	3.341	20.098
10	1	5	5	anoxic	seminal	0.904	0.481
11	1	5	5	anoxic	nodal	0.904	20.040
12	1	6	3	anoxic	seminal	2.701	6.204
13	1	6	3	anoxic	nodal	2.701	19.670
14	1	7	1	control	seminal	3.572	1.146
15	1	7	1	control	nodal	3.572	3.109
·							·
·							·
·							·
106	3	17	5	anoxic	seminal	2.562	2.469
107	3	17	5	anoxic	nodal	2.562	16.963
108	3	18	5	control	seminal	2.562	1.754
109	3	18	5	control	nodal	2.562	4.435

give any improvement in the precision with which the effects of the treatments are estimated?

(e) Interpret the results of your analysis. In particular, consider the following points:

(i) Is the porosity of roots affected by anoxia, and does it vary between genotypes and root types? If so, what is the direction of the effects of anoxia and root type? Which genotype has the most porous roots, and which the least?

(ii) Is there evidence of two-way interactions between anoxia, root type and genotype? If so, what is the nature of these interactions?

(iii) Is there evidence of a three-way interaction between these factors?

These data have also been analysed and interpreted by McDonald *et al.* (2001).

5. An experiment was performed to determine the effect of nutritionally inadequate diets on the quality of sheep's wool. The experiment consisted of a 21-day pre-experimental period (ending on 22 October 1997), a 12-week restricted-intake

period (ending on 14 January 1998) and a 6-week recovery period (ending on 25 February). Three experimental treatments were defined:

- Treatment 1. A control dietary regime.
- Treatment 2. An energy-deficient/protein-deficient nutritional regime.
- Treatment 3. An energy-deficient/protein-adequate nutritional regime.

Animals in the control group were fed at $1.3\times$ the level required for maintenance ($1.3\,\text{M}$) throughout the experiment. The energy-deficient/protein-deficient group were fed at $0.75\,\text{M}$ during the restricted-intake period, then *ad libitum* up to $2\,\text{M}$ during the recovery period. The energy-deficient/protein-adequate group received the same diet as the energy-deficient/protein-deficient group except that they received a supplement of 3.3 g/day of active methionine throughout the experimental period.

Twenty-one Merino wether hoggets were used in the experiment. Following the 21-day adjustment period the animals were divided into seven groups on the basis of live weight, and one randomly chosen animal from each group was allocated to each treatment. Measurements of the live weight and the fibre diameter of the wool of each animal were taken at the end of the pre-experimental period, and measurements of fibre diameter were also taken during the recovery period. The results are presented in Table 7.13. (Data reproduced by kind permission of Rachel Kirby.)

The experimental design includes a block factor which is not shown explicitly in Table 7.13.

(a) Assign an appropriate block number to each row of the data.

(b) Analysing measurements made on each occasion separately, and using the block factor and the treatment factor as model terms, perform analyses of variance on the measurements of fibre diameter made during the restricted-intake period and the recovery period. Interpret the results.

(c) Stack the measurements made on these three occasions into a single variate, and perform a new analysis by mixed modelling. Include the following variables in your initial model:

  (i) the time at which the measurement was made;

  (ii) the fibre diameter from the same sheep during the pre-experimental period;

  (iii) the live weight of the sheep during the pre-experimental period.

Determine whether any terms can reasonably be dropped from the model. If so, make the appropriate changes and re-fit the model.

(d) Interpret the results of your analysis. In particular, consider the following points:

  (i) Is the fibre diameter influenced by the sheep's pre-experiment live weight?

  (ii) Is the fibre diameter during the experiment related to the fibre diameter in the pre-experimental period?

  (iii) Is the fibre diameter influenced by the treatment, the date (i.e. the time elapsed since the restricted-intake period) and the treatment $\times$ date interaction? If so, what is the direction or nature of these effects?

Table 7.13 Live weight and fibre diameter of sheep on a range of nutritional regimes

Tag	Treatment	Live weight (kg), 20 October 1997	Fibre diameter (microns)			
			22 October 1997	07 January 1998	28 January 1998	18 February 1998
100	3	36.5	20.44	13.92	15.43	19.90
170	2	33.5	16.62	16.38	17.09	16.68
1519	1	34.0	18.12	16.30	16.74	19.22
162	1	35.0	18.55	15.74	15.25	18.02
1582	3	32.0	19.33	15.77	16.41	19.22
175	2	35.0	18.26	16.85	17.47	18.08
124	1	35.0	20.35	19.39	18.94	21.68
95	1	33.0	18.27	16.30	16.95	19.70
1497	2	35.0	17.95	18.55	19.78	20.01
1337	3	34.0	19.12	14.78	16.52	19.27
62	1	31.5	20.88	17.58	18.52	21.22
1420	3	33.5	17.05	13.18	14.35	17.12
171	2	33.0	18.21	17.77	18.44	19.14
1556	2	33.5	18.05	16.59	17.55	17.93
167	1	35.5	18.16	14.67	17.61	20.87
1048	1	33.0	20.63	14.93	15.58	19.40
1083	3	35.0	17.97	13.64	14.86	18.66
1422	2	35.5	17.70	16.69	17.74	18.64
1042	2	31.5	18.41	16.90	17.96	19.18
1065	3	32.5	18.57	14.64	15.86	19.09
142	3	34.0	20.08	16.20	18.48	21.71

(e) Obtain predicted values of fibre diameter for each treatment at representative times, and make a graphical display of the relationship between these variables.

6. The data on the yield of F_3 families of wheat presented in Exercise 3 in Chapter 3 are a subset from a larger investigation, in which two crosses were studied, and the F_3 families were grown with and without competition from ryegrass. The experiment had a split-split plot design, with the following relationship between the block and treatment structures:

- 'cross' was the treatment factor that varied only between main plots;
- 'family' was the treatment factor that varied between sub-plots within each main plot, but not between sub-sub-plots within the same sub-plot;
- presence or absence of ryegrass was the treatment factor that varied between sub-sub-plots within the same sub-plot.

The mean grain yield per plant was determined in each sub-sub-plot. (When the subset of the data is considered in isolation, it has a randomised block design, as described in the earlier exercise.) The first and last few rows of the spreadsheet holding the data are presented in Table 7.14: the full data set is held in the file 'wheat with ryegrass.xls' (www.wiley.com/go/mixed_modelling). (Data reproduced by kind permission of Soheila Mokhtari.)

Table 7.14　Yield per plant (g) of F_3 families from two crosses between inbred lines of wheat, grown with and without competition from ryegrass in a split-split plot design. 0 = ryegrass absent; 1 = ryegrass present.

	A	B	C	D	E	F	G	H
1	rep	mainplot	subplot	subsubplot	cross	family	ryegrass	yield
2	1	1	1	1	1	29	0	15.483
3	1	1	1	2	1	29	1	5.333
4	1	1	2	1	1	26	0	13.4
5	1	1	2	2	1	26	1	3.483
6	1	1	3	1	1	40	0	11.817
7	1	1	3	2	1	40	1	3.583
.								.
.								.
.								.
382	2	2	47	1	2	4	0	7.55
383	2	2	47	2	2	4	1	4.02
384	2	2	48	1	2	15	0	6.867
385	2	2	48	2	2	15	1	3.183

(a) Analyse the data according to the experimental design, both by analysis of variance and by mixed modelling.

(b) Modify your mixed model so that family-within-cross is regarded as a random-effect term.

(c) Interpret the results of your analysis. In particular, consider the following points:

　(i) Does the presence of ryegrass affect the yield of the wheat plants? If so, in which direction?

　(ii) Is there a difference in mean yield between the crosses?

　(iii) Do the families within each cross vary in yield?

　(iv) Do the effects of crosses and families interact with the effect of ryegrass?

(d) Compare the results with those that you obtained from the subset of the data in Chapter 3.

7. Return to the data set concerning the vernalisation of F_3 chickpea families, introduced in Exercise 2 in Chapter 5.

(a) Use analysis of variance methods to obtain the estimated mean number of days to flowering of vernalised plants of each F_3 family.

(b) Create a new variable, with one value for each control plant, giving the corresponding family mean for vernalised plants.

(c) Fit a mixed model in which the response variable is the number of days to flowering of the control plants, and the model terms are:

　• the corresponding family-mean value for vernalised plants,

　• family,

- plant group within family, and
- plant within group.

Interpret the results.

(d) Obtain estimates of the constant, and of the effect of days to flowering in vernalised plants.

(e) Display the results of your analysis graphically, showing the linear trend relating the number of days to flowering in the control plants to that in the vernalised plants and the mean values of these variables for each family.

(f) Obtain estimates of the amount of variation in the number of days to flowering of the control plants that is accounted for by each term in the mixed model.

(g) How can the analysis be interpreted to give an estimate of the effect of omitting the low-temperature stimulus on the subsequent development of plants in each family?

8

The use of mixed models for the analysis of unbalanced experimental designs

8.1 A balanced incomplete block design

In Chapter 2 we saw that the analysis of variance of a standard, balanced experimental design, the split plot design, could be viewed as the fitting of a mixed model, the treatment terms being the fixed-effect terms and the block terms being the random-effect terms. One of the major uses of mixed modelling is the analysis of experiments that cannot be tackled by anova, because it has not been possible to achieve exact balance during the design phase.

A *balanced incomplete block design* presented by Cox (1958, Chapter 11, Sections 11.1 and 11.2, pp 219–230) provides the starting point for a simple illustration of this problem. The data are displayed in a spreadsheet in Table 8.1. (Data reproduced by permission of John Wiley & Sons, Inc). This experiment comprises five treatments, T1 to T5. It is expected that there will be random variation in the response from day to day, and 'day' is therefore to be included in the analysis as a block term. However, each treatment does not occur on every day: though there are five treatments, there are only three observations per day, so this is clearly not an ordinary randomised complete block design. Each block is *incomplete*. Such a compromise is often necessary: nature does not always provide groups of experimental units (areas of land, litters of animals) that are the right size to permit the application of every treatment to one unit within every group. Despite this, the design of the experiment is *balanced*. Each treatment occurs exactly six times, and any pair of treatments occurs on the same day exactly three times: for example, Treatments T1 and T2 occur together on Days 3, 6 and 10, whereas Treatments T3 and T4 occur together on Days 5, 7 and 8. Therefore every possible comparison between two treatments is made with the same precision.

Introduction to Mixed Modelling: Beyond Regression and Analysis of Variance   N. W. Galwey
© 2006 John Wiley & Sons, Ltd

Table 8.1 Data from an experiment with a balanced incomplete block design. The conventions used in this spreadsheet are the same as those used in Table 1.1, Chapter 1.

	A	B	C
1	day!	T!	response
2	1	T4	4.43
3	1	T5	3.16
4	1	T1	1.40
5	2	T4	5.09
6	2	T2	1.81
7	2	T5	4.54
8	3	T2	3.91
9	3	T4	6.02
10	3	T1	3.32
11	4	T5	4.66
12	4	T3	3.09
13	4	T1	3.56
14	5	T3	3.66
15	5	T4	2.81
16	5	T5	4.66
17	6	T2	1.60
18	6	T3	2.13
19	6	T1	1.31
20	7	T3	4.26
21	7	T1	3.86
22	7	T4	5.87
23	8	T3	2.57
24	8	T5	3.06
25	8	T2	3.45
26	9	T2	3.31
27	9	T3	5.10
28	9	T4	5.42
29	10	T5	5.53
30	10	T1	4.46
31	10	T2	3.94

Consequently this design can be analysed by the ordinary methods of anova, as is done by the following GenStat statements:

```
IMPORT \
    'Intro to Mixed Modelling\\Chapter 8\\incmplt block design.xls'
```

```
BLOCKS day
TREATMENTS T
ANOVA [FPROB = yes] response
```

The output of the ANOVA statement is as follows:

Analysis of variance

Variate: response

Source of variation	d.f.	s.s.	m.s.	v.r.	F pr.
day stratum					
T	4	4.4503	1.1126	0.28	0.882
Residual	5	20.1348	4.0270	7.17	
day.*Units* stratum					
T	4	15.7533	3.9383	7.01	0.002
Residual	16	8.9915	0.5620		
Total	29	49.3298			

Information summary

Model term	e.f.	non-orthogonal terms
day stratum		
T	0.167	
day.*Units* stratum		
T	0.833	day

Message: the following units have large residuals.

day 5 *units* 2	−1.42	s.e.	0.55
day 8 *units* 3	1.18	s.e.	0.55

Tables of means

Variate: response

Grand mean 3.73

T	T1	T2	T3	T4	T5
	2.88	2.90	3.60	4.82	4.46

Standard errors of differences of means

Table	T
rep.	6
d.f.	16
s.e.d.	0.474

As in the analysis of the split plot design (Chapter 2, Section 2.2), the anova is divided into strata defined by the block term. However, whereas in the split plot design each treatment term was tested in only one stratum, in the present case the treatment term 'T' is tested both in the 'day' stratum and in the 'day.*Units*' stratum (the within-day stratum). This reflects the fact that each comparison between treatments is made partly between and partly within days. The value of $MS_{Residual}$ in the 'day' stratum is several times larger than that in the 'day.*Units*' stratum, confirming that the decision to treat days as blocks was justified. Consequently the comparisons among treatments are made with considerably more precision in the 'day.*Units*' stratum, and it is only in this stratum that the effect of 'T' is significant according to the F test. The *efficiency factor* (e.f.) in the *information summary* shows how the information concerning the effects of 'T' is distributed between the strata. In the present simple case, the proportion of the information on 'T' given by comparison among observations within each day is given by

$$\frac{\text{(no. of treatments)} \times \text{(no. of units per block} - 1)}{\text{(no. of treatments} - 1) \times \text{(no. of units per block)}} = \frac{5 \times (3-1)}{(5-1) \times 3} = 0.833. \quad (8.1)$$

The remaining proportion, 0.167, is given by comparison among the means for each day. The information summary also notes that 'T' is non-orthogonal to 'day' – that is, the effects of this term are estimated neither entirely within days, nor entirely between them.

The remainder of the output follows the pattern that we have seen in previous anovas. However, the methods for the calculation of the treatment means and of the $SE_{Difference}$ for comparisons between them are modified to take account of the incomplete block structure of the experiment. Each mean is not the simple mean of the observations for the treatment concerned, but is adjusted to allow for the effects of those blocks in which the treatment occurs (Cox, 1958). Thus the means are based only of the variation among observations within days, and the same is true of $SE_{Difference}$, which is given by

$$SE_{Difference} = \sqrt{\frac{2MS_{Residual, day.*Units*stratum}}{r \cdot \text{(efficiency factor, day.*Units*stratum)}}} = \sqrt{\frac{2 \times 0.5620}{6 \times 0.833}} = 0.474$$

$$(8.2)$$

where
$r = $ the number of replications of each treatment.

This approach to the treatment means of an incomplete block design was standard in the days when calculations were performed by hand, but it is preferable (and straightforward using statistical software) to obtain alternative estimates of the means and $SE_{Difference}$, taking account of the between-days variation. These estimates are produced by the statement

```
ANOVA [PRINT = cbmeans] response
```

The option setting 'PRINT = cbmeans' indicates that 'combined means', based on information from all strata in the anova, are to be presented (see Payne and Tobias, 1992). The output of this statement is as follows:

Tables of combined means					
Variate: response					
T	T1	T2	T3	T4	T5
	2.91	2.92	3.57	4.85	4.42

Standard errors of differences of combined means

Table	T
rep.	6
s.e.d.	0.463
effective d.f.	17.51

The treatment means are slightly different, though still not the same as the simple means. We will see later (Section 8.2) how this adjustment relative to the simple means is performed. The value of $SE_{Difference}$ is reduced from 0.474 to 0.463, showing that a gain in precision has been obtained by the fuller use of the information available.

8.2 Imbalance due to a missing block. Mixed-model analysis of the incomplete block design

Now suppose that it were only possible to conduct this experiment on nine days. The design would still be quite a good one: in the analysis without recovery of interblock information, only the comparisons between T1, T2 and T5 (the treatments applied on Day 10) would be made with slightly less precision. However, the design is no longer balanced, and no longer analysable by standard analysis of variance techniques. When the ANOVA statement is applied to the data with Day 10 omitted, it produces the following output:

Fault 11, code AN 1, statement 1 on line 11

Command: ANOVA [FPROB = yes] response
Design unbalanced - cannot be analysed by ANOVA
Model term T (non-orthogonal to term day) is unbalanced, in the day.*Units* stratum.

This unbalanced design can, however, be analysed by mixed-modelling methods. This is done by the following statements:

VCOMPONENTS [FIXED = T] RANDOM = day
REML [PRINT = Wald, means] response

The REML statement produces the following output:

Wald tests for fixed effects

Sequentially adding terms to fixed model

Fixed term	Wald statistic	d.f.	Wald/d.f.	chi pr
T	25.11	4	6.28	<0.001

Dropping individual terms from full fixed model

Fixed term	Wald statistic	d.f.	Wald/d.f.	chi pr
T	25.11	4	6.28	<0.001

Message: chi-square distribution for Wald tests is an asymptotic approximation (i.e. for large samples) and underestimates the probabilities in other cases.

Table of predicted means for Constant

3.593   Standard error: 0.2926

Table of predicted means for T

T	T1	T2	T3	T4	T5
	2.670	2.814	3.465	4.759	4.259

Standard errors of differences

Average:	0.5147
Maximum:	0.5399
Minimum:	0.4860

Average variance of differences: 0.2652

The Wald statistic indicates that the variation among the levels of 'T' is highly significant, as was indicated by the within-blocks F statistic in the anova of the complete experiment ($F_{4,16} = 7.01$, $P = 0.002$). The treatment means are similar, but not identical, to those given by analysis of variance on the complete experiment. Thus by using mixed modelling in place of analysis of variance, the requirement that the experiment should have exactly 10 blocks is overcome.

In order to compare the results of the anova and mixed-modelling approaches more closely, the complete (10-day) experiment can be analysed by mixed modelling. This is done with option settings in the REML statement that specify fuller output:

```
VCOMPONENTS [FIXED = T] RANDOM = day
REML [PRINT = model, components, Wald, means] response
```

The output of this statement is as follows:

REML variance components analysis

Response variate:	response
Fixed model:	Constant + T
Random model:	day
Number of units:	30

Residual term has been added to model

Sparse algorithm with AI optimisation

Estimated variance components

Random term	component	s.e.
day	0.6889	0.4227

Residual variance model

Term	Factor	Model(order)	Parameter	Estimate	s.e.
Residual		Identity	Sigma2	0.558	0.1966

Wald tests for fixed effects

Sequentially adding terms to fixed model

Fixed term	Wald statistic	d.f.	Wald/d.f.	chi pr
T	28.71	4	7.18	<0.001

Dropping individual terms from full fixed model

Fixed term	Wald statistic	d.f.	Wald/d.f.	chi pr
T	28.71	4	7.18	<0.001

Message: chi-square distribution for Wald tests is an asymptotic approximation (i.e. for large samples) and underestimates the probabilities in other cases.

Table of predicted means for Constant

3.733   Standard error: 0.2958

Table of predicted means for T

T	T1	T2	T3	T4	T5
	2.906	2.925	3.570	4.849	4.415

Standard error of differences: 0.4627

The value of (Wald statistic)/$DF_{Waldstatistic}$ in the mixed-model analysis (= 7.18) is slightly larger than the F statistic for the effect of 'T' in the 'day,*Units*' stratum of the anova (= 7.01), confirming that there has been a slight gain in statistical power from the incorporation of the between-days information into the test. The estimates of treatment means, and the value of $SE_{Difference}$ for comparisons between them, are the same as those obtained from the anova taking account of the between-days variation.

As noted above (Section 8.1), the treatment means are adjusted to take account of the incomplete block design of this experiment, and we can use the results of the mixed-modelling analysis to see how this adjustment is performed. The adjustment made to each mean is as shown in Table 8.2.

Table 8.2 Adjustment of treatment means to take account of an incomplete block design.

Treatment	Simple mean	Adjusted mean	Adjustment
T1	2.985	2.906	−0.079
T2	3.003	2.925	−0.078
T3	3.468	3.570	0.102
T4	4.940	4.849	−0.091
T5	4.268	4.415	0.147

These adjustments are based on the estimates of the day effects, which are printed by the following statement:

```
REML [PRINT = effects; PTERMS = day; METHOD = Fisher] response
```

The output of this statement is as follows:

Table of effects for day					
day	1	2	3	4	5
	−0.8348	−0.1965	0.6747	0.1097	−0.4473
day	6	7	8	9	10
	−1.1446	0.6995	−0.4803	0.6527	0.9668

Standard errors of differences

Average:	0.5591
Maximum:	0.5678
Minimum:	0.5547

Average variance of differences: 0.3126

The adjusted mean for each treatment is obtained from the formula

adjusted mean

$$= \text{simple mean} - \text{mean(effects of days on which the treatment occurs)}. \quad (8.3)$$

For example, Treatment T1 occurs on Days 1, 3, 4, 6, 7 and 10. The mean of the effects of these days is

$$\frac{-0.8348 + 0.6747 + 0.1097 - 1.1446 + 0.6995 + 0.9668}{6} = 0.07855$$

and

$$\text{adjusted mean}_{T1} = 2.985 - 0.07855 = 2.906.$$

If 'day' is treated as a fixed-effect term, i.e. if the VCOMPONENTS statement is changed from

```
VCOMPONENTS [FIXED = T] RANDOM = day
```

to

```
VCOMPONENTS [FIXED = T, day]
```

then the treatment means obtained are those in the original anova, based only on the within-day variation.

8.3   Use of R to analyse the incomplete block design

To prepare the incomplete block data for analysis by R, they are transferred from the Excel workbook 'Intro to Mixed Modelling\Chapter 8\incmplt block design.xls' to the text file 'incmplt block design.dat' in the same directory. Exclamation marks (!) are removed from the ends of headings in this file.

The following commands import the data, perform the appropriate anova and present the results:

```
incomplete.blk <- read.table(
    "Intro to Mixed Modelling\\Chapter 8\\incmplt block design.dat",
    header=TRUE)
attach(incomplete.blk)
fday <- factor(day)
fT <- factor(T)
incomplt.blk.modelaov <-
    aov(response ~ fT + Error(fday))
summary(incomplt.blk.modelaov)
model.tables(incomplt.blk.modelaov, type = "means", se = TRUE)
```

The output of these commands is as follows:

```
Error: fday
          Df  Sum Sq Mean Sq F value Pr(>F)
fT         4  4.4503  1.1126  0.2763 0.8817
Residuals  5 20.1348  4.0270

Error: Within
          Df  Sum Sq Mean Sq F value   Pr(>F)
fT         4 15.7533  3.9383  7.0081 0.001852 **
Residuals 16  8.9915  0.5620
---
Signif. codes:  0 '***' 0.001 '**' 0.01 '*' 0.05 '.' 0.1 ' ' 1

Tables of means
Grand mean

3.733


  fT
fT
    T1    T2    T3    T4    T5
 2.729 2.751 3.580 5.012 4.593
Warning message:
SEs for type  means  are not yet implemented in:
model.tables.aovlist(incomplt.blk.modelaov, type = "means", se =
    TRUE)
```

The anova agrees with that produced by GenStat. However, the estimated treatment means do *not* agree with those from GenStat, with or without taking account of the between-day information.

When the data from Day 10 are omitted, the output produced by R is as follows:

```
Error: fday
            Df  Sum Sq Mean Sq F value Pr(>F)
fT           4 10.3519  2.5880  0.9025 0.5384
Residuals    4 11.4707  2.8677

Error: Within
            Df  Sum Sq Mean Sq F value     Pr(>F)
fT           4 14.6689  3.6672  5.8599 0.005502 **
Residuals   14  8.7614  0.6258
---
Signif. codes:  0 '***' 0.001 '**' 0.01 '*' 0.05 '.' 0.1 ' ' 1
```

R does not give a warning that the design is unbalanced.

The following commands produce and present a mixed-model analysis of the data with Day 10 omitted:

```
incomplt.blk.modellme <- lme(response ~ fT,
    data = incomplete.blk, random = ~ 1|fday)
summary(incomplt.blk.modellme)
anova(incomplt.blk.modellme)
```

The output of these commands is as follows:

```
Linear mixed-effects model fit by REML
Fixed: response ~ fT
 Data: incomplete.blk
       AIC      BIC    logLik
  84.36937 93.44022 -35.18468

Random effects:
 Groups    Name        Variance Std.Dev.
 fday      (Intercept) 0.56135  0.74924
 Residual              0.62233  0.78888
# of obs: 27, groups: fday,  9

Fixed effects:
             Estimate Std. Error DF t value  Pr(>|t|)
(Intercept)   2.66997    0.44998 22  5.9335 5.691e-06 ***
fTT2          0.14413    0.53991 22  0.2669 0.7919969
fTT3          0.79496    0.50687 22  1.5684 0.1310666
fTT4          2.08950    0.50687 22  4.1224 0.0004475 ***
fTT5          1.58867    0.53991 22  2.9425 0.0075325 **
```

```
---
Signif. codes:  0 '***' 0.001 '**' 0.01 '*' 0.05 '.' 0.1 ' ' 1

Correlation of Fixed Effects:
      (Intr) fTT2    fTT3    fTT4
fTT2 -0.600
fTT3 -0.622   0.533
fTT4 -0.622   0.533   0.540
fTT5 -0.600   0.500   0.533   0.533

Analysis of Variance Table
     Df  Sum Sq Mean Sq   Denom F value    Pr(>F)
fT    4 15.6274  3.9069 22.0000  6.2778 0.001581 **
---
Signif. codes:  0 '***' 0.001 '**' 0.01 '*' 0.05 '.' 0.1 ' ' 1
```

The F value in the anova table agrees with the value of (Wald Statistic)/DF given by GenStat. In the estimates of the fixed effects, the mean for treatment T1 is used as the intercept, and its value, 2.66997, agrees with that given by GenStat. The effects of the other treatments are presented relative to this intercept, and the means are related to the effects by the formula

$$\text{mean(Treatment } i) = \text{intercept} + \text{effect(Treatment } i). \tag{8.4}$$

Thus
$$\text{mean(Treatment } 2) = 2.66997 + 0.14413 = 2.81410.$$

These values agree with those given by GenStat.

8.4 Relaxation of the requirement for balance: alpha designs

In the early literature on experimental design, exemplified by Cochran and Cox (1957), there was a strong emphasis on balance, which was necessary in order for the calculations required for statistical analysis to be feasible. However, the advent of electronic computers and the development of the mixed-modelling approach permitted the development of a range of designs in which the requirement for balance was relaxed, in order to allow the investigator to follow more closely the pattern of natural variation in his or her experimental material. A common problem, which occurs in many contexts, is that the number of experimental units that form a natural, homogeneous block is much smaller than the number of treatments to be compared. In this situation, a randomised block design will only take account of a small proportion of the natural variation, leaving an undesirably large residual variance. The merit of the balanced incomplete block design is that the number of experimental units per block is smaller than the number of treatments, which helps to overcome this problem. However, the possibility of finding an appropriate design of this type depends on the particular combination of:

- the number of treatments,

- the number of replications, and

- the number of units per block.

If the requirement for balance is relaxed, designs covering a much wider range of situations can be specified.

Hence the ability to apply mixed-model analyses to unbalanced designs is useful in many situations in which perfect balance is difficult to achieve. However, this does not exempt the investigator from trying to achieve balance. To illustrate this, consider two alternative designs for an experiment in which three treatments, A, B and C, are to be allocated to five blocks, each comprising three experimental units, as shown in Figure 8.1. Design 1 is a standard randomised block design, whereas in Design 2, the treatments have been randomly allocated to the units without regard to the block structure. If the block term is included in the model, Design 2 is unbalanced and cannot be analysed by standard anova methods; however, it can still be analysed by mixed modelling. In this design the treatment effects are estimated partly within, and partly between, blocks. For example, the difference between the means of Block 3 and Block 5 could be due either to natural variation among the blocks, or to a difference between the effects of Treatment B and Treatment C. If the variance component due to blocks is greater than zero – that is, if there is any real natural variation among the blocks – there will be a reduction in the precision with which the treatment effects are estimated, relative to that obtained from Design 1. No amount of sophistication in the analysis methods used can overcome this loss of efficiency, and the investigator should therefore seek an efficient experimental design, with balance or near-balance as one of the criteria, even though other designs are analysable.

Design 1				Design 2			
Block				Block			
1	B	A	C	1	C	B	A
2	C	B	A	2	B	C	A
3	C	A	B	3	B	B	A
4	B	C	A	4	B	A	C
5	B	C	A	5	A	C	C

Figure 8.1   Alternative designs for an experiment with three treatments.

Many extensions to the concept of the balanced incomplete block design have been devised to help the investigator in this task. One of these is *alpha designs*, which consist of incomplete blocks, each comprising only a small proportion of the treatments, but in which these blocks can be grouped together into complete replications. The treatments allocated to each block are chosen so that the design is nearly balanced. Alpha designs were originally devised for the analysis of plant breeding trials (Patterson and Williams, 1976), and a set of arrays for generating alpha designs with varying numbers of treatments, numbers of replications and block sizes was presented by Patterson *et al.* (1978). This task can be performed by several software systems, including GenStat. For example, the following statement produces a design with 23 treatments in three replicates, each replicate comprising five blocks:

```
AGALPHA [PRINT = design] LEVELS = 23; NREPLICATES = 3; \
    NBLOCKS = 5; SEED = 60594; \
    TREATMENTS = trtmnt; REPLICATES = rep; BLOCKS = blk; \
    UNITS = unt
```

The arbitrary value in the setting of the SEED parameter is used to initiate the randomisation of the design. The output of this statement is as follows:

Treatment combinations on each unit of the design					
unt	1	2	3	4	5
rep blk					
1 1	16	15	22	-	14
2	10	19	9	17	18
3	7	23	6	3	1
4	8	-	12	13	5
5	11	4	20	21	2
2 1	14	7	2	13	9
2	1	11	17	5	15
3	16	6	20	-	10
4	-	8	21	23	18
5	3	19	4	22	12
3 1	16	17	3	21	13
2	8	10	14	1	4
3	11	6	9	-	12
4	5	7	20	22	18
5	-	2	23	19	15

Treatment factors are listed in the order: trtmnt.

Because the number of treatments is not an exact multiple of the number of blocks, some blocks contain five units and others four. The 'missing' unit in a four-unit block is indicated by a hyphen (-). The design is nearly balanced: for example, Treatments 15 and 16 occur in the same block in Replicate 1, but not in any other replicate, and the same is true of many other pairs of treatments. However, the balance is not perfect: some pairs of treatments, e.g. Treatments 15 and 18, do not occur in the same block in any replication. The number of blocks in which each pairwise combination of treatments occurs is shown in the following matrix:

```
 1 | 0
 2 | 0 0
 3 | 1 0 0
 4 | 1 1 1 0
 5 | 1 0 0 0 0
 6 | 1 0 1 0 0 0
 7 | 1 1 1 0 1 1 0
 8 | 1 0 0 1 1 0 0 0
 9 | 0 1 0 0 0 1 1 0 0
10 | 1 0 0 1 0 1 0 1 1 0
11 | 1 1 0 1 1 1 0 0 1 0 0
12 | 0 0 1 1 1 1 0 1 1 0 1 0
13 | 0 1 1 0 1 0 1 1 1 0 0 1 0
14 | 1 1 0 1 0 0 1 1 1 1 0 0 1 0
15 | 1 1 0 0 1 0 0 0 0 0 1 0 0 1 0
16 | 0 0 1 0 0 1 0 0 0 1 0 0 1 1 1 0
17 | 1 0 1 0 1 0 0 0 1 1 1 0 1 0 1 1 0
18 | 0 0 0 0 1 0 1 1 1 1 0 0 0 0 0 0 1 0
19 | 0 1 1 1 0 0 0 0 1 1 0 1 0 0 1 0 1 1 0
20 | 0 1 0 1 1 1 1 0 0 1 1 0 0 0 0 1 0 1 0 0
21 | 0 1 1 1 0 0 0 1 0 0 1 0 1 0 0 1 1 1 0 1 0
22 | 0 0 1 1 1 0 1 0 0 0 0 1 0 1 1 1 0 1 1 1 0 0
23 | 1 1 1 0 0 1 1 1 0 0 0 0 0 0 1 0 0 1 1 0 1 0 0
   +--------------------------------------------------------------
     1 2 3 4 5 6 7 8 9 10 11 12 13 14 15 16 17 18 19 20 21 22 23
```

Each combination occurs either in one block or in none: no combination occurs in more than one block. The number of treatments with which each treatment is combined is as shown in Table 8.3. The numbers are nearly equal – each treatment is combined with 10, 11 or 12 others. Thus the experiment is as balanced as is possible given the constraints of the design, or nearly so.

In addition to being presented in the GenStat output, the treatment, block and unit for each observation in the experiment are stored in the factors 'trtmnt', 'rep' and 'unt' respectively. These can be retrieved when the experiment is complete, to analyse the data.

Table 8.3   Number of treatments with which each treatment is combined in the incomplete block design.

i = treatment number. N = number of treatments with which the ith treatment is combined.

i	N	i	N	i	N	i	N	i	N
1	12	6	10	11	11	16	10	21	11
2	11	7	12	12	10	17	12	22	11
3	12	8	10	13	11	18	11	23	10
4	12	9	11	14	11	19	11		
5	11	10	11	15	10	20	11		

If each replication of an experiment is to be laid out in an array of rows and columns (as in the GenStat output above), an alpha design will take account of the natural variation among rows, but not of that among columns: this will contribute to residual variation and reduce the precision with which treatment effects are estimated. However, the concepts of alpha designs can be extended further, to give approximate balance in both dimensions. Results from an experiment with such a design (known as an alphalpha design) were presented by Mead (1997). The experiment was a field trial of 35 wheat genotypes, laid out in two replicates, the plots in each replicate being arranged in an array of five rows and seven columns. The grain yield from each plot was measured (units not given). The data are presented in the spreadsheet in Table 8.4. (Data reproduced by kind permission of Mike Talbot, Biomathematics and Statistics Scotland.)

Table 8.4 Yields of wheat genotypes, investigated in an alphalpha design.

	A	B	C	D	E
1	replicate!	row!	column!	genotype!	yield
2	1	1	1	20	3.77
3	1	1	2	4	3.21
4	1	1	3	33	4.55
5	1	1	4	28	4.09
6	1	1	5	7	5.05
7	1	1	6	12	4.19
8	1	1	7	30	3.27
9	1	2	1	10	3.44
10	1	2	2	14	4.30
11	1	2	3	16	*
12	1	2	4	21	3.86
13	1	2	5	31	3.26
14	1	2	6	6	4.30
15	1	2	7	18	3.72
16	1	3	1	22	3.49
17	1	3	2	11	4.20
18	1	3	3	19	4.77
19	1	3	4	26	2.56
20	1	3	5	29	2.87
21	1	3	6	15	1.93
22	1	3	7	23	2.26
23	1	4	1	24	3.62
24	1	4	2	25	4.52
25	1	4	3	5	4.23

(Continued overleaf)

Table 8.4 (*continued*)

	A	B	C	D	E
26	1	4	4	32	3.76
27	1	4	5	2	3.61
28	1	4	6	27	3.62
29	1	4	7	8	4.01
30	1	5	1	17	3.81
31	1	5	2	9	3.75
32	1	5	3	3	4.81
33	1	5	4	34	3.69
34	1	5	5	13	4.61
35	1	5	6	35	2.68
36	1	5	7	1	4.15
37	2	1	1	31	4.70
38	2	1	2	19	7.37
39	2	1	3	25	5.03
40	2	1	4	34	5.33
41	2	1	5	20	5.73
42	2	1	6	8	4.70
43	2	1	7	6	5.63
44	2	2	1	24	4.07
45	2	2	2	21	5.66
46	2	2	3	12	4.98
47	2	2	4	4	4.04
48	2	2	5	23	4.27
49	2	2	6	13	4.10
50	2	2	7	3	4.75
51	2	3	1	11	5.66
52	2	3	2	7	6.43
53	2	3	3	26	4.59
54	2	3	4	5	5.20
55	2	3	5	35	4.83
56	2	3	6	10	4.70
57	2	3	7	30	4.23
58	2	4	1	33	5.71
59	2	4	2	9	6.13
60	2	4	3	17	4.63
61	2	4	4	18	5.48
62	2	4	5	32	5.47

Table 8.4 (*continued*)

	A	B	C	D	E
63	2	4	6	15	*
64	2	4	7	2	4.16
65	2	5	1	1	5.22
66	2	5	2	27	6.16
67	2	5	3	16	4.20
68	2	5	4	29	4.66
69	2	5	5	14	5.54
70	2	5	6	28	3.81
71	2	5	7	22	3.60

In this experiment, if two treatments occur in the same row *or column* in Replicate 1, then in most cases they occur in different rows *and columns* in Replicate 2. However, the balance is not perfect and there are exceptions: for example, Genotypes 4 and 12 both occur in Row 1 in Replicate 1, and both in Row 2 in Replicate 2. The following statements import and analyse these data:

```
IMPORT \
    'Intro to Mixed Modelling\\Chapter 8\\alphalpha design.xls'
VCOMPONENTS [FIXED = genotype] RANDOM = replicate /(row + column)
REML [PRINT = model, components, Wald, means] yield
```

The random-effects model,

$$\text{replicate}/(\text{row} + \text{column}) = \text{replicate} + \text{replicate.row} + \text{replicate.column},$$

indicates that both rows and columns are nested within replicates. That is:

- there is no main effect of 'row', and Row 1 in Replicate 1 is not the same row as Row 1 in Replicate 2;

- similarly, there is no main effect of 'column', and Column 1 in Replicate 1 is not the same column as Column 1 in Replicate 2.

The output of the REML statement is as follows:

REML variance components analysis

Response variate: yield
Fixed model: Constant + genotype
Random model: replicate + replicate.row + replicate.column
Number of units: 68 (2 units excluded due to zero weights or missing values)

Residual term has been added to model

Sparse algorithm with AI optimisation

Estimated variance components

Random term	component	s.e.
replicate	0.70356	1.05611
replicate.row	0.06387	0.04897
replicate.column	0.19265	0.09741

Residual variance model

Term	Factor	Model(order)	Parameter	Estimate	s.e.
Residual		Identity	Sigma2	0.0902	0.03685

Wald tests for fixed effects

Sequentially adding terms to fixed model

Fixed term	Wald statistic	d.f.	Wald/d.f.	chi pr
genotype	166.52	34	4.90	<0.001

Dropping individual terms from full fixed model

Fixed term	Wald statistic	d.f.	Wald/d.f.	chi pr
genotype	166.52	34	4.90	<0.001

Message: chi-square distribution for Wald tests is an asymptotic approximation (i.e. for large samples) and underestimates the probabilities in other cases.

Table of predicted means for Constant

4.363   Standard error: 0.6110

Table of predicted means for genotype

genotype	1	2	3	4	5	6	7	8
	4.814	3.915	5.098	3.521	4.395	5.409	5.085	4.603

genotype	9	10	11	12	13	14	15	16
	4.351	4.328	4.931	4.946	4.682	4.764	3.212	3.958

genotype	17	18	19	20	21	22	23	24
	4.154	4.565	5.669	4.320	4.593	4.011	3.423	3.888

genotype	25	26	27	28	29	30	31	32
	4.640	3.759	4.699	4.295	3.793	3.953	3.859	4.264

genotype	33	34	35
	4.914	4.298	3.602

Standard errors of differences

Average:	0.3854
Maximum:	0.5358
Minimum:	0.3511

Average variance of differences: 0.1498

The variance-component estimate for the term 'replicate' is smaller than its standard error, suggesting that the natural variation between the two replicates may be no greater than that between rows or columns within a replicate. However, the variance components for 'row' and 'column' are both larger than their respective standard errors, suggesting that there is real natural variation among the rows and columns, and hence that the treatment effects are estimated with greater precision than they would have been if each replicate had been laid out in a randomised complete block. The Wald test indicates that there are highly significant differences among the genotype means. As in the case of the balanced incomplete block design, these are slightly different from the corresponding simple means (e.g. the simple mean yield of Genotype 1 is $(4.15 + 5.22)/2 = 4.69$, whereas the value given by the mixed-model analysis is 4.814), and the adjustment is related to the effects of the rows and columns in which each treatment occurs.

Methods for producing two-dimensional incomplete block designs, and many other aspects of incomplete block designs, are discussed by John and Williams (1995). The software CycDesigN (distributed by CSIRO Forestry and Forest Products, Australia; see web site http://www.ffp.csiro.au/tigr/software/cycdesign/cycdes.htm) can be used to generate such designs.

8.5 Use of R to analyse the alphalpha design

To prepare the alphalpha-design data for analysis by R, they are transferred from the Excel workbook 'Intro to Mixed Modelling\Chapter 8\ alphalpha design.xls' to the text file 'alphalpha design.dat' in the same directory. Exclamation marks (!) are removed from the ends of headings in this file.

The following commands import the data, perform the appropriate mixed-model analysis and present the results:

```
alphalpha <- read.table(
    "Intro to Mixed Modelling\\Chapter 8\\alphalpha design.dat",
    header=TRUE)
attach(alphalpha)
freplicate <- factor(replicate)
frow <- factor(row)
fcolumn <- factor(column)
fgenotype <- factor(genotype)
alphalpha.modellme <- lmer(yield ~ fgenotype +
    (1|freplicate) + (1|freplicate:frow) + (1|freplicate:fcolumn),
    data = alphalpha)
summary(alphalpha.modellme)
anova(alphalpha.modellme)
```

The output of these commands is as follows:

```
Linear mixed-effects model fit by REML
Formula: yield ~ fgenotype + (1 | freplicate) + (1 |
    freplicate:frow) +      (1 | freplicate:fcolumn)
```

```
     Data: alphalpha
        AIC       BIC     logLik MLdeviance REMLdeviance
     149.2624 233.6037 -36.63122    34.65678      73.26244
     Random effects:
      Groups               Name       Variance Std.Dev.
      freplicate:fcolumn (Intercept) 0.192650 0.43892
      freplicate:frow    (Intercept) 0.063875 0.25273
      freplicate         (Intercept) 0.703558 0.83878
      Residual                       0.090172 0.30029
     # of obs: 68, groups: freplicate:fcolumn, 14; freplicate:frow,
        10; freplicate, 2


     Fixed effects:
                   Estimate Std. Error t value
     (Intercept)   4.814286   0.662550  7.2663
     fgenotype2   -0.898831   0.376579 -2.3868
     fgenotype3    0.283324   0.369033  0.7677
     fgenotype4   -1.293620   0.375756 -3.4427
     fgenotype5   -0.419103   0.381466 -1.0987
     fgenotype6    0.595013   0.385572  1.5432
     fgenotype7    0.270735   0.380741  0.7111
     fgenotype8   -0.211508   0.358782 -0.5895
     fgenotype9   -0.463420   0.367871 -1.2597
     fgenotype10  -0.486201   0.377974 -1.2863
     fgenotype11   0.116612   0.356312  0.3273
     fgenotype12   0.131470   0.390178  0.3369
     fgenotype13  -0.131799   0.366146 -0.3600
     fgenotype14  -0.049845   0.363807 -0.1370
     fgenotype15  -1.602178   0.486153 -3.2956
     fgenotype16  -0.856276   0.460194 -1.8607
     fgenotype17  -0.659906   0.368333 -1.7916
     fgenotype18  -0.249766   0.356286 -0.7010
     fgenotype19   0.854610   0.386472  2.2113
     fgenotype20  -0.494783   0.375268 -1.3185
     fgenotype21  -0.221370   0.383129 -0.5778
     fgenotype22  -0.802927   0.364750 -2.2013
     fgenotype23  -1.390965   0.353116 -3.9391
     fgenotype24  -0.925911   0.351951 -2.6308
     fgenotype25  -0.173866   0.382134 -0.4550
     fgenotype26  -1.055253   0.387952 -2.7201
     fgenotype27  -0.115191   0.375953 -0.3064
     fgenotype28  -0.519658   0.368918 -1.4086
     fgenotype29  -1.021030   0.363487 -2.8090
     fgenotype30  -0.861319   0.355089 -2.4256
     fgenotype31  -0.955346   0.357684 -2.6709
     fgenotype32  -0.550075   0.380358 -1.4462
     fgenotype33   0.099977   0.360267  0.2775
     fgenotype34  -0.515832   0.366712 -1.4066
     fgenotype35  -1.212487   0.371843 -3.2607


     Correlation of Fixed Effects:
```

```
      .
      .
      .
<Here follows a large correlation matrix.>
      .
      .
      .

Analysis of Variance Table
          Df   Sum Sq Mean Sq
fgenotype 34 15.0159  0.4416
```

The estimates of the variance components agree with those given by GenStat, and comparison of the mean square for 'fgenotype' with the estimate of the residual variance gives

$$F_{\text{fgenotype}} = \frac{0.4416}{0.090172} = 4.8973$$

which agrees with the value of (Wald statistic)/DF given by GenStat. As in the analysis of the balanced incomplete block design using R (Section 8.3), in the estimates of the fixed effects, the mean for the first level of the treatment factor is used as the intercept. Its value, 4.814286, agrees with that given by GenStat for the mean of Genotype 1. The effects of the other treatments are presented relative to this intercept, and the means are related to the effects by Equation 8.4. Thus

$$\text{mean(Genotype 2)} = 4.814286 - 0.898831 = 3.915455.$$

These values agree with those given by GenStat.

8.6 Summary

A major uses of mixed modelling is the analysis of experiments that cannot be tackled by analysis of variance, because it has not been possible to achieve exact balance during the design phase.

A balanced incomplete block design provides the starting point for a simple illustration of this problem.

Information concerning the treatment effects is distributed over two strata of this design, the among-blocks and within-block strata. The efficiency factor indicates the proportion of the information in each stratum.

If one block is omitted from a balanced incomplete block design it becomes unbalanced, and cannot be analysed by the ordinary methods of analysis of variance. However, it can still be analysed by mixed modelling.

If a balanced incomplete block design is analysed by the methods of analysis of variance, the significance of the treatment term is tested in both the among-blocks stratum and the within-blocks stratum. If it is analysed by mixed modelling, a single significance test, pooled over the two strata, is performed.

It is expected that this pooling will give a gain in statistical power.

The estimates of treatment means in an incomplete block design may be adjusted to take account of the block effects.

If this adjustment is made, it is effectively assumed that the block effects are random; if they are regarded as fixed, the unadjusted means are obtained.

The number of experimental units that form a natural, homogeneous block is often much smaller than the number of treatments to be compared. Balanced incomplete block designs can sometimes be used in such circumstances, but often no such design can be found that matches

- the number of treatments,

- the number of replications, and

- the number of units per block.

If the requirement for balance is relaxed, designs covering a much wider range of situations can be specified.

However, imbalance always carries a penalty in loss of efficiency, and the experimenter should therefore still attempt to specify a design that is as nearly balanced as possible.

Many extensions of this type to the concept of the balanced incomplete block design have been devised. One such extension is the alpha design, which consists of incomplete blocks, each comprising only a small proportion of the treatments, but in which these blocks can be grouped together into complete replications.

If each replication of an experiment is to be laid out in an array of rows and columns, an alpha design will take account of the natural variation among rows, but not of that among columns. A further extension is given by alphalpha designs, which give approximate balance over both rows and columns.

8.7 Exercises

1. Seven experimental treatments, A, B, C, D, E, F and G, are to be compared in a replicated experiment. However, the experimental units form natural groups of six, so one treatment must be omitted from each group.

   (a) Devise a balanced incomplete block design within these constraints. How many replications are required to achieve balance?

   (b) Invent values for the response variable from this experiment, and perform the appropriate analysis of variance.

   (c) Confirm that the efficiency factor in this analysis agrees with Equation 8.1, and that $SE_{Difference}$ for the treatment means agrees with Equation 8.2.

   (d) Show that if one block is omitted from the experiment it can no longer be simply analysed by analysis of variance, but that it can still be analysed by mixed modelling.

   (e) Obtain the treatment means from the analysis of variance of the complete experiment:

       (i) with recovery of interblock information

       (ii) without recovery of interblock information.

   (f) Obtain the same two sets of means from mixed-model analyses.

2(a) Produce an alpha design with 37 treatments in four replicates, each replicate comprising seven blocks.

(b) Determine the number of blocks in which each pairwise combination of treatments occurs. Confirm that each combination occurs either in one block or in none.

(c) Determine the number of treatments with which each treatment is combined. How closely do these values approach the ideal outcome?

(d) Invent values for the response variable from this experiment, and perform the appropriate mixed-model analysis.

9

Beyond mixed modelling

9.1 Review of the uses of mixed models

Just as mixed modelling is an extension of the linear modelling methods comprising regression analysis and anova, so mixed modelling itself can be further extended in several directions, to give even more versatile and realistic models. This chapter reviews the various contexts in which we have seen that mixed modelling is preferable to a simple regression or anova approach, then outlines the ways in which the concepts of mixed modelling can be developed further. Fuller accounts of such advanced uses of mixed modelling are given by Brown and Prescott (1999) and by Pinhero and Bates (2000). Brown and Prescott demonstrate the use of the statistical software SAS to fit the models, whereas Pinhero and Bates use the statistical computer language S (of which the software R is one implementation – see Chapter 1, Section 1.11). Both books place much more emphasis on the underlying mathematical theory than is given here.

A mixed-model analysis provides a fuller interpretation of the data than a simple regression or anova approach, and permits wider inferences about the observations to be expected in future, in the following situations:

- When one or more of the factors in a regression model is a random-effect term, and should therefore contribute to the SEs of estimates of effects of other terms. Examples include:

  - the variation in house prices among towns, which contributed to the SE of the estimated effect of latitude (Chapter 1);

  - the variation in bone mineral density among patients sampled at different hospitals, which contributed to the SEs of the estimated effects of gender, age, height and weight (Chapter 7, Section 7.2).

- Where variance components are of intrinsic interest, e.g. in the investigation of the sources of variation in the strength of a chemical paste (delivery, cask and sample – Chapter 3, Sections 3.2–3.6). The relative magnitude of the different sources

---

Introduction to Mixed Modelling: Beyond Regression and Analysis of Variance N. W. Galwey
© 2006 John Wiley & Sons, Ltd

of variation was estimated, with a view to their control by replication in subsequent investigation.

- When a preliminary screening process is used to identify candidates for further evaluation, e.g. in the identification of high-yielding breeding lines among the progeny of a cross between two barley varieties (Chapter 3, Sections 3.7–3.15). The best linear unbiased predictor (BLUP), obtained from the mixed-model analysis, provides a more realistic – and more conservative – prediction of the future performance of the selected candidates than is given by the simple mean performance of each candidate (or by the closely related best linear unbiased estimate (BLUE)).

- When it has not been possible to achieve exact balance in the design of an experiment. For example:

  - if one block has to be omitted from a balanced incomplete block design (Chapter 8, Sections 8.1–8.2);
  - in an alpha or alphalpha design (Chapter 8, Section 8.4).

9.2   The generalised linear mixed model. Fitting a logistic (sigmoidal) curve to proportions of observations

All the models that we have considered so far are *linear*: that is, they can be expressed in the form

$$y_k = \beta_0 + \beta_1 x_{1k} + \beta_2 x_{2k} + \cdots + \beta_p x_{pk} + \upsilon_1 z_{1k} + \upsilon_2 z_{2k} + \cdots + \upsilon_q z_{qk} + \varepsilon_k \quad (9.1)$$

where y_k = the kth observation of the response variable Y,

x_{ik} = the kth observation of the ith explanatory variable in the fixed-effect model, X_i,

p = the number of explanatory variables in the fixed-effect model,

z_{jk} = the kth observation of the jth explanatory variable in the random-effect model, Z_j,

q = the number of explanatory variables in the random-effect model,

ε_k = the kth value of the random variable E, which represents the residual variation in Y,

$\beta_0 \ldots \beta_p$ and $\upsilon_1 \ldots \upsilon_q$ are parameters to be estimated.

When one or more of the explanatory variables are factors, some ingenuity is needed to express the model in this form. For example, the model used in Chapter 1 to relate house prices to latitude and town can be expressed in this form by setting

$$Y = \log(\text{house price})$$

$$X_1 = \text{latitude}$$

$$Z_1 = 1 \text{ for observations from Bradford,}$$

$$0 \text{ otherwise}$$

$$Z_2 = 1 \text{ for observations from Buxton,}$$
$$0 \text{ otherwise}$$

.

.

.

$$Z_{11} = 1 \text{ for observations from Witney,}$$
$$0 \text{ otherwise.}$$

The variables Z_1 to Z_{11}, with their arbitrary values that indicate the category (the town) to which each observation belongs, are known as *dummy variables*. When the response variable and the explanatory variables are specified in this way, we find that

$$\beta_0 = \text{intercept}$$
$$\beta_1 = \text{effect of latitude}$$
$$\upsilon_1 = \text{effect of Bradford}$$
$$\upsilon_2 = \text{effect of Buxton}$$

.

.

.

$$\upsilon_{11} = \text{effect of Witney.}$$

The estimates of $\upsilon_1 \ldots \upsilon_{11}$ are also the estimates of the parameters T_1 to T_{11}, the deviations of the town means from the regression line relating log(house price) to latitude, defined in Chapter 1, Section 1.4. The decision to treat this model as a mixed model is equivalent to a decision to treat these parameters as values of a random variable, as described in Section 1.6.

In addition to being linear, all the models considered so far have had residuals that can reasonably be assumed to be normally distributed. There are many other regression models, relating a response variable to one or more explanatory variables, that do not have these properties. As an example of a situation in which neither a linear model nor a normal distribution of the residuals is adequate, we can consider the results of an experiment to determine the toxicity of ammonia to a species of beetle, *Tribolium confusum* (Finney, 1971, Chapter 9, Section 9.1, p 177). The experiment was performed in two batches, each comprising a series of samples. These will be represented by dummy variables, Z_1 and Z_2, and the explanatory variable X is $\log_{10}$(concentration of ammonia) applied to each sample. In any sample, the number of dead beetles, R, must be an integer between 0 and the number of beetles in the sample, N. The data are shown in the spreadsheet in Table 9.1. (Data reproduced by kind permission of Cambridge University Press.)

If the death of each individual is independent of that of every other individual, then the random variable R has a *binomial distribution*, the precise shape of which is

Table 9.1 Mortality of the beetle *Tribolium confusum* at different concentrations of ammonia.

$X = \log_{10}$(concentration of ammonia). $N =$ number of beetles in the sample. $R =$ number of dead beetles in the sample. The conventions used in this spreadsheet are the same as those used in Table 1.1, Chapter 1.

	A	B	C	D
		A	B	C	D
1	batch!	X	N	R
2	1	0.72	29	2
3	2	0.72	29	1
4	1	0.80	30	7
5	2	0.80	31	12
6	1	0.87	31	12
7	2	0.87	32	4
8	1	0.93	28	19
9	2	0.93	31	18
10	1	0.98	26	24
11	2	0.98	31	25
12	1	1.02	27	27
13	2	1.02	28	27
14	1	1.07	26	26
15	2	1.07	31	29
16	1	1.10	30	30
17	2	1.10	31	30

determined by the values of N and by the probability that an individual beetle dies, designated by π. This statement can be written in symbolic shorthand as

$$R \sim \text{binomial}(N, \pi).$$

The value of π may depend on the value of X under consideration and on the batch, i.e. π may be a function of X and 'batch'. This value can never be known, though it can be estimated from the data.

A brief digression on the binomial distribution is required here. This distribution is defined by the statement

$$P(R = r) = \frac{N!}{r!(N - r)!}\pi^r(1 - \pi)^{(N-r)} \tag{9.2}$$

where

$$N! = N \times (N - 1) \times (N - 2) \times \cdots \times 3 \times 2 \times 1.$$

For example, if

$$\pi = 0.3,$$

then in a sample of 30 beetles, the probability that 8 die is given by

$$P(R = 8) = \frac{30!}{8! \times 22!} \times 0.3^8 \times 0.7^{22} = 0.1501.$$

Substituting each possible value of r from 0 to 30 into Equation 9.2, we obtain the distribution illustrated in Figure 9.1.

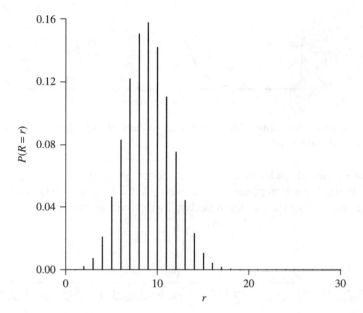

Figure 9.1   The binomial distribution with parameters $\pi = 0.3$ and $N = 30$.

For a fuller account of the binomial distribution, and why it occurs in such situations, see, for example, Snedecor and Cochran (1989, Chapter 7, Sections 7.1 to 7.5, pp 107–117) or Bulmer (1979, Chapter 6, pp 81–90).

In a system of this kind, the relationship between the combined value of the explanatory variables ('batch' and X in this case) and π is often *sigmoidal* (S-shaped – see Figure 9.2). At one extreme of the range of the explanatory variables, the probability of the event under consideration (death in this case) is close to zero; at the other extreme it is close to one.

There are two commonly used functions that specify a relationship of this form. One of these, the integral of the normal distribution, is the basis of a method called probit analysis (Finney, 1971). This method of fitting a sigmoid curve probably has the clearer conceptual basis: it is based on the assumption that an underlying variable, in this case the tolerance of the beetles to the toxin, is normally distributed, and that

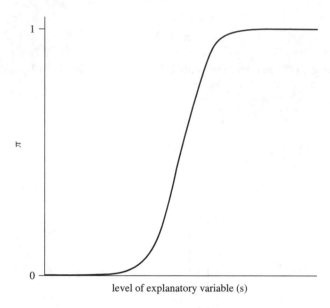

Figure 9.2 A sigmoidal relationship between an explanatory variable (or variables) and the probability (π) of a response.

at any given dose, all individuals up to a certain level of tolerance are killed. The alternative is the logistic function, which we will use here as it has the advantage of being rather easier to express algebraically. In the present case it is

$$p_{ij} = \frac{r_{ij}}{n_{ij}}$$

$$= \frac{1}{1+\exp(-[\beta_0+\beta_1(x_j - \bar{x})+\upsilon_1 z_{1i} + \upsilon_2 z_{2i}+\upsilon_3 z_{1i}(x_j - \bar{x})+\upsilon_4 z_{2i}(x_j - \bar{x})])}+\varepsilon_{ij}$$

(9.3)

where

r_{ij} = the value of R at the jth level of X in the ith batch, i.e. the number of dead beetles in this sample (the ijth sample),

n_{ij} = the value of N, i.e. the total number of beetles, in the ijth sample,

x_j = the jth level of X,

z_{1i} = the value of the dummy variable Z_1 in the ith batch, indicating whether each observation was obtained from Batch 1,

z_{2i} = the value of the dummy variable Z_2 in the ith batch, indicating whether each observation was obtained from Batch 2,

ε_{ij} = the value of the random variable E in the ijth sample, which represents the residual effect on p_{ij},

$\bar{x} = \dfrac{\displaystyle\sum_{j=1}^{7} x_j}{7}$ = the mean value of X over all samples,

and the β and υ are parameters to be estimated, namely

β_0 = constant,
β_1 = effect of X,
υ_1 = effect of Batch 1,
υ_2 = effect of Batch 2,
υ_3 = effect of interaction between X and Batch 1,
υ_4 = effect of interaction between X and Batch 2.

The relationship between the batches and the dummy variables is as shown in Table 9.2.

The value p_{ij} gives an estimate of π based on the information from the ijth sample, applicable to the batch and the level of X in question. However, the fitting of the logistic function will give estimates based on all the data, for any combination of levels of 'batch' and X.

The relationship between Y and the explanatory variables in this equation is not linear. However, if we ignore the residual term ε_{ij} for the time being, we can *transform* the equation to the familiar linear model form

$$\log_e \left(\frac{p_{ij}}{1 - p_{ij}} \right) = \beta_0 + \beta_1(x_j - \bar{x}) + \upsilon_1 z_{1i} + \upsilon_2 z_{2i} + \upsilon_3 z_{1i}(x_j - \bar{x}) + \upsilon_4 z_{2i}(x_j - \bar{x})$$

(9.4)

The function $\log_e(p_{ij}/(1 - p_{ij})$ is known as the *logit function*. Models of this type, which can be expressed in linear form by dropping the residual term and applying a suitable transformation to the response variable, are known as *generalised linear*. Every generalised linear model is characterised by a probability distribution and a *link function*: in the present case, the binomial distribution and the logit function. For each probability distribution, there is a particular link function known as the *canonical link* which has special mathematical properties, including the fact that, when the residual term is the only random-effect term in the model, it always gives a unique set of parameter estimates, the *sufficient statistic* (McCullagh and Nelder, 1989, Chapter 2, Sections 2.2.2 to 2.2.4, pp 28–32). In the case of the binomial distribution, the logit function is the canonical link – another reason for preferring it to the *probit function*, the corresponding function used in probit analysis.

In the notation of Wilkinson and Rogers (1973; see Chapter 2, Section 2.2), the model specified here is

$$X * batch.$$

It is reasonable to specify 'X' as a fixed-effect term, and 'batch' as a random-effect term, from which it follows that 'X.batch' is also a random-effect term (see the rules

Table 9.2 The relationship between batches and dummy variables in the model fitted to the data on beetle mortality.

Batch	Z_1	Z_2
1	$z_{11} = 1$	$z_{21} = 0$
2	$z_{12} = 0$	$z_{22} = 1$

for making these decisions given in Chapter 6, Section 6.3). We have then specified a *generalised linear mixed model* (GLMM).

This GLMM can be fitted to the data by the following GenStat statements:

```
IMPORT \
    'Intro to Mixed Modelling\\Chapter 9\\ammonia Tribolium.xls'
GLMM [PRINT = model, monitoring, components, \
    vcovariance, effects; \
    DISTRIBUTION = binomial; LINK = logit; DISPERSION = *; \
    RANDOM = batch + X.batch; FIXED = X; \
    PTERMS = 'constant' + X*batch] Y = R; NBINOMIAL = N
```

In the GLMM statement, the DISTRIBUTION option specifies that the response variate 'R' follows the general form of the binomial distribution, though its variance may be greater or less than that of a binomial variable, as will be specified in a moment. The LINK option specifies that the logit function has been chosen as the link function. If the assumptions that each death is independent and that R follows the binomial distribution in every respect were correct, then it would follow that

$$\text{var}(R) = N\pi(1 - \pi) \tag{9.5}$$

and there would be no need to estimate the residual variance from the data. We could indicate that we were willing to make this assumption by setting the option 'DISPERSION = 1'; instead, by setting this option to a missing value ('*') we indicate that the residual variance is to be estimated. The options RANDOM and FIXED specify the random-effect and fixed-effect model terms respectively. The parameter Y specifies the response variate, and the parameter NBINOMIAL specifies the variate that holds the number of observations in each sample, i.e. the maximum value that each value of the response variate might take.

The output of the GLMM statement is as follows:

Generalized linear mixed model analysis

Method:	c.f. Schall (1991) Biometrika
Response variate:	R
Distribution:	binomial
Link function:	logit
Random model:	batch + (batch.X)
Fixed model:	Constant + X

Dispersion parameter estimated

Monitoring information

Iteration	Gammas		Dispersion	Max change
1	0.02903	0.0000004987	2.137	1.6173E+01
2	0.02653	4.9867E-09	2.238	1.0117E-01
3	0.02659	4.9867E-11	2.237	8.3595E-04
4	0.02659	0.0000001000	2.237	1.8842E-04
5	0.02659	1.0000E-09	2.237	1.0463E-05

Estimated variance components

Random term	component	s.e.
batch	0.059	0.203
batch.X	0.000	bound

Residual variance model

Term	Factor	Model(order)	Parameter	Estimate	s.e.
Dispersn		Identity	Sigma2	2.237	0.877

Estimated variance matrix for variance components

batch	1	0.0411		
batch.X	2	0.0000	0.0000	
Dispersn	3	−0.0282	0.0000	0.7700
		1	2	3

Table of effects for Constant

0.3314 Standard error: 0.26543

Table of effects for X

17.84 Standard error: 2.305

Table of effects for batch

batch	1	2
	0.1118	−0.1118

Standard errors of differences are not available.

Table of effects for batch.X

batch	1	2
	4.6529E-10	−4.6529E-10

Standard errors of differences are not available.

The output first specifies the fitting method used and the model fitted. Next comes some monitoring information, indicating the estimated values of certain model parameters at successive iterations of the model-fitting process, leading to convergence, i.e. successful fitting (see Section 9.3). Next come estimates of the variance components for the random-effect terms, with their SEs. The estimate for 'batch' is smaller than its SE, and that for 'batch.X' is zero, indicating that these terms could probably be dropped from the model. The estimate of the residual variance is presented in terms of the *dispersion*, which is given by the ratio

$$\frac{\text{observed residual variance}}{\text{residual variance expected if the response variate follows the distribution specified in the model}}.$$

In the present case the dispersion is substantially larger than 1, indicating that there is more residual variation from sample to sample than would be expected if the distribution were truly binomial. Next comes the matrix of covariances among these variance estimates. The values on the diagonal are simply the squares of the SEs above – the variances of the variance estimates. For example,

$$\mathrm{var}(\hat{\sigma}^2_{\mathrm{batch}}) = 0.203^2 \approx 0.04.$$

The off-diagonal values indicate the extent to which the estimate of one variance component is associated with that of another.

Next come the estimated effects of each model term. These can be substituted into the original model (Equation 9.3), together with the value

$$\bar{x} = 0.9363,$$

to provide an estimate of π, the true probability that an insect dies, for any combination of batch and X. Thus:

$$\hat{\pi}_{ij}$$

$$= \frac{1}{1+\exp\left[-\left(\begin{array}{l}0.3314+17.83\times(x_j-0.9363)+0.1109\times z_{1i}-0.1109\times z_{2i}+\\4.6528\times 10^{-7}\times z_{1i}(x_j-0.9363)-4.6528\times 10^{-7}\times z_{2i}(x_j-0.9363)\end{array}\right)\right]}.$$

This function is displayed over the range of the data, together with the observed values, in Figure 9.3.

Overall, the curves give a reasonable fit to the data, permitting a realistic estimate of the proportion of insects that would be killed by a particular concentration of ammonia. The effect of batch corresponds to the horizontal displacement between the two curves. This is small relative to the scatter of the data, confirming that the batch has little if any effect. The 'X.batch' interaction effect corresponds to the difference in average slope between the two curves, which is too slight to be detected by eye. The fact that the dispersion is larger than 1 indicates that the scatter of the observations about the curves is wider than would be expected if the deaths of individuals within each sample were independent: that is, there is evidence of some *heterogeneity* in the conditions of each sample, even after allowing for the effects of ammonia and batch. Because the 'X.batch' interaction effect is negligible, the decision to use a mixed model has had little effect on the precision of the parameter estimates in this case. However, if this effect were substantial, it would be important to take it into account in order to obtain realistic values for the SEs of $\hat{\beta}_0$, $\hat{\beta}_1$ and $\hat{\beta}_2$.

9.3 Fitting a GLMM to a contingency table. Trouble-shooting when the mixed-modelling process fails

There are several other types of data that cannot be realistically represented by an ordinary linear model with normally distributed residual variation, but which do fulfil the criteria for fitting a GLMM namely:

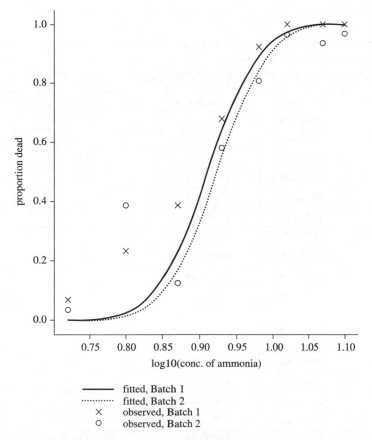

Figure 9.3 Fitted curves from a logistic model relating the proportion of *T. confusum* individuals killed to the concentration of ammonia applied.

- They can be represented by a model that can be converted to the general linear form by omitting the residual term and applying an appropriate transformation.

- An appropriate probability distribution can be specified for the response variable.

Hence the use of GLMMs permits a wide extension to the range of situations in which the concepts of mixed modelling can be applied. Another important case in which generalised linear models can be used is the analysis of *contingency tables*. These are data sets in which events of a particular type are counted, and are classified by factors that indicate the combination of circumstances, or contingency, in which each event occurred. This type of data set is illustrated by an example concerning the frequency of damage caused by waves to the forward sections of cargo-carrying ships (McCullagh and Nelder, 1989, Section 6.3.2, pp 204–208). Each occurrence of damage is classified by the type of ship to which it occurred (A to E), the year of construction of the ship and its period of operation. For each category defined by these factors, i.e. each contingency, the number of incidents of damage observed was recorded, and also the number of months of service over which observations were available. The first and last few rows of the data are shown in the spreadsheet in Table 9.3. The null hypothesis to

Table 9.3   Number of incidents of damage to ships, classified by ship type, year of construction and period of operation.

Only the first and last few rows of the data are shown. The conventions used in this spreadsheet are the same as those used in Table 1.1, Chapter 1.

	A	B	C	D	E
1	type!	constrctn_y!	operatn_period!	service_months	damage_incidents
2	A	1960–64	1960–74	127	0
3	A	1960–64	1975–79	63	0
4	A	1965–69	1960–74	1095	3
5	A	1965–69	1975–79	1095	4
6	A	1970–74	1960–74	1512	6
7	A	1970–74	1975–79	3353	18
8	A	1975–79	1960–74		
9	A	1975–79	1975–79	2244	11
10	B	1960–64	1960–74	44882	39
11	B	1960–64	1975–79	17176	29

34	E	1960–64	1960–74	45	0
35	E	1960–64	1975–79	0	0
36	E	1965–69	1960–74	789	7
37	E	1965–69	1975–79	437	7
38	E	1970–74	1960–74	1157	5
39	E	1970–74	1975–79	2161	12
40	E	1975–79	1960–74		
41	E	1975–79	1975–79	542	1

be tested (H_0) is that none of the factors type of ship, year of construction or period of operation influenced the frequency of incidents, and in this case the number of incidents in each category is expected to be proportional to the number of months of service. (Data reproduced by kind permission of Chapman and Hall.)

We can represent the number of incidents of damage to ships of the ith type, constructed during the jth range of years, during the kth period of operation (the ijkth category) by the symbol r_{ijk}. If each damage incident is independent of all the others, then it can be shown that r_{ijk} is an observation of a random variable R_{ijk} which has a *Poisson distribution*, the mean of which is given by

$$N\pi_{ijk}$$

where

$N =$ the mean total number of incidents of damage in a hypothetical infinite population of data sets, each based on the same total number of months of observation as the present data set,

π_{ijk} = the true probability that an individual damage incident falls in the ijkth
      category.

This statement can be written in symbolic shorthand as

$$R_{ijk} \sim \text{Poisson}(N\pi_{ijk})$$

Again a brief digression is required, this time on the Poisson distribution. This is
defined by the statement that if

$$R \sim \text{Poisson}(\mu),$$

where
μ = the mean number of incidents in an observation period,

then

$$P(R = r) = \frac{\mu^r}{r!}e^{-\mu} \tag{9.6}$$

For example, if the mean number of incidents of damage in a particular category (over
an infinite hypothetical population of data sets similar to the present data set) is 8,
then the observed number of incidents will be distributed as shown in Figure 9.4. For
a fuller account of the Poisson distribution, and why it occurs in this context, see,
for example, Snedecor and Cochran (1989, Chapter 7, Section 7.14, pp 130–133) or
Bulmer (1979, Chapter 6, pp 90–97).

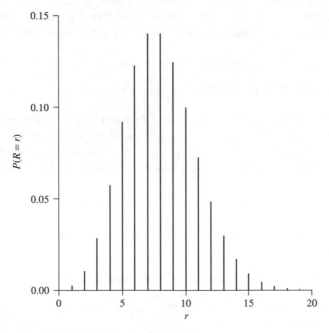

Figure 9.4   The Poisson distribution with parameter (mean value) $\mu = 8$.

In the present case, if H_0 is true, then

$$\pi_{ijk} = a_{ijk} \tag{9.7}$$

where

a_{ijk} = the proportion of months of service that fall in the ijkth category.

In this case the expected number of incidents in the ijkth category is Na_{ijk}, and

$$r_{ijk} = Na_{ijk} + \varepsilon_{ijk} \tag{9.8}$$

where

ε_{ijk} = the value of the random variable E in the ijkth category, which represents the residual effect on r_{ijk}.

We can add terms to this model to represent the possibility that the factors type of ship, year of construction and period of operation, and their interactions, influence the probability of a damage incident, i.e.

$$r_{ijk} = Na_{ijk}q_{i..}q_{.j.}q_{..k}q_{ij.}q_{i.k}q_{.jk}q_{ijk} + \varepsilon_{ijk} \tag{9.9}$$

where

$q_{i..}$ = the main effect of the ith type of ship,
$q_{.j.}$ = the main effect of the jth range of years of construction,
$q_{..k}$ = the main effect of the kth period of operation,
$q_{ij.}$ = the effect of interaction between the ith type and the jth range of years, etc.

N and the q are parameters to be estimated from the data. This model, like that in Equation 9.3, can be transformed to the linear model form by ignoring the residual term ε_{ijk} and applying a suitable transformation, in this case the logarithmic transformation:

$$\log(r_{ijk}) = \log(N) + \log(a_{ijk}) + \log(q_{i..}) + \log(q_{.j.}) + \log(q_{..k}) + \log(q_{ij.})$$
$$+ \log(q_{i.k}) + \log(q_{.jk}) + \log(q_{ijk}). \tag{9.10}$$

However, the term $\log(a_{ijk})$ does not have to be estimated: it is a separate variable supplied with the data. It is equivalent to a term $\beta \log(a_{ijk})$ when the parameter β is required to take the value 1. Such a term is called an *offset*. In the notation of Wilkinson and Rogers (1973), the model specified here (excluding the offset term) is

type ∗ constrctn_y ∗ operatn_period.

The following statements import the data and fit this model, treating all terms as fixed-effect terms for the time being:

```
IMPORT 'Intro to Mixed Modelling\\Chapter 9\\ship damage.xls'
CALCULATE logservice = LOG(service_months/SUM(service_months))
MODEL [DISTRIBUTION=poisson; LINK=log; DISPERSION = 1; \
    OFFSET=logservice] damage_incidents
TERMS type * constrctn_y * operatn_period
FIT [PRINT = *] type
ADD [PRINT = *] constrctn_y
ADD [PRINT = *] operatn_period
ADD [PRINT = *] type.constrctn_y
```

```
ADD [PRINT = *] type.operatn_period
ADD [PRINT = model, accumulated; FPROBABILITY = yes]\
   constrctn_y.operatn_period
```

The CALCULATE statement obtains the natural logarithm of the proportion of months of service in each category, transforming this variable to the scale on which it will be required in the model. A message in the output (not shown here) warns that in the case of Unit 34 (row 35 of the spreadsheet) an attempt has been made to obtain the logarithm of zero, and that the result is a missing value. Consequently this unit is omitted from the analysis, which is appropriate, as it represents a category of ship that spent no time at sea, and therefore was not exposed to the risk of damage. In the MODEL statement, the DISTRIBUTION option specifies that the response variate ('damage_incidents') follows the general form of the Poisson distribution, though its variance may be greater or less than that of a Poisson variable, unless we constrain it using the DISPERSION option (see below). The LINK function specifies the function required to transform the model to the linear form, just as the same option did in the GLMM statement in the previous example (Section 9.2). If the assumption that each damage incident is independent and that R follows the Poisson distribution in every respect is correct, then it follows that

$$\text{var}(R) = \mu \qquad (9.11)$$

and there is no need to estimate the residual variance from the data. (It is a peculiarity of the Poisson distribution that its variance is equal to its mean.) The option setting 'DISPERSION = 1' indicates that we are willing to make this assumption. The offset term is specified by the OFFSET option.

In order to obtain an analysis in which the deviance accounted for by each model term is shown separately, it is necessary to specify each of the terms in a separate statement. The FIT statement specifies the model on H_0, comprising only the constant term $\log(N)$ and the offset term $\log(a_{ijk})$. The option setting 'PRINT = *' indicates that no output is to be produced from this initial model. The main-effect and interaction terms are added by a succession of ADD statements, again with no printing, until the final ADD statement specifies that the complete model and the *accumulated analysis of deviance* are to be printed. The *analysis of deviance* is a method closely related to anova: in the special case where the residual variation is normally distributed (i.e. in all the models considered prior to the present chapter), the two are equivalent. The term *accumulated* indicates that the deviance accounted for by adding each term to the model successively is to be presented (cf. Chapter 1, Section 1.4, where an anova constructed on the same basis is presented). The FPROBABILITY option specifies that the analysis of deviance table is to include a P value for the significance of each term in the model. Note that it is not necessary to fit the three-way interaction term 'type.constrctn_y.operatn_period' explicitly, as there is only one observation for each combination of these factors, and this term is therefore the residual term.

The output of the final ADD statement is as follows:

Message: term constrctn_y.operatn_period cannot be fully included in the model because 1 parameter is aliased with terms already in the model.

(constrctn_y 1975–79 .operatn_period 1975–79) = (constrctn_y 1975–79)

Regression analysis

Response variate:	damage_incidents
Distribution:	Poisson
Link function:	Log
Offset variate:	logservice
Fitted terms:	Constant + type + constrctn_y + operatn_period+

type.constrctn_y + type.operatn_period + constrctn_y.operatn_period

Accumulated analysis of deviance

Change	d.f.	deviance	mean deviance	deviance ratio	approx chi pr
+ type	4	55.4391	13.8598	13.86	<.001
+ constrctn_y	3	41.5341	13.8447	13.84	<.001
+ operatn_period	1	10.6601	10.6601	10.66	0.001
+ type.constrctn_y	12	24.1079	2.0090	2.01	0.020
+ type.operatn_period	4	6.0661	1.5165	1.52	0.194
+ constrctn_y.operatn_period					
	2	1.6643	0.8322	0.83	0.435
Residual	7	6.8568	0.9795		
Total	33	146.3283	4.4342		

Message: ratios are based on dispersion parameter with value 1

A message first warns that the term 'constrctn_y.operatn_period' cannot be fully included in the model. This is because of the presence of missing values, as a result of which not all combinations of levels of the three factors are represented in the data. The model fitted and the fitting method used are then specified. The accumulated analysis of deviance then shows how the variation among the values of 'damage_incidents' (after adjusting for 'logservice') is distributed among the terms in the model. Roughly speaking, the deviance per degree of freedom – the mean deviance – gives a measure of the amount of variation accounted for by each term, when the terms are added successively to the model. Thus the mean deviance corresponds to the mean square in an anova. It is divided by the dispersion parameter to give the deviance ratio: since the dispersion parameter has been fixed at 1 (i.e. 'damage_incidents' has been assumed to follow a Poisson distribution in every respect, including the value of the variance), the two are identical. If the dispersion parameter had not been specified, it would be estimated by the residual mean deviance; hence the deviance ratio is equivalent to the variance ratio (the F statistic) in an anova. If the assumption that 'damage_incidents' follows a Poisson distribution is correct, and if H_0 is true, then the deviance for each term is distributed approximately as χ^2 with the degrees of freedom indicated. Thus each deviance provides a significance test for the term in question, and the column headed 'approx chi pr' gives the corresponding P value. These P values indicate that the main effects of the three factors are highly significant, and that the 'type.constructn_y' interaction is also significant, but that the other interactions are not.

It is of interest to obtain estimates of the mean frequency of damage to ships of each type, but it is not possible to do so from the model fitted above, because some pairwise combinations of factor levels are not represented in the data. However, we can omit the non-significant two-way interaction terms from the model, and obtain estimates based on the simpler model comprising only the main effects of the three factors. This is done by the following statements:

```
FIT [PRINT = *] type + constrctn_y + operatn_period
PREDICT [PRINT = description, prediction, se, sed] type
```

The output of the PREDICT statement is as follows:

Predictions from regression model

These predictions are estimated mean values, formed on the scale of the response variable, adjusted with respect to some factors as specified below.

The predictions have been formed only for those combinations of factor levels for which means can be estimated without involving aliased parameters.

The predictions are based on the mean of the offset variate: logservice −4.956

The predictions have been standardized by averaging over the levels of some factors:

Factor	Weighting policy	Status of weights
operatn_period	Marginal weights	Constant over levels of other factors
constrctn_y	Marginal weights	Constant over levels of other factors

The standard errors are appropriate for interpretation of the predictions as summaries of the data rather than as forecasts of new observations.

Response variate: damage_incidents

type	Prediction	s.e.
A	4.211	0.665
B	2.446	0.168
C	2.118	0.613
D	3.903	0.975
E	5.831	1.057

Standard errors of differences of predictions

type A	1	*				
type B	2	0.700	*			
type C	3	0.901	0.637	*		
type D	4	1.160	1.003	1.155	*	
type E	5	1.227	1.082	1.214	1.430	*
		1	2	3	4	5

Message: s.e.'s, variances and lsd's are approximate, since the model is not linear.

Message: s.e.'s are based on dispersion parameter with value 1

The note that the estimated means are formed on the scale of the response variable indicates that they must be back-transformed, using the inverse of the link function, in order to obtain them on the original scale, namely number of incidents of damage. For example, the mean value for ships of Type A corresponds to

$$e^{4.211} = 67.42 \text{ incidents.}$$

The accompanying SE indicates that the true value is likely to lie in the range from

$$e^{(4.211-0.665)} = 34.67$$

to

$$e^{(4.211+0.665)} = 131.10 \text{ incidents.}$$

Another note states that these estimates are based on a value of -4.956 for the offset variate 'logservice'. This means that they are based on the value

$$\log_e A = -4.959, \tag{9.12}$$

where

$A =$ the proportion of months of service that is assumed to fall in the category under consideration.

Although A varies from category to category in the data, it is held constant for the purpose of prediction, in order to provide a valid basis for comparison between risks in the different categories. From this decision, it follows that

$$\begin{gathered} \text{number of months of service assumed to fall in the category under} \\ \text{consideration for the purpose of prediction} = TA \end{gathered} \tag{9.13}$$

where

$T =$ total number of months of service.

Totalling the values of 'service_months' in the data we obtain

$$T = 163574, \tag{9.14}$$

rearranging Equation 9.12 we obtain

$$A = e^{-4.959}, \tag{9.15}$$

and substituting Equations 9.14 and 9.15 into Equation 9.13, we find that the predicted numbers of incidents are based on

$$163574e^{-4.956} = 1151.7 \text{ months of service.}$$

Another note indicates that the calculation of this mean has required averaging over the levels of the other factors, and that in this process *marginal weights* have been applied: that is, operation periods and construction years that are more heavily represented in the data contribute more heavily to the mean. However, it is noted that the status of weights is constant over levels of other factors; this means that a *combination of* 'operatn_period' and 'constrctn_y' that is more heavily represented in the data than the marginal weights would lead one to predict does *not* contribute more heavily to the mean. Another note states that the SEs are not appropriate for the forecasting of new observations: that is, they indicate the precision *of the means themselves*, not the amount of variation among individual observations. As noted earlier (Chapter 4, Section 4.6), SEs of differences between means are usually of more interest than SEs of the means themselves, and these are provided, for all pairwise comparisons, on the transformed scale.

If 'constrctn_y' and 'operatn_period' are representative of a broader population of construction and operation dates, it may be reasonable to specify these factors as random-effect terms. However, it may be reasonable to retain 'type' as a fixed-effect term representing particular methods of construction that are of individual interest, and that may be amenable to choice or control: shipbuilders can decide what type of hull to construct, and insurers can decide what premiums to offer on each type of ship. According to the rules given in Chapter 6, Section 6.3, it follows that the remaining terms in the model should then be specified as random-effect terms. The following statements specify the fitting of this model:

```
GLMM [DISTRIBUTION = poisson; LINK = log; DISPERSION = 1; \
   OFFSET = logservice; \
   RANDOM = constrctn_y * operatn_period + type.constrctn_y + \
   type.operatn_period; FIXED = type] Y = damage_incidents
```

The structure of this statement corresponds to that of the GLMM statement in Section 9.2: the only new feature is the OFFSET option, which has the same meaning as the OFFSET option in the MODEL statement earlier in this section. However, the output from this statement includes a number of warning messages, including the following:

Warning 7, code VC 38, statement 134 in procedure GLMM

Command: REML [PRINT = *] TRANS
Value of deviance at final iteration larger than at previous iteration(s)
Minimum deviance = 26.1747: value at final iteration = 43.4902

Warning 8, code VD 12, statement 134 in procedure GLMM

Command: REML [PRINT = *] TRANS
REML algorithm diverged/parameters out of bounds – output not available
Results may be unreliable. Printed estimates of variance parameters/monitoring information are available from REML or VDISPLAY and will indicate which parameters are
unstable. Redefine the model or use better initial values.

Warning 9, code VC 86, statement 134 in procedure GLMM

Command: REML [PRINT = *] TRANS
Parameter(s) out of bounds

.

.

.

> *Message: negative variance components present. Tables of effects/means will be produced for random model terms but should be used with caution.*

Whereas the methods of ordinary regression analysis and analysis of variance can be applied without arithmetic problems to almost any data set, the same is not true of mixed modelling, and certainly not of GLMM. We will now examine the kinds of problems that can be encountered, in order to interpret these warning messages.

In ordinary regression analysis and analysis of variance, analytical formulae are applied which give, in a single step, the best estimates, according to a certain criterion, of the parameters of the model being fitted. The criterion used to identify the 'best' parameter estimates is that they should be the values that maximise the probability of the observed data, and hence minimise the value of the deviance (a concept introduced in Chapter 3, Section 3.12). These are said to be the parameter estimates with the highest *likelihood*: as noted in the GenStat output presented in Section 3.12,

$$\text{deviance} = -2\log_e(\text{likelihood}). \tag{9.16}$$

When all the random-effect terms are assumed to be normally distributed (as is the case in all the examples considered in earlier chapters), these estimates are also those that minimise the estimate of the residual variance. However, in mixed modelling, the maximum likelihood estimates cannot generally be obtained in a single step: a *search* must be made for them. In broad terms, the search strategy is to start from an arbitrary set of parameter values, then determine a set of changes that can be made to these that will increase the value of the likelihood. This process is repeated, the change made getting smaller at each iteration, until no change that produces a further increase can be found. At this point the model-fitting process is said to have *converged* on the maximum likelihood estimates. The process is analogous to trying to climb a mountain in fog, following the rule 'always walk uphill'. The process will lead to the right solution, provided that:

- one starts on the flank of the mountain, not in some other part of the landscape,

- one takes steps that are not too large, and

- the mountain has one, and only one, peak.

If these criteria are not met, one may walk uphill indefinitely (failure of convergence), or the process may lead to a small peak on the flank of the likelihood 'mountain', not its true summit (convergence to a local maximum). The latter problem may be recognised by unrealistic parameter estimates and a poor fit to the data. Sometimes these problems can be overcome by a careful choice of initial parameter values for the fitting process, giving the process a better chance of 'climbing the right peak'.

In the present case the model-fitting process has encountered severe difficulties, and it is not advisable to rely on the results of such an analysis. Generally, such problems occur because the model being fitted does not represent the pattern of variation in the data well: a review of the model may suggest terms that can be dropped, or others that should be added. In the output above, the message that negative variance components are present suggests that the difficulties may have occurred because some of the model terms have little or no effect: if such terms are dropped, model fitting may be more successful. Inspection of the results produced when all terms were specified as fixed-effect terms indicates that the non-significant terms 'type.operatn_period' and 'constrctn_y.operatn_period' are candidates for omission. Fitting of the resulting model is specified by the following statement:

```
GLMM [DISTRIBUTION = poisson; LINK = log; DISPERSION = 1; \
    OFFSET = logservice; \
    RANDOM = constrctn_y + operatn_period + type.constrctn_y; \
    FIXED = type] Y = damage_incidents
```

The output produced by this statement is as follows:

Generalized linear mixed model analysis

Method: c.f. Schall (1991) Biometrika
Response variate: damage_incidents
Distribution: poisson
Link function: logarithm
Random model: constrctn_y + operatn_period + (constrctn_y.type)
Fixed model: Constant + type

Dispersion parameter fixed at value 1.000

******** Warning from GLMM:
missing values generated in weights/working variate.

Monitoring information

Iteration	Gammas			Dispersion	Max change
1	0.06317	0.07271	0.1455	1.000	2.5130E-01

******** Warning from GLMM:

missing values generated in weights/working variate.

2	0.07559	0.07117	0.1366	1.000	1.2420E-02

.
.
.

\<similar messages from subsequent iterations\>

.
.
.

6	0.07661	0.07151	0.1438	1.000	1.0626E-04

******** Warning from GLMM:
missing values generated in weights/working variate.

7	0.07661	0.07151	0.1439	1.000	7.1114E-05

Estimated variance components

Random term	component	s.e.
constrctn_y	0.077	0.129
operatn_period	0.072	0.111
constrctn_y.type	0.144	0.130

Residual variance model

Term	Factor	Model(order)	Parameter	Estimate	s.e.
Dispersn		Identity	Sigma2	1.000	fixed

Estimated variance matrix for variance components

constrctn_y	1	0.016543			
operatn_period	2	−0.000168	0.012291		
constrctn_y.type	3	−0.004545	−0.000003	0.016841	
Dispersn	4	0.000000	0.000000	0.000000	0.000000
		1	2	3	4

Table of effects for Constant

6.349   Standard error: 0.3638

Table of effects for type

type	A	B	C	D	E
	0.0000	−0.5856	−0.7391	−0.2174	0.4039

Standard errors of differences

Average:	0.4279
Maximum:	0.5056
Minimum:	0.3450

Average variance of differences: 0.1853

Tables of means with standard errors

Table of predicted means for type

type	A	B	C	D	E
	6.349	5.763	5.610	6.131	6.753

Standard errors of differences

Average:	0.4279
Maximum:	0.5056
Minimum:	0.3450

Average variance of differences: 0.1853

Back-transformed Means (on the original scale)

type
A	571.8
B	318.3
C	273.0
D	460.0
E	856.3

This simpler model has been successfully fitted. The output follows the same general form as that from the previous GLMM. The statement of the fitting method used and the model fitted is followed by a warning about the missing values in the data. Next comes the monitoring information on the fitting process, which shows that convergence has been achieved: this is indicated by the values of the *gammas*, statistics closely related to the variance components for the three random-effect terms (see Section 9.6), which hardly change between the fourth and fifth iteration. Next come the estimates of the variance components. Although the main effects of 'constrctn_y' and 'operatn_period' were significant in the fixed-effects-only analysis, the variance-component estimates for these terms are smaller than their respective SEs. It should be remembered that in the fixed-effects-only analysis, each term was tested against the residual deviance: the significance of these terms may have been due to other variance components that contribute to their deviance. Several variance components can contribute to the deviance for a particular model term, just as several variance components can contribute to an expected mean square in an anova (see Chapter 3, Section 3.3). Next come the tables of effects for the fixed-effect terms, which are presented relative to level 'A' of the factor 'type', so that the effect for this level is zero. Next come the estimated means for each level of type, on the transformed scale, i.e. the scale after the link function has been applied to the model: in this case, the logarithmic scale. The means are related to the effects by the formula

$$\text{mean}(\text{Level } i \text{ of 'type'}) = \text{constant} + \text{effect}(\text{Level } i \text{ of 'type'}).$$

Thus

$$\text{mean}(\text{Level A}) = 6.349 + 0.0000 = 6.349$$

$$\text{mean}(\text{Level B}) = 6.349 - 0.5856 = 5.763.$$

These means are somewhat different from those obtained from the fixed-effect model: consistently larger, but less variable. However, the ranking of the five types of ship is the same, Type C giving the fewest incidents of damage and Type E the most. The SEs of differences between these means are consistently smaller than those from the fixed-effect model. Overall, the variation among the means is somewhat less, relative to the SEs of differences, than when the fixed-effect model is used. This is perhaps to be expected, as the mixed model recognises 'constrctn_y.type' as a random-effect term, which contributes to the SEs of differences between levels of 'type'. Each mean can be back-transformed to the original scale (count of incidents) using the inverse of the link function, namely the exponential function: for example,

$$\text{back-transformed mean(Level A)} = e^{6.349} = 571.8.$$

9.4 The hierarchical generalised linear model

In the GLMMs fitted above (Sections 9.2 and 9.3), it is assumed that although the residual variation may not be normally distributed, the other random-effect terms do follow the normal distribution. However, this is not always the most natural assumption. For example, it has been suggested (Lee and Nelder, 2001) that when a Poisson distribution and a logarithmic link function are specified for the residual variation (as in the model fitted in Section 9.3), a more appropriate assumption for the other random-effect terms might be a gamma distribution and a logarithmic link function. Such models, in which a probability distribution and a link function can be specified for each random-effect term, have been called hierarchical generalised linear models (HGLMs). However, this name is slightly misleading, as the random-effect terms do not have to be nested to form a hierarchy: they may be crossed, as in the model in Section 9.3. An alternative name, taking account of this possibility, would be stratified generalised linear models.

A system to fit HGLMs has been developed (Lee and Nelder, 1996, 2001), and is available in GenStat. The following statements use this system to fit the same GLMM as was fitted in Section 9.3, with no change to the distributions or link functions:

```
HGFIXEDMODEL [DISTRIBUTION = poisson; LINK = logarithm; \
    OFFSET = logservice]  type
HGRANDOMMODEL [DISTRIBUTION = normal; LINK = identity] \
    constrctn_y + operatn_period + type.constrctn_y
HGANALYSE damage_incidents
HGPLOT METHOD = histogram, fittedvalues, normal, halfnormal
HGPLOT [RANDOMTERM = type.constrctn_y] \
    METHOD = histogram, fittedvalues, normal, halfnormal
```

The HGFIXEDMODEL and HGRANDOMMODEL statements specify the fixed-effect and random-effect models respectively. The HGFIXEDMODEL statement also specifies the distribution and link function for the residual term, and the offset variable, as was done in the equivalent GLMM statement (Section 9.3). The HGRANDOMMODEL statement further indicates that the other random-effect terms are to have a normal distribution and the *identity* link function, i.e. no transformation, as was assumed implicitly in the GLMM statement. The HGANALYSE statement indicates that the response variable is 'damage_incidents'. The output from this statement (not shown) is numerically equivalent to that from the GLMM statement. The two HGPLOT statements produce diagnostic plots of the random effects, for the term 'type.constrctn_y' and the residual term respectively. The other two random-effect terms do not have enough levels to justify the production of diagnostic plots. These plots are presented in Figure 9.5.

The distributions used in the model fitted here are not both normal, and the reference to normal plots and normal quantiles in the labelling of these diagnostic plots is

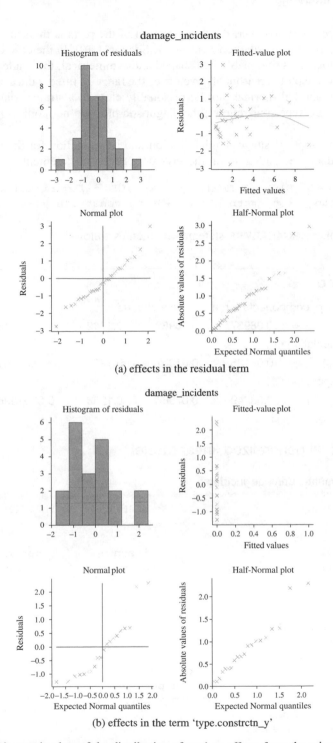

(a) effects in the residual term

(b) effects in the term 'type.constrctn_y'

Figure 9.5 Diagnostic plots of the distribution of random effects from the mixed model relating damage to ships to their type, year of construction and period of operation, fitted with the normal distribution and the identity link function.

therefore imprecise. However, the interpretation of the plots is the same as in models that assume a normal distribution (Chapter 1, Section 1.10). In the case of the residual term, the histogram is reasonably bell shaped and symmetrical, the scatter of the points in the fitted-value plot is reasonably even over the range of fitted values, and the points in the normal and half-normal plot lie reasonably close to a straight diagonal line. In the case of term 'type.constrctn_y', the diagnostic plots do not conform so closely to these ideals.

In order to adopt the suggested distribution and link function for the random-effect terms other than the residual term, the HGRANDOMMODEL statement is modified to

```
HGRANDOMMODEL [DISTRIBUTION = gamma; LINK = logarithm] \
   constrctn_y + operatn_period + type.constrctn_y
```

The output of the HGANALYSE statement is then as follows:

Monitoring

cycle no., disp. components & max. relative change

3	0.08967	0.06988	0.1712	0.8215

Aitken extrapolation OK

7	0.08135	0.07067	0.1270	0.02001

Aitken extrapolation OK

11	0.08249	0.07064	0.1258	0.0001896

 converged

Hierarchical generalized linear model

Response variate: damage_incidents

```
                                      MEAN        DISPERSION
    ------------------------------------------------------------
    DISTRIBUTIONS
        fixed                         Poisson        gamma
        random                        gamma          gamma
    LINKS
        fixed                            log            log
        random                           log            log
    LIN. PREDICTORS
        fixed                           type
        random
                               constrctn_y
                              operatn_period
                              constrctn_y.type
    dispersion fixed
    lmat matrices unset
    offset on
    ------------------------------------------------------------
```

Estimates from the mean model

		estimate	s.e.	t
1	Constant	6.381	0.3538	18.036
2	type B	-0.578	0.3280	-1.761
3	type C	-0.731	0.4470	-1.636
4	type D	-0.160	0.4114	-0.388
5	type E	0.448	0.3900	1.148

Note: s.e.s assume dispersion =   1.000

Estimates on the log scale from the dispersion model

		estimate	s.e.	t
1	lambda 1	-2.495	1.093	-2.283
2	lambda 2	-2.650	1.479	-1.791
3	lambda 3	-2.073	0.624	-3.323

Note: s.e.s assume dispersion =   1.000

The monitoring information at the beginning of the output confirms that the model-fitting process has converged. The model fitted is then specified. Next come the estimates of the constant and of the effects of 'type', which are very little different from those obtained from the GLMM. The parameter estimates from the dispersion model, $\hat{\lambda}_1$, $\hat{\lambda}_2$ and $\hat{\lambda}_3$, indicate the deviance due to the random-effect terms 'constrctn_y', 'operatn_period' and 'constrctn_y.type' respectively. These estimates are given relative to the dispersion: that is,

$$\hat{\lambda}_i = \frac{\text{deviance}(\text{term } i)}{\text{dispersion}}. \tag{9.17}$$

In the present case we have specified that

$$\text{dispersion} = 1, \tag{9.18}$$

so rearranging Equation 9.17 and substituting the numerical values from the output and from Equation 9.18, we obtain, for example,

$$\text{deviance}(\text{constrctn\_y}) = -2.495 \times 1 = -2.495.$$

The deviances for the random-effect terms can be interpreted and compared in roughly the same way as estimates of variance components. For example, each is compared in the output with its own SE, to obtain a t statistic that gives a tentative indication of whether any variation is accounted for by the term (cf. Chapter 3, Section 3.11, where the analogous statistic based on estimates of variance components is referred to as a z statistic).

The diagnostic plots obtained from this model are shown in Figure 9.6. The plot for the residual term is very little changed, but that for term 'type.constrctn_y' is considerably *less* satisfactory than when the normal distribution and the identity link function were used. Thus in this particular case, the use of the HGLM system has not

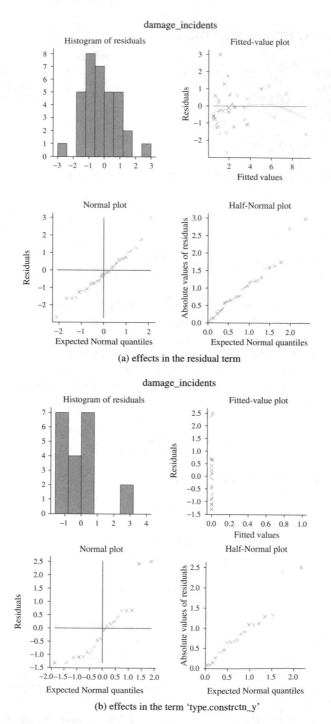

Figure 9.6   Diagnostic plots of the distribution of random effects from the mixed model relating damage to ships to their type, year of construction and period of operation, fitted with the gamma distribution and the logarithmic link function.

improved the outcome of the modelling process. This may be because the parameters of a non-normal distribution are more difficult to fit, even when such a distribution is more appropriate on theoretical grounds (Duncan Hedderly, personal communication).

9.5 The role of the covariance matrix in the specification of a mixed model

In all the mixed models we have examined so far, each random-effect term has consisted of a set of factor levels. Any two observations either come from the same level of the factor, in which case they have the same random effect for the term in question, or come from different levels, in which case their random effects are independent. In the residual stratum, each observation comprises a separate level, and its random effect is independent of those of all other observations. However, the relationship between the random effects in each term need not be so simple. In order to explore more complicated relationships among the random effects, we need first to establish a notation in which to display them. Such a notation is provided by the *covariance matrix*, which is specified as follows.

Consider a simple data set comprising four groups of three observations of a response variable (Table 9.4). The natural model to fit to these data is

$$y_{ij} = \mu + \gamma_i + \varepsilon_{ij} \tag{9.19}$$

Table 9.4 A simple data set comprising a grouping factor and a response variable. The conventions used in this spreadsheet are the same as those used in Table 1.1, Chapter 1.

	A	B
	A	B
1	group!	y
2	1	1.97
3	1	1.01
4	1	4.53
5	2	0.76
6	2	1.60
7	2	3.52
8	3	4.37
9	3	4.05
10	3	5.17
11	4	6.94
12	4	7.50
13	4	5.62

Table 9.5   Anova of a simple data set comprising a grouping factor and a response variable.

Source of variation	DF	MS	F	P
Group	3	13.875	8.44	0.007
Residual	8	1.644		
Total	11			

where

y_{ij} = the jth observation of the response variable Y in the ith group (the ijth observation),

μ = the overall mean,

γ_i = the effect of the ith group,

ε_{ij} = the residual effect on the ijth observation of Y.

We will make the usual assumption that the ε_{ij} are values of a random, normally distributed variable E, i.e.

$$E \sim N(0, \sigma_E^2).$$

The anova corresponding to this model is shown in Table 9.5, and the estimate of σ_E^2 is given by MS_{Residual}

$$\hat{\sigma}_E^2 = 1.644.$$

In order to express the idea that the 12 values ε_{ij}, $i = 1 \ldots 4$, $j = 1 \ldots 3$, are independent, we need to think of each of them as an value of a different variable, E_{ij}. Similarly, each value of the response variable, y_{ij}, is thought of as an observation of a different variable, Y_{ij}. The relationships among the 12 variables E_{ij} can then be expressed in a covariance matrix:

i	j	j'	i' 1 / 1	1 / 2	1 / 3	2 / 1	2 / 2	2 / 3	3 / 1	3 / 2	3 / 3	4 / 1	4 / 2	4 / 3
1	1		σ_E^2	0	0	0	0	0	0	0	0	0	0	0
1	2		0	σ_E^2	0	0	0	0	0	0	0	0	0	0
1	3		0	0	σ_E^2	0	0	0	0	0	0	0	0	0
2	1		0	0	0	σ_E^2	0	0	0	0	0	0	0	0
2	2		0	0	0	0	σ_E^2	0	0	0	0	0	0	0
2	3		0	0	0	0	0	σ_E^2	0	0	0	0	0	0
3	1		0	0	0	0	0	0	σ_E^2	0	0	0	0	0
3	2		0	0	0	0	0	0	0	σ_E^2	0	0	0	0
3	3		0	0	0	0	0	0	0	0	σ_E^2	0	0	0
4	1		0	0	0	0	0	0	0	0	0	σ_E^2	0	0
4	2		0	0	0	0	0	0	0	0	0	0	σ_E^2	0
4	3		0	0	0	0	0	0	0	0	0	0	0	σ_E^2

where

i = the group to which the first value in a particular comparison belongs,

j = the position within group i of the first value,

i' = the group to which the second value in the comparison belongs,

j' = the position within group i' of the second value.

The value of $\mathrm{cov}(E_{ij}, E_{i'j'})$, the covariance between E_{ij} and $E_{i'j'}$, is given in the cell in the ijth row and the $i'j'$th column of this matrix. Since E_{ij} is the only random term that contributes to Y_{ij},

$$\mathrm{cov}(Y_{ij}, Y_{i'j'}) = \mathrm{cov}(E_{ij}, E_{i'j'}) \tag{9.20}$$

The variance of each of the variables E_{ij} – its covariance with itself – is given along the *leading diagonal* of the matrix (the 12 cells from top left to bottom right). All other covariances are zero, reflecting the assumption that all the variables are mutually independent. Because

$$\mathrm{cov}(E_{ij}, E_{i'j'}) = \mathrm{cov}(E_{i'j'}, E_{ij}) \tag{9.21}$$

by definition, the matrix is symmetrical about the leading diagonal; hence to improve clarity, the top right hand half can be omitted.

Now suppose that we decide to treat the groups as a random-effect term. We then make the assumption that the $\gamma_i, i = 1 \ldots 4$, are values of a variable Γ, and that

$$\Gamma \sim N(0, \sigma_\Gamma^2).$$

The anova can now be interpreted as shown in Table 9.6, and the estimate of σ_Γ^2 is

$$\hat{\sigma}_\Gamma^2 = \frac{MS_{\mathrm{Group}} - MS_{\mathrm{Residual}}}{3} = \frac{13.875 - 1.644}{3} = 4.077.$$

In order to express the idea that the four values $\gamma_i, i = 1 \ldots 4$, are independent, we think of each of them as a value of a different variable, Γ_i. The variable representing the random part of Y_{ij} is now no longer E_{ij} but $\Gamma_i + E_{ij}$, and the covariances among these 12 variables are as follows:

i	j \ i' j'	1 1	1 2	1 3	2 1	2 2	2 3	3 1	3 2	3 3	4 1	4 2	4 3
1	1	$\sigma_\Gamma^2+\sigma_E^2$											
1	2	σ_Γ^2	$\sigma_\Gamma^2+\sigma_E^2$										
1	3	σ_Γ^2	σ_Γ^2	$\sigma_\Gamma^2+\sigma_E^2$									
2	1	0	0	0	$\sigma_\Gamma^2+\sigma_E^2$								
2	2	0	0	0	σ_Γ^2	$\sigma_\Gamma^2+\sigma_E^2$							
2	3	0	0	0	σ_Γ^2	σ_Γ^2	$\sigma_\Gamma^2+\sigma_E^2$						
3	1	0	0	0	0	0	0	$\sigma_\Gamma^2+\sigma_E^2$					
3	2	0	0	0	0	0	0	σ_Γ^2	$\sigma_\Gamma^2+\sigma_E^2$				
3	3	0	0	0	0	0	0	σ_Γ^2	σ_Γ^2	$\sigma_\Gamma^2+\sigma_E^2$			
4	1	0	0	0	0	0	0	0	0	0	$\sigma_\Gamma^2+\sigma_E^2$		
4	2	0	0	0	0	0	0	0	0	0	σ_Γ^2	$\sigma_\Gamma^2+\sigma_E^2$	
4	3	0	0	0	0	0	0	0	0	0	σ_Γ^2	σ_Γ^2	$\sigma_\Gamma^2+\sigma_E^2$

Table 9.6 Expected mean squares in the anova of a simple data set comprising a grouping factor and a response variable.

Source of variation	DF	Expected MS
Group	3	$3\sigma_\Gamma^2 + \sigma_E^2$
Residual	8	σ_E^2
Total	11	

The variance of each observation of the response variable is now $\sigma_\Gamma^2 + \sigma_E^2$. The covariance between observations from different groups – for example, between Y_{12} and Y_{23} – is zero, as before. However, the covariance between observations from the same group – for example, between Y_{12} and Y_{13} – is now σ_Γ^2, because they share the same group effect (Γ_1) though they have different individual observation effects (E_{12} and E_{13}). This covariance matrix can be expressed in the notation of *matrix algebra* as

$$
\begin{pmatrix}
1 & & & & & & & & & & & \\
1 & 1 & & & & & & & & & & \\
1 & 1 & 1 & & & & & & & & & \\
0 & 0 & 0 & 1 & & & & & & & & \\
0 & 0 & 0 & 1 & 1 & & & & & & & \\
0 & 0 & 0 & 1 & 1 & 1 & & & & & & \\
0 & 0 & 0 & 0 & 0 & 0 & 1 & & & & & \\
0 & 0 & 0 & 0 & 0 & 0 & 1 & 1 & & & & \\
0 & 0 & 0 & 0 & 0 & 0 & 1 & 1 & 1 & & & \\
0 & 0 & 0 & 0 & 0 & 0 & 0 & 0 & 0 & 1 & & \\
0 & 0 & 0 & 0 & 0 & 0 & 0 & 0 & 0 & 1 & 1 & \\
0 & 0 & 0 & 0 & 0 & 0 & 0 & 0 & 0 & 1 & 1 & 1
\end{pmatrix} \sigma_\Gamma^2
$$

$$
+ \begin{pmatrix}
1 & & & & & & & & & & & \\
0 & 1 & & & & & & & & & & \\
0 & 0 & 1 & & & & & & & & & \\
0 & 0 & 0 & 1 & & & & & & & & \\
0 & 0 & 0 & 0 & 1 & & & & & & & \\
0 & 0 & 0 & 0 & 0 & 1 & & & & & & \\
0 & 0 & 0 & 0 & 0 & 0 & 1 & & & & & \\
0 & 0 & 0 & 0 & 0 & 0 & 0 & 1 & & & & \\
0 & 0 & 0 & 0 & 0 & 0 & 0 & 0 & 1 & & & \\
0 & 0 & 0 & 0 & 0 & 0 & 0 & 0 & 0 & 1 & & \\
0 & 0 & 0 & 0 & 0 & 0 & 0 & 0 & 0 & 0 & 1 & \\
0 & 0 & 0 & 0 & 0 & 0 & 0 & 0 & 0 & 0 & 0 & 1
\end{pmatrix} \sigma_E^2 .
$$

In the first term in this expression, all rows and all columns representing observations in the same group are identical. (To see this, it may be helpful to fill in the top right-hand half of the matrix.) Hence the information in this term can be expressed more concisely in the form

$$\text{covariance matrix} = \begin{pmatrix} 1 & & & \\ 0 & 1 & & \\ 0 & 0 & 1 & \\ 0 & 0 & 0 & 1 \end{pmatrix} \sigma_\Gamma^2. \qquad (9.22)$$

9.6 A more general pattern in the covariance matrix. Analysis of pedigree data

In all the mixed models we have considered so far, the covariance matrix for each term is of the form shown in Equation 9.22 – an *identity matrix*, having ones along the leading diagonal and zeros elsewhere, multiplied by the variance component for the term in question. However, the covariance matrix need not be so simple. For example, consider the pedigree in Figure 9.7, which represents three generations of animals. Earlier generations are represented in higher rows of the pedigree: animals represented in the bottom row are the most recent generation, those in the row above are their parents, and those in the top row their grandparents.

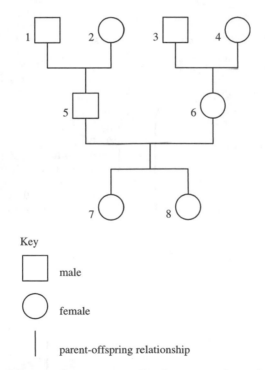

Key

☐ male

◯ female

│ parent-offspring relationship

Figure 9.7 A pedigree representing three generations of animals.

Suppose that a continuous variable Y (e.g. weight) is measured on each animal in this pedigree, and that this variable has a genetic component Γ and an environmental component E, so that

$$Y = \mu + \Gamma + E \tag{9.23}$$

where
μ = the mean value of Y.

Then in the population from which this pedigree is randomly sampled, the variance of Y among unrelated animals (such as the grandparents 1, 2, 3 and 4) has two components, namely σ_Γ^2, the genetic component, and σ_E^2, the environmental component, the total variance of Y being $\sigma_\Gamma^2 + \sigma_E^2$. The values of Γ in any two unrelated animals are independent, as are the values of E, so the covariance of Y between such animals is zero. However, Animal 7 shares half her genes with her sister, Animal 8, so the values of Γ in these two animals are not independent: the covariance of Y between them is $\frac{1}{2}\sigma_\Gamma^2$. Conversely, the variance among siblings is less than that among unrelated animals: it is $\frac{1}{2}\sigma_\Gamma^2 + \sigma_E^2$. Animal 7 also shares a quarter of her genes with her paternal grandmother, Animal 2, so that the covariance of Y between these animals is $\frac{1}{4}\sigma_\Gamma^2$. Similar arguments can be applied to every pair of animals, giving the following covariance matrix:

$$
\begin{array}{c|cccccccc}
i\ i' & 1 & 2 & 3 & 4 & 5 & 6 & 7 & 8 \\
\hline
1 & \sigma_\Gamma^2+\sigma_E^2 \\
2 & 0 & \sigma_\Gamma^2+\sigma_E^2 \\
3 & 0 & 0 & \sigma_\Gamma^2+\sigma_E^2 \\
4 & 0 & 0 & 0 & \sigma_\Gamma^2+\sigma_E^2 \\
5 & \frac{1}{2}\sigma_\Gamma^2 & \frac{1}{2}\sigma_\Gamma^2 & 0 & 0 & \sigma_\Gamma^2+\sigma_E^2 \\
6 & 0 & 0 & \frac{1}{2}\sigma_\Gamma^2 & \frac{1}{2}\sigma_\Gamma^2 & 0 & \sigma_\Gamma^2+\sigma_E^2 \\
7 & \frac{1}{4}\sigma_\Gamma^2 & \frac{1}{4}\sigma_\Gamma^2 & \frac{1}{4}\sigma_\Gamma^2 & \frac{1}{4}\sigma_\Gamma^2 & \frac{1}{2}\sigma_\Gamma^2 & \frac{1}{2}\sigma_\Gamma^2 & \sigma_\Gamma^2+\sigma_E^2 \\
8 & \frac{1}{4}\sigma_\Gamma^2 & \frac{1}{4}\sigma_\Gamma^2 & \frac{1}{4}\sigma_\Gamma^2 & \frac{1}{4}\sigma_\Gamma^2 & \frac{1}{2}\sigma_\Gamma^2 & \frac{1}{2}\sigma_\Gamma^2 & \frac{1}{2}\sigma_\Gamma^2 & \sigma_\Gamma^2+\sigma_E^2
\end{array}
$$

$$
= \begin{pmatrix}
1 \\
0 & 1 \\
0 & 0 & 1 \\
0 & 0 & 0 & 1 \\
\frac{1}{2} & \frac{1}{2} & 0 & 0 & 1 \\
0 & 0 & \frac{1}{2} & \frac{1}{2} & 0 & 1 \\
\frac{1}{4} & \frac{1}{4} & \frac{1}{4} & \frac{1}{4} & \frac{1}{2} & \frac{1}{2} & 1 \\
\frac{1}{4} & \frac{1}{4} & \frac{1}{4} & \frac{1}{4} & \frac{1}{2} & \frac{1}{2} & \frac{1}{2} & 1
\end{pmatrix} \sigma_\Gamma^2
+ \begin{pmatrix}
1 \\
0 & 1 \\
0 & 0 & 1 \\
0 & 0 & 0 & 1 \\
0 & 0 & 0 & 0 & 1 \\
0 & 0 & 0 & 0 & 0 & 1 \\
0 & 0 & 0 & 0 & 0 & 0 & 1 \\
0 & 0 & 0 & 0 & 0 & 0 & 0 & 1
\end{pmatrix} \sigma_E^2
$$

where
i = the first animal in a particular pairwise comparison,
i' = the second animal in the comparison.

The matrix

$$
\begin{pmatrix}
1 \\
0 & 1 \\
0 & 0 & 1 \\
0 & 0 & 0 & 1 \\
\frac{1}{2} & \frac{1}{2} & 0 & 0 & 1 \\
0 & 0 & \frac{1}{2} & \frac{1}{2} & 0 & 1 \\
\frac{1}{4} & \frac{1}{4} & \frac{1}{4} & \frac{1}{4} & \frac{1}{2} & \frac{1}{2} & 1 \\
\frac{1}{4} & \frac{1}{4} & \frac{1}{4} & \frac{1}{4} & \frac{1}{2} & \frac{1}{2} & \frac{1}{2} & 1
\end{pmatrix}
$$

is called the *relationship matrix* among the animals in the pedigree, and is said to hold the *variance structure* among them. The covariance matrix can be applied to the observations of Y, designated as y_i, $i = 1 \ldots 8$, using the methods of mixed modelling, to obtain estimates of the variance components, $\hat{\sigma}_\Gamma^2$ and $\hat{\sigma}_E^2$. From these, an estimate of the heritability of Y,

$$h^2 = \frac{\hat{\sigma}_\Gamma^2}{\hat{\sigma}_\Gamma^2 + \hat{\sigma}_E^2}, \tag{9.24}$$

(see also Equation3.22), can be obtained. For a fuller account of genetic covariance between relatives and heritability, see Falconer and Mackay (1996, Chapters 9 and 10, pp 145–183).

This method of analysis can be applied to a set of simulated data described by Goldin *et al.* (1997). (Data reproduced by kind permission of John Blangero, Tom Dyer, Jean MacCluer and Marcy Speer. Data presented at the 10th Genetic Analysis Workshop, supported by grant R01 GM31575 from the National Institute of General Medical Sciences.) The first and last few lines of a spreadsheet holding the pedigree data and the phenotypic data (i.e. the observations on the individuals in the pedigrees) are shown in the spreadsheet in Table 9.7. The first three columns specify the relationships among the individuals, assumed here to belong to a species of domesticated animal. Thus in row 19 we are told that the father and mother of Individual 1378 are Individuals 922 and 924 respectively. The information on these two parents is held in rows 5 and 7, where it is confirmed that 922 is a male (sex $= 1$) and 924 is a female (sex $= 2$). Q4 is a quantitative trait, the inheritance of which is to be studied.

These data are imported and analysed by the following statements:

```
IMPORT 'Intro to Mixed Modelling\\Chapter 9\\GAW10 ped phen.xls'
VPEDIGREE INDIVIDUALS = ID; MALE = father; FEMALE = mother; \
     INVERSE = ainv
VCOMPONENTS [FIXED = sex * age] RANDOM = ID
VSTRUCTURE [TERM = ID] MODEL = fixed; INVERSE = ainv
REML [PRINT = model, monitor, component, Wald, effects] Q4
```

In the VPEDIGREE statement, the parameter INDIVIDUALS specifies the factor that identifies each individual to be considered, and the parameters MALE and FEMALE specify factors that identify the parents of each individual. From this information is produced, not the relationship matrix described above, but its *inverse* – a matrix derived from it which contains the same information. (The details of the relationship between a matrix and its inverse need not concern us here.) The parameter INVERSE indicates the data structure in which this inverse matrix is to be stored, namely 'ainv'. The VCOMPONENTS statement specifies the terms in the fixed-effect and random-effect parts of the model in the usual way. In the present case sex, age and their interaction are to be fitted as fixed-effect terms. This indicates that sex and age may have effects on the phenotypic trait studied, which should be taken into account when assessing the degree of similarity between relatives. The 'sex.age' interaction term indicates that the effect of age may differ between the sexes. The only random term is the individual observations. However, this term does not merely comprise a set of independent residual effects: it is structured by the pedigree relationships among the individuals, and this is indicated in the VSTRUCTURE statement that follows.

Table 9.7 Pedigree data indicating the relationships among a set of individuals, with observations of a quantitative phenotypic trait.

Only the first and last few rows of the data are shown. The conventions used in this spreadsheet are the same as those used in Table 1.1, Chapter 1. The numbering of the individuals has been changed from that in the original version of this data set, in order to meet GenStat's requirement that every individual, including those who appear under 'father!' or 'mother!' but not under 'ID!', should have a unique number.

	A	B	C	D	E	F
1	ID!	father!	mother!	sex!	age	Q4
2	919	1	460	1	56	
3	920	2	461	2	59	
4	921	3	462	2	80	
5	922	4	463	1	80	
6	923	5	464	1	67	
7	924	6	465	2	77	
8	925	7	466	2	61	
9	926	8	467	2	63	
10	927	9	468	1	64	
11	928	10	469	2	37	13.1086
12	929	11	470	1	40	10.1638
13	930	12	471	1	40	10.3412
14	931	13	472	1	49	10.9942
15	932	14	473	1	64	10.685
16	933	15	474	1	65	11.4764
17	934	16	475	2	66	11.3533
18	935	17	476	1	67	10.1688
19	1378	922	924	2	80	
20	1379	922	924	1	80	
21	1380	922	924	1	80	
22	1381	927	926	2	63	
23	1382	927	926	1	72	
24	1596	919	1378	2	64	11.6905
25	1597	923	1381	1	58	10.5241
26	1598	923	1381	1	53	
27	1599	923	1381	2	64	11.3787
.						.
.						.
.						.
1484	2035	1588	1585	1	49	11.6033
1485	2329	2024	1590	2	37	12.1533

Table 9.7 (*continued*)

	A	B	C	D	E	F
1486	2330	2024	1590	2	47	10.9051
1487	2331	2024	1590	2	40	9.9029
1488	2332	2024	1590	1	42	12.1348
1489	2333	2024	1590	2	46	10.8755
1490	2334	2024	1590	2	45	12.3281
1491	2335	2025	1373	2	34	10.8796
1492	2409	1365	2331	1	19	10.5293
1493	2410	1370	2334	2	24	10.7028
1494	2411	1370	2334	1	16	11.9798
1495	2412	1374	2329	1	16	11.0855
1496	2413	2035	2333	2	26	9.3804
1497	2414	2035	2333	2	24	11.3734
1498	2415	2035	2333	1	22	9.6838

The option setting 'TERM = ID' indicates the term in the random-effect model for which a variance structure is to be specified. In the present case there is only one term, ID, but if the random-effect model contained several terms, a different variance structure (or none) might be specified for each of them. The parameter setting 'MODEL = fixed' indicates that the variance structure is fully specified prior to the model-fitting process – in this case, by the VPEDIGREE statement. In the next example (Section 9.7) we will see a situation in which some aspects of the variance structure are estimated during the model-fitting process. The parameter setting 'INVERSE = ainv' points to the data structure that holds the information defining the variance structure. Finally, the REML statement specifies that the response variable to which the model is to be fitted is Q4. The PRINT option of this statement specifies that the output should comprise a statement of the model fitted, information monitoring the fitting process, and estimates of the variance components.

The output of the REML statement is as follows:

```
6. . . . . . . . . . . . . . . . . . . . . . . . . . . . . . . . . . . . . . . . . . . . . . . . . . . . . . . . .
***** REML Variance Components Analysis *****


Response Variate : Q4

Fixed model       : Constant+sex+age+sex.age
Random model      : ID

Number of units   : 1000 (497 units excluded due to zero weights
or missing values)

* Residual term has been added to model
```

```
* Sparse algorithm with AI optimisation
* All covariates centred

*** Covariance structures defined for random model ***

Covariance structures defined within terms:

Term     Factor      Model                        Order Nrows
ID       ID          Fixed matrix ainv (inverse)      1 2415




*** Convergence monitoring ***

Cycle  Deviance  Current variance parameters: gammas, sigma2,
   others
    0      *        1.00000  1.00000
    1    877.367   0.468573  1.02253
    2    877.202   0.464133  1.08776
    3    876.839   0.451823  1.17101
    4    876.592   0.437181  1.23839
    5    876.538   0.426116  1.23548
    6    876.538   0.426581  1.23569
    7    876.538   0.426546  1.23568




*** Estimated Parameters for Covariance Models ***

Random term(s) Factor  Model(order) Parameter  Estimate     S.e.

ID               ID     Fixed matrix Scalar       1.236     0.298

Note: the covariance matrix for each term is calculated as G or
      R where var(y) = Sigma2( ZGZ'+R ), ie. relative to the
      residual variance, Sigma2


*** Residual variance model ***

Term      Factor  Model(order)  Parameter   Estimate     S.e.

Residual          Identity      Sigma2         0.427     0.0513


*** Wald tests for fixed effects ***


   Fixed term    Wald statistic   d.f.   Wald/d.f.  Chi-sq prob

* Sequentially adding terms to fixed model
```

```
    sex                    0.71           1       0.71        0.399
    age                    6.09           1       6.09        0.014
    sex.age                1.76           1       1.76        0.185

*  Dropping individual terms from full fixed model

    sex.age                1.76           1       1.76        0.185
    sex                    0.99           1       0.99        0.321

*  Message: chi-square distribution for Wald tests is an
   asymptotic approximation (i.e. for large samples) and
   underestimates the probabilities  in other cases.


*** Table of effects for Constant ***

            11.44      Standard error:    0.056



*** Table of effects for sex ***

         sex              1               2
                     0.00000        -0.05684

Standard error of differences:        0.05725



*** Table of effects for age ***

         0.006357      Standard error:    0.0023813



*** Table of effects for sex.age ***

         sex              1               2
                     0.000000       -0.004416

Standard error of differences:        0.003333
```

First comes the usual information about the model fitted – the response variable, the fixed-effect and random-effect models, and so on – with some additional information about the variance structure specified (here referred to as the covariance structure). It is noted that the inverse of the relationship matrix, 'ainv', has 2415 rows, corresponding to the 2415 individuals (including parents) represented in the data set. Next comes the information on the convergence of the fitting process, which shows that after 5 cycles of parameter estimation there is little or no change in the deviance from the fitted model. Convergence has been achieved, and the process terminates successfully after 7 cycles. Next come the estimates of the variance components. The first of these is the variance accounted for by the covariance structure specified for factor ID – that is, the genetic component that follows the pattern of relationships among the individuals, σ_Γ^2. This is followed by the residual variance component that affects each individual independently, σ_E^2. The estimate of the residual variance component is expressed straightforwardly as

$$\hat{\sigma}_E^2 = 0.427.$$

However, it is noted that the genetic component is calculated *relative to* the residual variance: that is, if we define

$$\gamma_\Gamma = \frac{\sigma_\Gamma^2}{\sigma_E^2} \tag{9.25}$$

then the estimate of γ_Γ is

$$\hat{\gamma}_\Gamma = 1.236$$

and

$$\hat{\sigma}_\Gamma^2 = \hat{\gamma}_\Gamma \hat{\sigma}_E^2 = 1.236 \times 0.427 = 0.5278. \tag{9.26}$$

(Note that the ratio $\gamma_\Gamma = \sigma_\Gamma^2/\sigma_E^2$ is not the same thing as the variable Γ introduced in Equations 9.19 and 9.23: the letter 'gamma' is used for both purposes.) Substituting the expression for $\hat{\sigma}_\Gamma^2$ in Equation 9.26 into Equation 9.24,

$$h^2 = \frac{\hat{\gamma}_\Gamma \hat{\sigma}_E^2}{\hat{\gamma}_\Gamma \hat{\sigma}_E^2 + \hat{\sigma}_E^2} = \frac{\hat{\gamma}_\Gamma}{\hat{\gamma}_\Gamma + 1}, \tag{9.27}$$

and in the present case,

$$h^2 = \frac{1.236}{1.236 + 1} = 0.553.$$

This tells us that the trait Q4 is heritable, but only moderately so. This estimate of heritability may be an overestimate, as it depends upon the assumption that all similarity between relatives is due to shared genes, not shared environment. This assumption may be reasonable in the case of domestic animals in an experimental or commercial environment. (It is never likely to be so in the case of human beings.) More elaborate methods of estimation can be used to attempt to correct for the effect of shared environment. Knowledge of the heritability of traits provides the basis for selective breeding of domestic animals: if Q4 were a trait of commercial importance in an animal species, it would be amenable to improvement by selection, though not easily so.

The Wald tests show that trait Q4 is significantly influenced by the animal's age, though not by its sex nor by a 'sex.age' interaction effect. The coefficient for the effect of age, 0.006357, shows that the value of this trait tends to increase with age.

In order to choose the animals from which to breed in a genetic improvement programme, it is desirable to estimate the *breeding value* of each animal, the amount by which the mean value of its progeny will differ from the population mean. The simplest estimate of an animal's breeding value is its phenotypic value: the observed value of the trait in question. However, a better estimate is provided by the estimated *genetic effect* on the trait for each animal, the estimated value of the variable Γ. In the present case, this is obtained by the following statement:

```
REML [PTERMS = ID; PRINT = effects; PSE = none] Q4
```

The output of this statement is as follows:

```
7. . . . . . . . . . . . . . . . . . . . . . . . . . . . . . . . . . . . . . . . . . . . . . . . .


*** Table of effects for ID ***

    ID          1          2          3          4          5
            0.0508     0.0016    -0.1929    -0.0702     0.0065


    ID          6          7          8          9         10
           -0.0702    -0.1748    -0.0842    -0.0842     0.6077
    .
    .
    .


<similar output for other individuals>
    .
    .
    .
    ID        2406       2407       2408       2409       2410
           -0.0190     0.3380     0.1222    -0.6363    -0.2330


    ID        2411       2412       2413       2414       2415
            0.2872     0.0768    -1.0287    -0.2661    -0.8934
```

In the present case the genetic effect is an improvement on the phenotypic value as a predictor not only because it is adjusted for the sex and age of the animal (the fixed-effect terms in the mixed model), but also because it incorporates information from the individual's relatives. If it seems counter-intuitive that these can provide more reliable information concerning an animal's breeding value than observation of the animal's own phenotype, consider the extreme case of the selection of dairy bulls. Each of these has the potential to produce many daughters, so the choice of the best animal is crucial, yet no observation of milk yield can be made on any of the bulls available. They are compared entirely on the basis of the yield of their female relatives. The relationship between the phenotypic values and the genetic effects in the present data set (for those individuals for which both values are available) is shown in Figure 9.8. A breeder seeking to increase the value of trait Q4 would choose many of the same

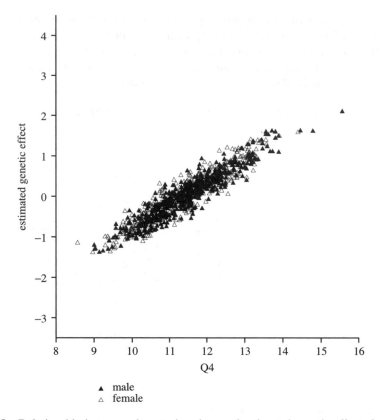

Figure 9.8   Relationship between phenotypic values and estimated genetic effects for trait Q4, obtained from pedigree data.

animals for mating on either basis, but there would be some discrepancies, and hence some increase in the progress made if selection were based on the estimated genetic effect.

Because the genetic effect of each animal is a random effect – an estimated value of the random variable Γ – it is a BLUP or shrunk estimate (see Chapter 5). This partly accounts for the narrow range of values of the estimated genetic effect (max − min = $2.11 - (-1.39) = 3.50$) relative to the range of Q4 (max − min = $15.56 - 8.57 = 6.99$). (The adjustment for age and sex also narrows the range of values of the estimated genetic effect.) This application of BLUPs is also considered by Robinson (1991, Section 6.2).

If information from genetic markers is available, indicating the distribution of genetic variation among the individuals in the pedigrees studied, the techniques of mixed modelling can be further used to address the problem of genetic mapping. A genetic map estimates the positions on the chromosomes (the linear structures in each cell of an organism that carry its hereditary information) of the genes that influence the phenotypic trait under consideration. Such a map is constructed by using genetic markers as explanatory variables and the phenotypic trait as the response variable. A clear account of the concepts involved is given by Sham (1998, Chapter 5, Section 5.3, pp 197–219). There are several variations on this basic approach. Among these, the variance component method is widely used and is implemented in several pieces of software, notably Merlin (described by Abecasis *et al.* (2002) and available from the University of Michigan's Center for Statistical Genetics via the web site http://csg.sph.umich.edu/). The variance-component method includes genetic markers in the model solely as random-effect terms, but an extension of it, the Quantitative Transmission Disequilibrium Test (QTDT), includes them as both random- and fixed-effect terms, making fuller use of the available information (Abecasis *et al.*, 2000a; Abecasis *et al.* 2000b). This test is implemented in the software QTDT (available from the same web site).

9.7 Estimation of parameters in the covariance matrix. Analysis of temporal and spatial variation

There are many other contexts in which it is possible, and valuable, to specify the structure of a covariance matrix. We will next consider the situation in which observations are obtained from a regular array of experimental units. This may be a one-dimensional array, such as a series of observations at regular time intervals, or a two-dimensional array, such as the spatial arrangement of plots in a field experiment on a crop species. In either case, a structured covariance matrix can be used to model the natural variation over the array, in order to improve the precision with which treatment effects are estimated. Consider a very small field experiment, comprising a 4×3 array of plots (Figure 9.9).

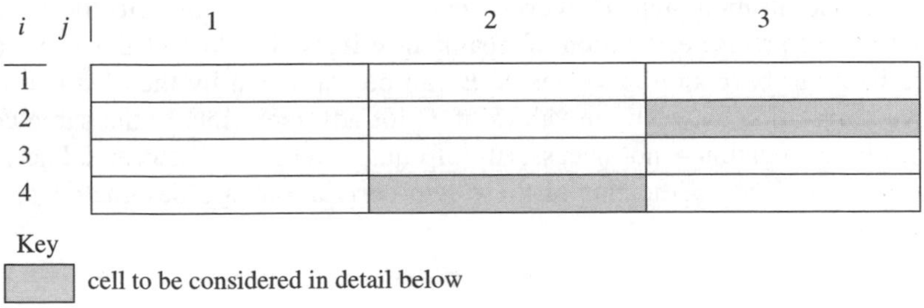

Key

cell to be considered in detail below

Figure 9.9   Lay-out of a small field experiment.

Provided that the experimental treatments are randomly allocated to the plots, the contribution of natural variation among plots to the response variable Y can be treated as a set of independent values of a random variable, E, which is usually assumed to be normally distributed: that is,

$$E \sim N(0, \sigma_E^2)$$

and the covariance matrix among the 12 values of E, ε_{ij}, $i = 1 \ldots 4$, $j = 1 \ldots 3$, is assumed to be

i	j	j'	$i'=1$ $j'=1$	1 2	1 3	2 1	2 2	2 3	3 1	3 2	3 3	4 1	4 2	4 3
1	1		σ_E^2											
1	2		0	σ_E^2										
1	3		0	0	σ_E^2									
2	1		0	0	0	σ_E^2								
2	2		0	0	0	0	σ_E^2							
2	3		0	0	0	0	0	σ_E^2						
3	1		0	0	0	0	0	0	σ_E^2					
3	2		0	0	0	0	0	0	0	σ_E^2				
3	3		0	0	0	0	0	0	0	0	σ_E^2			
4	1		0	0	0	0	0	0	0	0	0	σ_E^2		
4	2		0	0	0	0	0	0	0	0	0	0	σ_E^2	
4	3		0	0	0	0	0	0	0	0	0	0	0	σ_E^2

where

i = the row in which the first observation in a particular pairwise comparison is located,

j = the column in which the first observation is located,

i' = the row in which the second observation in the comparison is located,

j' = the column in which the second observation is located.

However, experience of natural variation in fields tells us that adjacent plots are likely to have similar values of E. This does not invalidate the simple analysis of a simple experimental design: the similarity between neighbouring observations is taken care of by the randomisation step. However, recognition of this similarity does open up the way to more precise estimation of treatment effects. The knowledge that adjacent plots are likely to have similar values of E can be expressed by the idea that there is a positive correlation between the values of E for adjacent plots in the same column, and a similar correlation – not necessarily of equal strength – between adjacent plots in the same row. The coefficients of these two correlations are designated ρ_1 and ρ_2 respectively. If

$$\rho_1 = \rho_2 = 0$$

then the values of E in adjacent plots are independent, but if

$$\rho_1 > 0 \text{ and } \rho_2 > 0$$

then plots that are close together will generally have more similar values of E than those that are further apart. The closer the plots, the greater the similarity between them, and the larger the values of ρ_1 and ρ_2, the more extensive the regions of similar plots (in a trial with many rows and columns). If the correlation coefficient between adjacent plots in the same column is ρ_1, it follows that that between plots separated by an intervening plot is ρ_1^2, that between plots separated by two intervening plots is ρ_1^3, and so on. Thus the pattern of correlations between plots in the same column can be expressed by the following matrix:

$$
\begin{array}{c|cccc}
i\ i' & 1 & 2 & 3 & 4 \\
\hline
1 & 1 & & & \\
2 & \rho_1 & 1 & & \\
3 & \rho_1^2 & \rho_1 & 1 & \\
4 & \rho_1^3 & \rho_1^2 & \rho_1 & 1
\end{array}.
$$

Likewise, the following matrix expresses the pattern of correlations between plots in the same row:

$$
\begin{array}{c|ccc}
j\ j' & 1 & 2 & 3 \\
\hline
1 & 1 & & \\
2 & \rho_2 & 1 & \\
3 & \rho_2^2 & \rho_2 & 1
\end{array}.
$$

These results can be generalised to give the correlation coefficient between two plots that are neither in the same column nor in the same row. This is determined by the number of columns and rows that separate them: for plots separated by $|i - i'|$ rows and $|j - j'|$ columns it is $\rho_1^{|i-i'|}\rho_2^{|j-j'|}$. Thus the pattern of correlations among all plots in the experiment is

$$
\begin{pmatrix}
1 & & & \\
\rho_1 & 1 & & \\
\rho_1^2 & \rho_1 & 1 & \\
\rho_1^3 & \rho_1^2 & \rho_1 & 1
\end{pmatrix}
\otimes
\begin{pmatrix}
1 & & \\
\rho_2 & 1 & \\
\rho_2^2 & \rho_2 & 1
\end{pmatrix}
=
$$

i'		1	1	1	2	2	2	3	3	3	4	4	4
i	j j'	1	2	3	1	2	3	1	2	3	1	2	3
1	1	1											
1	2	ρ_2	1										
1	3	ρ_2^2	ρ_2	1									
2	1	ρ_1	$\rho_1\rho_2$	$\rho_1\rho_2^2$	1								
2	2	$\rho_1\rho_2$	ρ_1	$\rho_1\rho_2$	ρ_2	1							
2	3	$\rho_1\rho_2^2$	$\rho_1\rho_2$	ρ_1	ρ_2^2	ρ_2	1						
3	1	ρ_1^2	$\rho_1^2\rho_2$	$\rho_1^2\rho_2^2$	ρ_1	$\rho_1\rho_2$	$\rho_1\rho_2^2$	1					
3	2	$\rho_1^2\rho_2$	ρ_1^2	$\rho_1^2\rho_2$	$\rho_1\rho_2$	ρ_1	$\rho_1\rho_2$	ρ_2	1				
3	3	$\rho_1^2\rho_2^2$	$\rho_1^2\rho_2$	ρ_1^2	$\rho_1\rho_2^2$	$\rho_1\rho_2$	ρ_1	ρ_2^2	ρ_2	1			
4	1	ρ_1^3	$\rho_1^3\rho_2$	$\rho_1^3\rho_2^2$	ρ_1^2	$\rho_1^2\rho_2$	$\rho_1^2\rho_2^2$	ρ_1	$\rho_1\rho_2$	$\rho_1\rho_2^2$	1		
4	2	$\rho_1^3\rho_2$	ρ_1^3	$\rho_1^3\rho_2$	$\rho_1^2\rho_2$	ρ_1^2	$\rho_1^2\rho_2$	$\rho_1\rho_2$	ρ_1	$\rho_1\rho_2$	ρ_2	1	
4	3	$\rho_1^3\rho_2^2$	$\rho_1^3\rho_2$	ρ_1^3	$\rho_1^2\rho_2^2$	$\rho_1^2\rho_2$	ρ_1^2	$\rho_1\rho_2^2$	$\rho_1\rho_2$	ρ_1	ρ_2^2	ρ_2	1

For example, consider the shaded plots in the plan of this field experiment:

- the cell in row 4, column 2 ($i = 4$, $j = 2$), and
- the cell in row 2, column 3 ($i' = 2$, $j' = 3$).

These are separated by $|i - i'| = |4 - 2| = 2$ rows and $|j - j'| = |2 - 3| = 1$ column. Hence the correlation coefficient between the values of E for these plots is $\rho_1^2 \rho_2$, as indicated in the shaded cell in the matrix above. The combination of matrices in this way is called the *direct product* (note the operator $\otimes$, which specifies this combination).

The covariance matrix among the values of E is then given by

$$\text{covariance matrix} = \begin{pmatrix} 1 & & & \\ \rho_1 & 1 & & \\ \rho_1^2 & \rho_1 & 1 & \\ \rho_1^3 & \rho_1^2 & \rho_1 & 1 \end{pmatrix} \otimes \begin{pmatrix} 1 & & \\ \rho_2 & 1 & \\ \rho_2^2 & \rho_2 & 1 \end{pmatrix} \sigma_E^2. \tag{9.28}$$

In all the mixed models considered previously, the covariance matrix for each random-effect term has been fully specified by the design of the experiment (the specification of treatments and blocks, or of the pedigree structure), except for the variance component, which was estimated during the model-fitting process. However, in the present case, not only the variance component σ_E^2, but also the correlation coefficients ρ_1 and ρ_2, are to be estimated from the data. Provided that the true values of either ρ_1 or ρ_2, or both, are greater than zero, the treatment effects will consequently be estimated with greater precision.

These ideas can be illustrated in the analysis of data from a field trial of wheat breeding lines, conducted in South Australia in 1994 (reproduced by kind permission of Gil Hollamby). A total of 107 lines and varieties were sown in a randomised complete block design, with three replications (blocks). In each replication, 3 standard varieties were each sown in 2 plots, namely varieties 82 (Tincurran), 89 (VF655) and 104 (WW1477). The remaining 104 varieties were each sown in a single plot. Thus each replication comprised 110 plots, which were arranged in 22 rows and 5 columns. The layout and randomisation of the experiment are shown in Figure 9.10. The variety sown in each plot is indicated.

An area of 4.2 m × 0.75 m in each plot was harvested, and the grain yield (kg) was measured. The first and last few rows of a spreadsheet holding the results are shown in Table 9.8.

The model that corresponds to the design of this experiment is the standard randomised complete block model:

Fixed-effect model (treatment model): variety
Random-effect model (block model): replication.

These data are imported, and this model fitted using anova, by the following statements:

```
IMPORT 'Intro to Mixed Modelling\\Chapter 9\\SA wheat yield.xls'
BLOCKSTRUCTURE replicate
TREATMENTSTRUCTURE variety
ANOVA [FPROB = yes] yield
```

Alternatively, the same model can be fitted by mixed-modelling methods, replacing the BLOCKSTRUCTURE, TREATMENTSTRUCTURE and ANOVA statements by the following statements:

Block		1						2					3		
Column	1	2	3	4	5	6	7	8	9	10	11	12	13	14	15
Row															
1	4	14	79	59	89	104	62	4	74	15	5	2	27	79	47
2	10	30	80	60	90	18	92	67	71	81	4	42	89	9	29
3	17	15	97	61	86	70	91	19	100	106	104	81	28	66	100
4	16	18	102	62	91	22	36	103	27	61	46	91	56	82	93
5	21	19	40	63	92	56	57	35	42	69	63	57	37	82	89
6	32	20	42	64	93	41	68	14	98	66	3	25	99	59	107
7	33	23	43	65	94	37	23	107	12	38	48	26	97	101	39
8	34	24	44	66	95	53	102	11	77	9	98	72	68	83	32
9	72	25	45	67	96	86	33	73	72	24	60	80	64	62	40
10	74	26	46	68	7	40	47	30	13	55	8	73	49	90	58
11	75	27	47	69	99	50	105	6	64	60	51	45	74	106	13
12	81	28	48	70	100	16	2	29	59	46	71	52	24	104	34
13	83	35	49	88	101	80	104	75	8	93	15	53	95	65	61
14	106	36	50	103	29	78	28	97	5	31	22	102	21	31	20
15	107	37	51	104	104	26	90	82	85	63	14	17	103	76	43
16	3	38	52	98	105	95	32	10	45	43	86	54	75	78	36
17	5	39	53	2	22	96	21	25	101	54	18	19	94	50	44
18	6	71	54	1	31	17	39	7	89	48	12	96	55	70	41
19	8	73	55	89	41	82	76	52	65	1	69	92	11	88	105
20	9	76	56	87	12	87	84	3	99	49	87	7	67	77	6
21	11	77	57	84	82	44	79	83	34	20	84	10	30	23	1
22	13	78	58	85	82	58	89	88	51	94	35	38	85	33	16

Figure 9.10   Arrangement of a field trial of wheat breeding lines in South Australia.

```
VCOMPONENTS [FIXED = variety] RANDOM = replicate
REML [PRINT = model, components, deviance, Wald] yield
```

The output of the REML statement is as follows:

REML variance components analysis

Response variate: yield
Fixed model: Constant + variety
Random model: replicate
Number of units: 330

Residual term has been added to model

Sparse algorithm with AI optimisation

Estimated variance components

Random term	component	s.e.
replicate	12642.	12764.

Table 9.8 Yields of wheat varieties in a field trial in South Australia, with a randomised block design and with information on the spatial location of each plot.

Only the first and last few rows of the data are shown. The conventions used in this spreadsheet are the same as those used in Table 1.1, Chapter 1.

	A	B	C	D	E
1	replicate!	row!	column!	variety!	yield
2	1	1	1	4	483
3	1	1	2	14	400
4	1	1	3	79	569
5	1	1	4	59	734
6	1	1	5	89	571
7	2	1	6	104	642
8	2	1	7	62	665
9	2	1	8	4	738
.					.
.					.
.					.
329	3	22	13	85	442
330	3	22	14	33	423
331	3	22	15	16	551

Residual variance model

Term	Factor	Model(order)	Parameter	Estimate	s.e.
Residual		Identity	Sigma2	13432.	1278.

Deviance: −2*Log-Likelihood

Deviance	d.f.
2471.63	221

Note: deviance omits constants which depend on fixed model fitted.

Wald tests for fixed effects

Sequentially adding terms to fixed model

Fixed term	Wald statistic	d.f.	Wald/d.f.	chi pr
variety	153.02	106	1.44	0.002

Dropping individual terms from full fixed model

Fixed term	Wald statistic	d.f.	Wald/d.f.	chi pr
variety	153.02	106	1.44	0.002

Message: chi-square distribution for Wald tests is an asymptotic approximation (i.e. for large samples) and underestimates the probabilities in other cases.

The estimated variance component for the term 'replicate', $\hat{\sigma}^2_{\text{replicate}} = 12642$, is smaller than its own SE, suggesting that there may be no real variation among the replicates. However, the F value for this term in the anova (not shown) is highly significant ($F_{2,106} = 104.53$, $P < 0.001$). The Wald test indicates that there is also highly significant variation among the varieties.

A model that allows for the possibility that there are correlations between the residual values in neighbouring plots, both within rows and within columns, is specified by the following statements:

```
VCOMPONENTS [FIXED = variety] RANDOM = row.column
VSTRUCTURE [TERMS = row.column] FACTOR = row, column; \
    MODEL = AR, AR; ORDER = 1, 1
REML \
    [PRINT = model, components, deviance, Wald, covariancemodels] \
    yield
```

The random-effect model 'replicate' has been replaced by 'row.column'. It might be thought that the term 'replicate' would be retained, so that the new model would be 'replicate + row.column'. However, the term 'replicate' is redundant when the term 'row.column' is included: any variation among replicates can be absorbed into the variation among columns. As in the analysis of pedigree data, the VSTRUCTURE statement specifies the structure of the covariance matrix among the random effects. The TERMS option indicates that such structure is to be specified for the term 'row.column'. (There might be other terms in the random-effect model for which a different structure, or none, was to be specified.) The parameter FACTOR indicates the factors within this model term for which a covariance matrix is to be specified, and over which the direct product is to be formed in order to specify the overall covariance matrix. (Some factors in the term might be excluded from this specification.) The parameter MODEL specifies that an *autoregressive* (AR) model is to be fitted for both 'row' and 'column': that is, a model in which the value in each plot is correlated with that in the neighbouring plot, as described above. The parameter ORDER indicates that for both 'row' and 'column' the AR model is to be first order: that is, only a correlation between immediately neighbouring plots is to be directly specified. Correlations between more distant neighbours arise only as a consequence of this. A first-order AR model for both rows and columns is designated as AR1 × AR1.

The output of the REML statement is as follows:

REML variance components analysis

Response variate: yield
Fixed model: Constant + variety
Random model: row.column
Number of units: 330

row.column used as residual term with covariance structure as below
Sparse algorithm with AI optimisation

Covariance structures defined for random model

Covariance structures defined within terms:

Term	Factor	Model	Order	No. rows
row.column	row	Auto-regressive (+ scalar)	1	22
	column	Auto-regressive	1	15

Residual variance model

Term	Factor	Model(order)	Parameter	Estimate	s.e.
row.column	Sigma2	19794.		3819.	
	row	AR(1)	phi_1	0.9039	0.0185
	column	AR(1)	phi_1	0.4288	0.0640

Estimated covariance models

Variance of data estimated in form:

$V(y) = Sigma2.R$

where: V(y) is variance matrix of data
   Sigma2 is the residual variance
   R is the residual covariance matrix

Residual term: row.column

Sigma2: 19794.

R uses direct product construction

Factor: row
Model : Auto-regressive

Covariance matrix (first 10 rows only):

```
1    1.000
2    0.904   1.000
3    0.817   0.904   1.000
4    0.739   0.817   0.904   1.000
5    0.668   0.739   0.817   0.904   1.000
6    0.603   0.668   0.739   0.817   0.904   1.000
7    0.545   0.603   0.668   0.739   0.817   0.904   1.000
8    0.493   0.545   0.603   0.668   0.739   0.817   0.904   1.000
9    0.446   0.493   0.545   0.603   0.668   0.739   0.817   0.904   1.000
10   0.403   0.446   0.493   0.545   0.603   0.668   0.739   0.817   0.904   1.000
       1       2       3       4       5       6       7       8       9      10
```

Factor: column
Model : Auto-regressive

Covariance matrix (first 10 rows only):

```
1    1.000
2    0.429   1.000
3    0.184   0.429   1.000
4    0.079   0.184   0.429   1.000
5    0.034   0.079   0.184   0.429   1.000
6    0.014   0.034   0.079   0.184   0.429   1.000
```

7	0.006	0.014	0.034	0.079	0.184	0.429	1.000			
8	0.003	0.006	0.014	0.034	0.079	0.184	0.429	1.000		
9	0.001	0.003	0.006	0.014	0.034	0.079	0.184	0.429	1.000	
10	0.000	0.001	0.003	0.006	0.014	0.034	0.079	0.184	0.429	1.000
	1	2	3	4	5	6	7	8	9	10

Deviance: −2*Log-Likelihood

Deviance	d.f.
2207.29	220

Note: deviance omits constants which depend on fixed model fitted.

Wald tests for fixed effects

Sequentially adding terms to fixed model

Fixed term	Wald statistic	d.f.	Wald/d.f.	chi pr
variety	772.51	106	7.29	<0.001

Dropping individual terms from full fixed model

Fixed term	Wald statistic	d.f.	Wald/d.f.	chi pr
variety	772.51	106	7.29	<0.001

Message: chi-square distribution for Wald tests is an asymptotic approximation (i.e. for large samples) and underestimates the probabilities in other cases.

The specification of the model at the beginning of the output now includes some information about the structure in the covariance matrix, including the fact that the experiment has 22 rows and 15 columns. The annotation '(+ scalar)' indicates that the model relating the residual value in each row to that in the next includes a constant term, which is subtracted from each residual before using the correlation coefficient ρ_1 to specify the relationship between them. This constant is required for only one of the two terms that define the covariance structure. The residual variance model relates entirely to the term 'row.column', as this term uniquely specifies every observation in the experiment. The model indicates that the estimate of residual variance is now $\hat{\sigma}^2_{\text{Residual}} = 19794$, somewhat higher than the value of $\hat{\sigma}^2_{\text{Residual}} = 13432$ given by the randomised complete block model. However, the estimates for the parameter 'phi_1' for the factors 'row' and 'column' indicate that the AR1 × AR1 model gives a substantially better fit to the data. These parameters are the estimates of the correlation coefficients ρ_1 and ρ_2 defined earlier. Both are substantially greater than their SEs, indicating that the true correlation coefficient is not zero in either case. The estimated correlation between adjacent rows, $\hat{\rho}_1 = 0.9039$, is much stronger than that between adjacent columns, $\hat{\rho}_2 = 0.4288$, probably because each row was only 0.75 m wide, whereas each column was 6 m wide. (The plots were trimmed to give the 4.2 m section harvested.) The covariance matrix for factor 'row' indicates the correlation between the first 10 plots in any column. The correlation between rows 1 and 2 is $\hat{\rho}_1 = 0.904$, that between rows 1 and 3 is $\hat{\rho}_1^2 = 0.817$, and so on. Similarly, the covariance matrix for factor 'column' indicates the correlation between the first 10 plots in any row. Note that these plots occupy the first 10 columns of the field experiment: the note 'first 10 rows only' refers to the covariance matrix and is the same regardless of the name of the factor in question.

The difference between the residual deviance from the randomised complete block model and that from the AR1 × AR1 model cannot be compared formally by a chi-square test as described in Chapter 3 (Section 3.12), because neither model is a reduced form of the other. However, the two deviances can be compared informally, i.e.

$$\text{Deviance}_{\text{randomisedcompleteblockmodel}} = 2471.63$$

$$\text{Deviance}_{\text{AR1} \times \text{AR1model}} = 2207.29.$$

The deviance from the AR1 × AR1 model is much lower, confirming that this model fits the data better, and suggesting that the variety means will be estimated with considerably more precision when this model is used. The two sets of estimates are compared in Figure 9.11: the correlation between them is $r = 0.656$. If the agreement between the two sets were perfect, the correlation would be $r = 1$ and the points in the figure would all lie on the diagonal line. The discrepancy is substantial, and would have a considerable effect on the decisions made by a plant breeder: the set of varieties chosen for further evaluation or commercial release would be very different, depending on which estimates were used to choose them. All the evidence from the analyses indicates that the estimates from the AR1 × AR1 model are closer to the true values, and give a better indication of the potential of each variety.

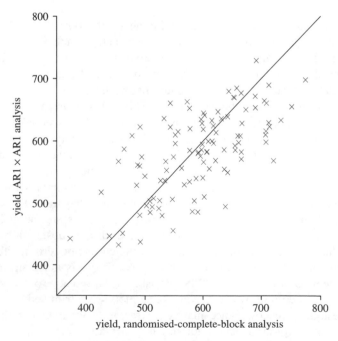

Figure 9.11   Comparison between estimates of mean yields of wheat varieties in South Australia from a randomised complete block analysis and an AR1 × AR1 analysis

There is a good deal more that can be done to improve the estimates of variety means from these data. Further terms can be added to the mixed model, representing additional aspects of spatial variation (for a full account, see Gilmour *et al.*, 1997). In brief:

- The value of each plot may have a direct dependence on the value of non-adjacent plots, in addition to its indirect dependence via the values of adjacent plots – a second-order autoregressive (AR2) model.

- A term can be added to the model to represent random variation among plots that is independent of the values in neighbouring plots: this is known as a *nugget* term.

- An irregular trend over the field may be represented by a fluctuating curve known as a *smoothing spline*, the precise shape of which is specified by a set of random effects (Verbyla *et al.*, 1999).

- Terms can be added to represent the effect of known husbandry factors, e.g. the direction in which the tractor was driven when trimming each plot, which affects the area harvested.

9.8 Summary

The basic concepts of mixed modelling can be extended in several directions. The following are illustrated in this chapter:

- Non-normal distribution of residuals:
  - sigmoid curve, binomial distribution of the response variable, logistic link function applied to the response variable (Section 9.2)
  - contingency table, Poisson distribution of the response variable, logarithmic link function applied to the response variable (Section 9.3).

- Non-normal distribution of effects in terms other than the residual term:
  - contingency table, gamma distribution of factorial terms, logarithmic link function (Section 9.4).

- Specification of the variance structure via a covariance matrix:
  - covariance between relatives in a pedigree structure (Section 9.6).

- Estimation of aspects of the variance structure not fully specified by the investigator:
  - correlation between observations in a regular array (Section 9.7).

9.9 Exercises

1. The seeds of some species of clover (*Trifolium* spp.) will not germinate immediately after ripening, but must undergo a period of 'softening', by exposure to fluctuating

high and low temperatures, usually on the soil surface. In an investigation of this phenomenon, seeds of eight clover species were sown into strips of fine mesh cotton bags, 100 seeds of a single species in each bag. Each strip comprised one bag of each species. At the start of summer, the bags were laid on the soil surface in full sun at a site cleared of vegetation. At intervals during the summer and autumn, two or four strips were removed from the field, and the seeds in each bag were tested to determine how many were still 'hard' (i.e. would not germinate when moistened in favourable conditions). The first and last few rows of the spreadsheet holding the data are presented in Table 9.9: the full data set is held in the file 'clover seed softening.xls' (www.wiley.com/go/mixed_modelling). (Data reproduced by kind permission of Hayley Norman.)

Table 9.9   Numbers of 'hard' and 'soft' seeds of several species of clover after exposure on the soil surface for varying lengths of time.

Observations on the same replication at the same time come from a single strip of bags: time = interval in days from the start of exposure on the soil surface; nhard = number of hard seeds in a single bag; nsoft = number of soft seeds in a single bag.

	A	B	C	D	E
1	species!	replication!	time	nhard	nsoft
2	T. glomeratum	1	0	98	2
3	T. glomeratum	2	0	99	1
4	T. glomeratum	3	0	99	1
5	T. glomeratum	4	0	99	1
6	T. spumosum	1	0	94	6
7	T. spumosum	2	0	93	7
8	T. spumosum	3	0	95	5
9	T. spumosum	4	0	96	4
.					.
.					.
.					.
234	T. scutatum	1	174	4	96
235	T. scutatum	2	174	11	89
236	T. scutatum	3	174	5	95
237	T. scutatum	4	174	5	95
238	T. pilulare	1	174	17	83
239	T. pilulare	2	174	28	72
240	T. pilulare	3	174		
241	T. pilulare	4	174		

Assume that the species studied are a representative sample from a large population of species potentially available for study.

(a) Specify a regression model for this experiment. Which term(s) in the model should be regarded as fixed, and which as random? What is the response variable?

(b) Fit your mixed model to the data, specifying an appropriate error distribution for the response variable, and an appropriate link function to relate the response variable to the linear model.

(c) Consider whether there is evidence that any terms can be omitted from the model. If so, fit the modified model to the data.

(d) Make a graphical display, showing the fitted relationship between the proportion of hard seed remaining for each species and the time elapsed since the start of exposure, and showing the scatter of the observed values around this relationship.

2. Return to the data set concerning the relationship between the distance from bushland and the level of predation on seeds, introduced in Exercise 3 in Chapter 7. In the earlier analysis of these data, it was assumed that the percentage of predation could be regarded as a normally distributed variable.

(a) Using the information on the number of seeds of each species per cage, convert each value of the percentage of predation ('%pred') to the actual number of seeds removed by predation.

(b) Refit your mixed model to the data, using the number of seeds removed by predation as the response variable, and specifying an appropriate error distribution for this response variable, and an appropriate link function to relate the response variable to the linear model.

(c) Obtain diagnostic plots of the residuals. Do these plots indicate that the assumptions underlying the analysis are more nearly fulfilled as a result of the changes to the response variable, the error distribution and the link function?

(d) Make a graphical display, showing the fitted relationship between the proportion of seeds removed by predation and the distance from the bushland, taking into account any other model terms (i.e. residue, cage type, species and/or interaction terms) that your analysis indicates are important. Your plot should also show the scatter of the observed values around this relationship.

3. Records of matings and of phenotypic variables were kept in a pedigree of the shrimp *Penaeus vannamei*. The variables recorded were as shown in Table 9.10, and the first and last few rows of the spreadsheet holding the data are presented in Table 9.11: the full data set is held in the file 'shrimp pedigree.xls' (www.wiley.com/go/mixed_modelling). (Data reproduced by kind permission of Dr Shaun Moss, Director, Shrimp Department, Oceani Institute, Hawaii, USA.) The variable of economic interest is the weekly growth rate.

(a) Consider whether any of the variables studied should be modelled as fixed-effect terms when estimating the genetic and residual components of variance of the weekly growth rate.

(b) Fit a mixed model to the data, taking account of the pedigree structure. Obtain estimates of the genetic and residual components, and estimate the heritability of the weekly growth rate.

Table 9.10   Variables recorded in a pedigree of the shrimp *Penaeus vannamei*.

Name of variable	Description
Kid	identifier of the shrimp
Ksire	identifier of the male parent
Kdam	identifier of the female parent
SEX	M = male; F = female
F	coefficient of inbreeding, relative to the founders of the pedigree
WeekGro	weekly growth rate, post-larva to harvest, (grams)
Brood	brood identifier, useful for the recognition of maternal effects when a female had more than one brood of full-sibs
CGU	contemporaneous group
seq	a database reference
DD	degree days accumulated during the growth period; this will be the same for all members of a brood and contemporaneous group

Table 9.11   Pedigree and phenotype data from shrimps of the species *Penaeus vannamei*.

	A	B	C	D	E	F	G	H	I	J
1	Kid	Ksire	Kdam	SEX	F	WeekGro	Brood	CGU	seq	DD
2	17	1	6	M					5120	
3	18	2	7	M					5121	
4	19	3	9	M					5122	
5	28	4	10	M					5473	
.										.
.										.
.										.
268	68533	63411	10166	M	0.02	1.52	101723411	34	22904	972
269	68547	63411	10166	F	0.02	1.22	101723411	34	22917	972
270	68571	63411	10166	F	0.02	1.72	101723411	34	22939	972
271	69500	11966	63406	M	0.03	1.54	634071966	34	25533	972
272	69528	11966	63406	F	0.03	1.45	634071966	34	25559	972
273	69545	11966	63406	F	0.03	1.43	634071966	34	25574	972
274	69676	11966	63391	M	0.02	1.29	633921966	34	25691	972
275	69712	11966	63391	F	0.02	1.38	633921966	34	25724	972
276	69718	11966	63391	F	0.02	1.41	633921966	34	25730	972

(c) Compare graphically the phenotypic value of the weekly growth rate for each individual and the estimated genetic effect on this variable.

4. Return to the data on yields of wheat genotypes, investigated in an alphalpha design, presented in Table 8.4 in Chapter 8.

(a) Explore the possibility of analysing these data using an autoregressive model, instead of the model based on the alphalpha design fitted in Chapter 8, Section 8.4.
(b) How well does your autoregressive model fit the data, relative to the model fitted earlier? Compare graphically the estimates of the genotype mean yields obtained by the two methods. How much effect will the choice of method have on the decisions made by a breeder seeking genetic improvement?

10

Why is the criterion for fitting mixed models called residual maximum likelihood?

10.1 Maximum likelihood and residual maximum likelihood

In the preceding chapters we have established the need for mixed models, i.e. statistical models with more than one random-effect term, and have seen how to construct such models, and how to interpret the results obtained when they are fitted to data. We have noted that the criterion used to fit a mixed model – that is, to obtain the best estimates of its parameters – is called residual maximum likelihood (REML), but we have not so far examined the meaning of this term. In this chapter we will explore the concept of maximum likelihood, and its use as a criterion for the estimation of model parameters. We will then show how the criterion for parameter estimation used in the earlier chapters can be viewed as *residual* or *restricted* maximum likelihood. The argument will proceed as follows:

- Consideration of a model comprising only the random-effect term E. The estimation of its variance σ^2 using the maximum likelihood criterion.

- Consideration of the simplest linear model, comprising the fixed-effect term μ and the random-effect term E. The simultaneous estimation of μ and σ^2 using the maximum likelihood criterion. An alternative estimate of σ^2 using the REML criterion.

- Extension of the REML criterion to the general linear model.

- Extension to models with more than one random-effect term.

Introduction to Mixed Modelling: Beyond Regression and Analysis of Variance   N. W. Galwey
© 2006 John Wiley & Sons, Ltd

10.2 Estimation of the variance σ^2 from a single observation using the maximum likelihood criterion

In the analysis of real data, it is usually necessary to estimate fixed-effect parameters, such as the mean, and random-effect parameters, such as variance components, simultaneously. However, we will start with a simplified, hypothetical situation in which the mean of a variable is known, and only its variance has to be estimated. Consider a variable E such that

$$E \sim N(0, \sigma^2).$$

Suppose that a sample of values of E, namely ε_i, $i = 1 \ldots n$, is to be taken, and that the variance σ^2 is to be estimated from this sample. We define the *maximum likelihood estimate* of the standard deviation σ as the value of σ that maximises the probability of the data. (For a full discussion of the merits of maximum likelihood as an estimation criterion, see Edwards, 1972.) We will denote this maximum likelihood estimate by $\hat{\sigma}$. It follows that the square of this estimate, $\hat{\sigma}^2$, is the maximum likelihood estimate of σ^2.

What value of σ has the required property? We will first address this question for a sample comprising a single observation of E, denoted by ε: that is, the case where

$$n = 1.$$

We will not obtain a formal algebraic solution to the problem (for such an account see, for example, Bulmer, 1979, Chapter 11, pp 197–200), but will explore graphically the probability distributions of E that are produced by a range of possible values that might be proposed for σ, designating such a proposed value by the symbol $\hat{\sigma}$, and we will note the probability of the observed value ε for each of these values (Figure. 10.1). These probabilities are infinitesimal: they are probability densities, designated by f(E) (see Chapter 1, Section 1.2). Nevertheless, their relative magnitudes can be compared. In this case, the value that maximises f(E) is

$$\hat{\sigma} = \varepsilon. \tag{10.1}$$

The observed value of E is itself the maximum likelihood estimate of the standard deviation: any other value of $\hat{\sigma}$ makes the observed value less probable. A smaller value of $\hat{\sigma}$ gives a probability distribution that is too narrow, so that ε lies relatively far from its maximum and f(E) is low. A larger value of $\hat{\sigma}$ gives a distribution that is too flat, so that although ε lies relatively close to its maximum, f(E) is again low.

10.3 Estimation of σ^2 from more than one observation

Note that probability density is represented in Figure. 10.1 by two methods: by the height on the vertical axis and by contour shading. The second method is useful when

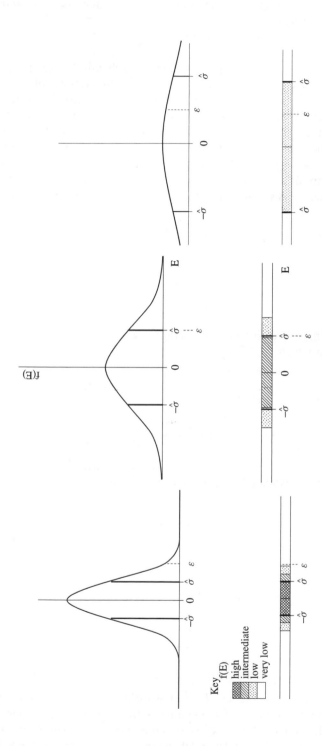

Figure 10.1 Comparison of possible estimates of the standard deviation of a variable E, when a single observation of E has been made. The upper part of the figure shows the probability density of E, $f(E)$, on the vertical axis; the lower part shows this value by contour shading. The heavy lines indicate the values of E at a distance $\pm\hat{\sigma}$ from the origin.

we move on to consider a sample comprising two observations of E, denoted by ε_1 and ε_2: that is, the case where

$$n = 2.$$

Figure: 10.2 extends the graphical representation of probability distributions to this *bivariate* case. It is necessary to regard ε_1 and ε_2 as observations of *two different variables*, designated E_1 and E_2. These are represented by two dimensions in the figure, and the probability density of every possible combination of values of E_1 and E_2 is represented in the third dimension. In the upper part of the figure, $f(E_1, E_2)$ is represented on the vertical axis; in the lower part, it is represented by contour shading. Once again, the figure explores a range of possible values for σ, and shows the probability density of the observed values $(\varepsilon_1, \varepsilon_2)$ for each proposed value of σ. The figure shows that in this case, the value of σ that maximises the value of $f(E_1, E_2)$, $\hat{\sigma}$, is such that

$$\sqrt{2}\hat{\sigma} = \sqrt{\varepsilon_1^2 + \varepsilon_2^2}$$

and hence

$$\hat{\sigma} = \sqrt{\frac{\varepsilon_1^2 + \varepsilon_2^2}{2}}. \tag{10.2}$$

The coefficient 2 in this equation reflects the sample size, n. To help justify its presence, consider a sample in which

$$(\varepsilon_1, \varepsilon_2) = (\hat{\sigma}, \hat{\sigma}).$$

Such a 'typical' observation lies at a distance

$$\sqrt{\hat{\sigma}^2 + \hat{\sigma}^2} = \sqrt{2}\hat{\sigma}$$

from the origin.

This argument can be extended to show that in general,

$$\hat{\sigma} = \sqrt{\frac{\sum\limits_{i=1}^{n} \varepsilon_i^2}{n}}. \tag{10.3}$$

Because $\hat{\sigma}$ is the maximum likelihood estimate of σ, it follows that

$$\hat{\sigma}^2 = \frac{\sum\limits_{i=1}^{n} \varepsilon_i^2}{n} \tag{10.4}$$

is the maximum likelihood estimate of σ^2: the estimate of the variance from the whole sample is simply the mean of the estimates from the individual observations. The value $\hat{\sigma}^2$ has the additional merit of being an *unbiased* estimate of σ^2: that is, if an infinite population of small samples is taken, and the value $\hat{\sigma}^2$ is obtained in each

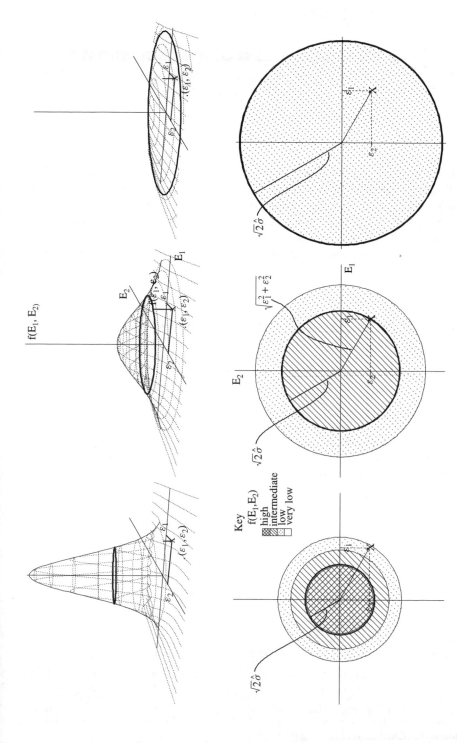

Figure 10.2 Comparison of possible estimates of the standard deviation of a variable E, when two observations of E have been made. The heavy line indicates the values of E_1 and E_2 at a distance $\sqrt{2}\hat{\sigma}$ from the origin.

small sample, the mean of all these estimates will be the true value, σ^2. However, it does *not* follow that $\hat{\sigma}$ is an unbiased estimate of σ.

10.4   The μ-effects axis as a dimension within the sample space

Now consider the simplest linear model,

$$Y = \mu + E \qquad\qquad (10.5)$$

where Y is the observed variable. It is no longer possible to observe E, but estimates of its values can be obtained from a sample of observations y_i, $i = 1 \dots n$. Figure. 10.3 shows a graphical representation of this model in the case where $n = 2$, the simplest case in which both μ and σ can be estimated. The observations y_1 and y_2 are regarded as values of two different variables, Y_1 and Y_2, and these variables provide axes that define a two-dimensional space with its origin at $\mathbf{0} = (0, 0)'$.[1] This is known as the

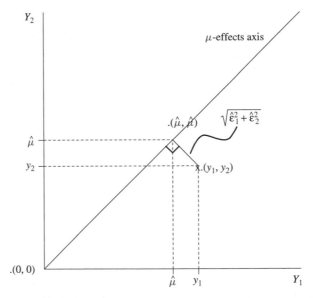

Figure 10.3   Graphical representation of the model $Y = \mu + E$, when two observations of Y have been made.

----

[1] Any point in n-dimensional space can be represented by a *vector* of values – a vertical column, which can in turn be represented by a single value – in this case $\mathbf{0} = \begin{pmatrix} 0 \\ 0 \end{pmatrix}$. This vector can also be represented as $\mathbf{0} = (0, 0)'$, using the prime symbol (') to indicate that a row of values is to be transposed to the corresponding column. For the rest of this chapter, the coordinates of a point in n-dimensional space will be represented in this way. In the figures, the prime symbol will be omitted to avoid clutter.

sample space. The observations, y_1 and y_2, define the point $\mathbf{y} = (y_1, y_2)'$. According to the model, the observations are related to μ and E by the following equations:

$$y_1 = \mu + \varepsilon_1 \qquad\qquad (10.6)$$

$$y_2 = \mu + \varepsilon_2. \qquad\qquad (10.7)$$

Any value of μ that may be proposed specifies a point, $\boldsymbol{\mu} = (\mu, \mu)'$, which represents the contribution of μ to y_1 and y_2. Any such value defines a line, passing through the origin and the point $\boldsymbol{\mu}$, which may be infinitely extended and which we will call the μ-effects axis. Any estimate of μ, which we will call $\hat{\mu}$, defines a point $\hat{\boldsymbol{\mu}} = (\hat{\mu}, \hat{\mu})'$ which lies on this axis. Because μ contributes equally to y_1 and y_2, the μ-effects axis makes the same angle ($45°$) with both the Y_1 axis and the Y_2 axis. Whatever value of $\hat{\mu}$ is chosen, Pythagoras's theorem can be used to show that

$$(\text{distance}(\mathbf{0}\hat{\boldsymbol{\mu}}))^2 = \hat{\mu}^2 + \hat{\mu}^2$$

and hence

$$\text{distance}(\mathbf{0}\hat{\boldsymbol{\mu}}) = \sqrt{2}\hat{\mu}. \qquad\qquad (10.8)$$

When the estimate $\hat{\mu}$ has been chosen, the estimates of the residual effects are given by rearranging Equations 10.6 and 10.7 and substituting $\hat{\mu}$ for μ to give

$$\hat{\varepsilon}_1 = y_1 - \hat{\mu} \qquad\qquad (10.9)$$

$$\hat{\varepsilon}_2 = y_2 - \hat{\mu}, \qquad\qquad (10.10)$$

and Pythagoras's theorem shows that the length of the line from $\mathbf{y}$ to $\hat{\boldsymbol{\mu}}$ is

$$\text{distance}(\hat{\boldsymbol{\mu}}\mathbf{y}) = \sqrt{\hat{\varepsilon}_1^2 + \hat{\varepsilon}_2^2}. \qquad\qquad (10.11)$$

10.5 Simultaneous estimation of μ and σ² using the maximum likelihood criterion

A natural estimate of μ is obtained by dropping a perpendicular from the point $\mathbf{y}$ to the μ-effects axis, as has been done in Figure. 10.3. Some more geometry (Figure. 10.4) shows that in this case,

$$\text{distance}(\mathbf{0}\hat{\boldsymbol{\mu}}) = \frac{y_1 + y_2}{\sqrt{2}}. \qquad\qquad (10.12)$$

Combining Equations 10.8 and 10.12, we obtain

$$\sqrt{2}\hat{\mu} = \frac{y_1 + y_2}{\sqrt{2}}$$

and hence

$$\hat{\mu} = \frac{y_1 + y_2}{2}. \qquad\qquad (10.13)$$

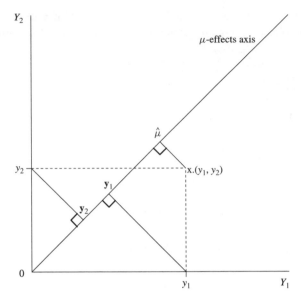

Figure 10.4 Demonstration of the relationship between y_1, y_2 and $\hat{\mu}$.

From Pythagoras's theorem,

$$\text{distance}(\mathbf{0y_1})^2 + \text{distance}(\mathbf{y_1}(0, y_1)')^2 = y_1.$$

But

$$\text{distance}(\mathbf{y_1}(0, y_1)') = \text{distance}(\mathbf{0y_1}).$$

Substituting and rearranging,

$$\text{distance}(\mathbf{0y_1}) = \frac{y_1}{\sqrt{2}}.$$

Similarly,

$$\text{distance}(\mathbf{0y_2}) = \frac{y_2}{\sqrt{2}}.$$

But

$$\text{distance}(\mathbf{0y_2}) = \text{distance}(\mathbf{y_1}\hat{\mu}).$$

(The demonstration of this is straightforward but is not given here.) Hence

$$\text{distance}(\mathbf{0}\hat{\mu}) = \text{distance}(\mathbf{0y_1}) + \text{distance}(\mathbf{y_1}\hat{\mu})$$

$$= \text{distance}(\mathbf{0y_1}) + \text{distance}(\mathbf{0y_2})$$

$$= \frac{y_1}{\sqrt{2}} + \frac{y_2}{\sqrt{2}} = \frac{y_1 + y_2}{\sqrt{2}}.$$

Once again, the coefficient 2 in this formula reflects the sample size. Again the argument can be extended to show that in general

$$\hat{\mu} = \frac{\sum\limits_{i=1}^{n} y_i}{n}. \tag{10.14}$$

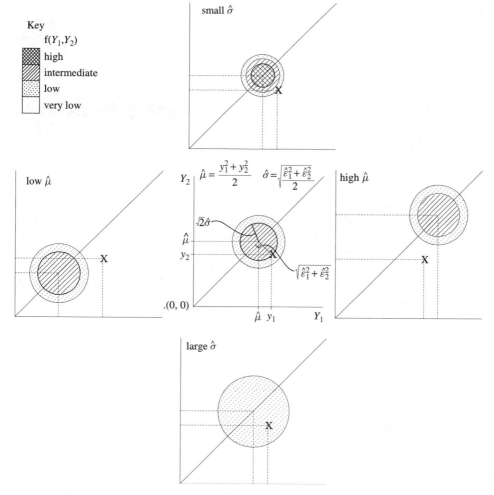

Figure 10.5 Comparison of possible estimates of the mean and standard deviation of a variable Y, when two observations of Y have been made.

The heavy line indicates the values of Y_1 and Y_2 at a distance $\sqrt{2}\hat{\sigma}$ from the point $(\hat{\mu}, \hat{\mu})'$.

The estimate of the population mean obtained by this geometrical argument is simply the familiar mean of the observations in the sample.

Figure. 10.5 shows the bivariate probability distributions of Y_1 and Y_2 that are produced by a range of possible values that might be proposed for the parameters μ and σ, and shows the probability density of the observed values y_1 and y_2 for each proposed combination of parameter values. The central frame in the figure shows the relationship of the data to the probability distribution when the estimates

$$\hat{\mu} = \frac{y_1 + y_2}{2} \tag{10.15}$$

and

$$\hat{\sigma} = \sqrt{\frac{\hat{\varepsilon}_1^2 + \hat{\varepsilon}_2^2}{2}} \tag{10.16}$$

are proposed. The four surrounding frames show that, when a higher or lower estimate of μ is proposed, or a larger or smaller estimate of σ, the probability density of the data is reduced: that is, the values $\hat{\mu}$ and $\hat{\sigma}$ given by Equations 10.15 and 10.16 are the maximum likelihood estimates of μ and σ respectively.

10.6   An alternative estimate of σ^2 using the REML criterion

As in the case where only the variance was to be estimated (Sections 10.2 and 10.3), if $\hat{\sigma}$ is the maximum likelihood estimate of σ, it follows that $\hat{\sigma}^2$ is the maximum likelihood estimate of σ^2. However, when the mean and the variance must both be estimated from the data, the maximum likelihood estimate $\hat{\sigma}^2$ is no longer unbiased, as it was in the hypothetical case where the mean was known. The values $\hat{\varepsilon}_i$ are calculated as deviations from $\hat{\mu}$, and they cluster around this value rather more closely than the unknown values ε_i cluster around the true mean μ (Figure. 10.6). Therefore, in an infinite population of small samples, the values of $\hat{\sigma}^2$ will somewhat underestimate the true value, σ^2.

We usually make an adjustment to overcome this deficiency, replacing

$$\hat{\sigma} = \sqrt{\frac{\sum\limits_{i=1}^{n} \hat{\varepsilon}_i^2}{n}} \tag{10.17}$$

by the familiar formula

$$\hat{\sigma} = \sqrt{\frac{\sum\limits_{i=1}^{n} \hat{\varepsilon}_i^2}{n - 1}}. \tag{10.18}$$

In the present case,

$$\hat{\sigma} = \sqrt{\frac{\hat{\varepsilon}_1^2 + \hat{\varepsilon}_2^2}{2}}$$

is replaced by

$$\hat{\sigma} = \sqrt{\frac{\hat{\varepsilon}_1^2 + \hat{\varepsilon}_2^2}{1}}. \tag{10.19}$$

But this, of course, is not the maximum likelihood estimate of σ. How can this adjusted formula be brought back within the framework of maximum likelihood?

If the estimate

$$\hat{\mu} = \frac{\sum\limits_{i=1}^{n} y_i}{n} \tag{10.20}$$

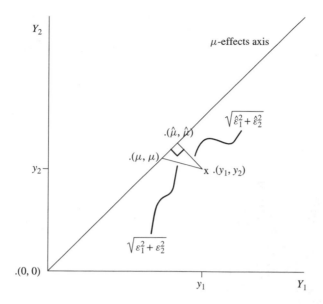

Figure 10.6 Demonstration that the observed values of a variable cluster more closely about the estimated mean than around the true mean.

The probability that the estimated mean coincides exactly with the true mean is infinitesimal. Unless this occurs, $\sqrt{\dfrac{\hat{\varepsilon}_1^2 + \hat{\varepsilon}_2^2}{2}}$ is always smaller than $\sqrt{\dfrac{\varepsilon_1^2 + \varepsilon_2^2}{2}}$.

is obtained first, and is then regarded as fixed, and if σ is subsequently estimated within this constraint, a new picture emerges. Figure 10.5 shows that when $\hat{\mu}$ is restricted to this particular function of the data, the observation **y** must lie on a line perpendicular to the μ-effects axis, passing through $\hat{\mu}$. Thus the estimation of σ is reduced from the two-dimensional problem illustrated in Figure. 10.3 to the one-dimensional problem illustrated in Figure. 10.7. This figure shows, using contour shading, the probability density of Y_1 and Y_2 that is produced by a range of possible values that might be proposed for σ, *within the constraint imposed by Equation 10.20.* Within this constraint, the value that maximises $f(y_1,\ y_2)$ is

$$\hat{\sigma} = \sqrt{\frac{\hat{\varepsilon}_1^2 + \hat{\varepsilon}_2^2}{1}}.$$

This is known as the *residual* or *restricted* maximum likelihood estimate, and is the familiar adjusted formula given in Equations 10.18 and 10.19.

This approach to the estimation of μ and σ can be extended to higher values of n by defining additional dimensions, Y_3 to Y_n, the μ-effects axis making the same angle with each of them. The maximum likelihood estimate of μ is given by

$$\hat{\mu} = \frac{\displaystyle\sum_{i=1}^{n} y_i}{n},$$

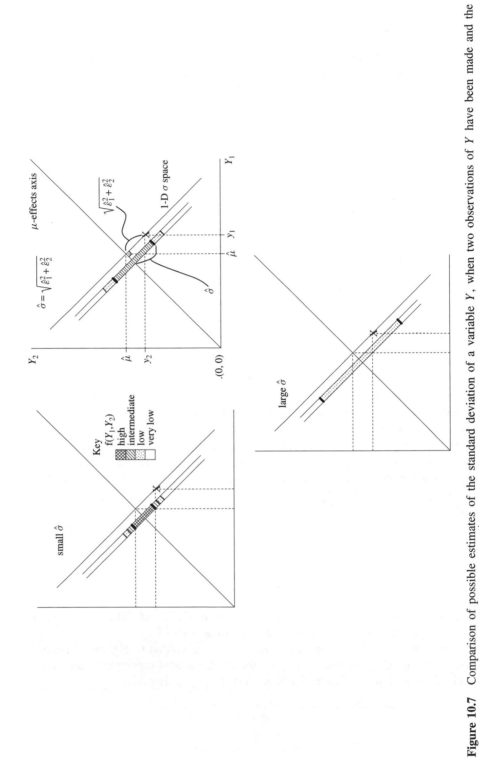

Figure 10.7 Comparison of possible estimates of the standard deviation of a variable Y, when two observations of Y have been made and the estimate of the mean is restricted to $\hat{\mu} = \dfrac{y_1 + y_2}{2}$.
The heavy lines indicate the values of Y_1 and Y_2 at a distance $\hat{\sigma}$ from the point $(\hat{\mu}, \hat{\mu})'$.

and the point

$$\hat{\boldsymbol{\mu}} = (\hat{\mu}, \hat{\mu} \ldots, \hat{\mu})' \qquad (10.21)$$

(a vector comprising n values) lies on the μ-effects axis. In the case of this simplest linear model, there is almost no distinction between obtaining the parameter estimate $\hat{\mu}$ and the fitted values $\hat{y}_1$ to $\hat{y}_n$: these are given by

$$\hat{\mathbf{y}} = (\hat{y}_1, \hat{y}_2 \ldots \hat{y}_n)' = (\hat{\mu}, \hat{\mu} \cdots \hat{\mu})' = \hat{\boldsymbol{\mu}} \qquad (10.22)$$

The estimated residual effects,

$$\hat{\boldsymbol{\varepsilon}} = (\hat{\varepsilon}_1, \hat{\varepsilon}_2 \ldots \hat{\varepsilon}_n)',$$

must then lie in an $(n - 1)$ dimensional subspace perpendicular to the μ-effects axis and passing through $\hat{\boldsymbol{\mu}}$, and the residual maximum likelihood estimate of σ is given by the familiar formula

$$\hat{\sigma} = \sqrt{\frac{\displaystyle\sum_{i=1}^{n} \hat{\varepsilon}_i^2}{n - 1}}. \qquad (10.23)$$

10.7 Extension to the general linear model. The fixed-effect axes as a subspace of the sample space

The residual maximum likelihood approach can be further extended to models that have more than one fixed-effect term, such as the general linear model (which encompasses all ordinary regression models),

$$Y = \beta_1 X_1 + \beta_2 X_2 + \ldots + \beta_p X_p + E, \qquad (10.24)$$

fitted to the data

$$y_i, x_{ij}, \quad i = 1 \ldots n, j = 1 \ldots p.$$

The observations of the response variable are represented by a point in the n-dimensional sample space $Y_1 \ldots Y_n$, and the explanatory variables, $X_1 \ldots X_p$, define axes that specify a p-dimensional subspace of this sample space, just as the model term μ was represented by the single dimension of the μ-effects axis.

The orientation of these *fixed-effects-model axes* in relation to the dimensions $Y_1 \ldots Y_n$ is determined by the values x_{ij}. We will first consider a very simple example of this type, in which there is a single explanatory variable and just two data points (Table 10.1). Figure. 10.8 shows that these two points lie exactly on the line

$$Y = 6 + 0.6X \qquad (10.25)$$

Table 10.1   Two observations of an explanatory variable X and a response variable Y

i	X	Y
1	$x_1 = 10$	$y_1 = 12$
2	$x_2 = 20$	$y_2 = 18$

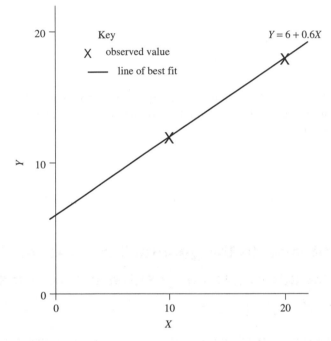

Figure 10.8   The line of best fit for the relationship between the explanatory variable X and the response variable Y when two pairs of observations have been made.

–there can be no residual variation when a model with two parameters (the constants 6 and 0.6) is fitted to two data points.

In order to apply the methods outlined above to these data, we first re-express the relationship between the variables in the form of the general linear model, thus:

$$Y = \beta_1 X_1 + \beta_2 X_2 \tag{10.26}$$

where
$\hat{\beta}_1 = 6$
$\hat{\beta}_2 = 0.6$

and the values of X_1, X_2 and Y are as given in Table 10.2. Note that X_1 is a *dummy variable*, all the values of which are 1, used to express the constant term in general linear model form.

The relationships between the sample space Y_1, Y_2, the dimensions represented by X_1 and X_2 and the data points in this model are represented graphically in Figure. 10.9.

Table 10.2   The two observations of an explanatory variable
and a response variable presented in Table 10.1, expressed in
the notation of the general linear model

i	X_1	X_2	Y
1	$x_{11} = 1$	$x_{21} = 10$	$y_1 = 12$
2	$x_{12} = 1$	$x_{22} = 20$	$y_2 = 18$

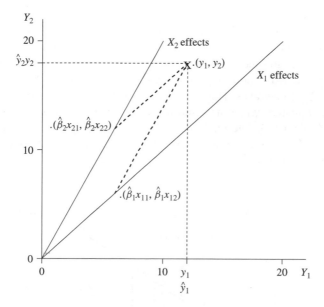

Figure 10.9   The relationship between the explanatory variables X_1 and X_2 and the response
variable Y, expressed by representing the effects of the explanatory variables as dimensions in
the sample space defined by two observations of Y.

The directions of the X-effects axes, relative to the Y_1 and Y_2 axes, are determined as
follows. The fitted values of Y are given by the following equations:

$$\hat{y}_1 = \hat{\beta}_1 x_{11} + \hat{\beta}_2 x_{21} \tag{10.27}$$

$$\hat{y}_2 = \hat{\beta}_1 x_{12} + \hat{\beta}_2 x_{22}. \tag{10.28}$$

(Because the two observations lie exactly on the line of best fit, the fitted values
are, in this case, the same as the observed values.) These equations show that the
contribution of X_1 to $\hat{y}_1$ is $\hat{\beta}_1 x_{11}$, and the contribution of X_1 to $\hat{y}_2$ is $\hat{\beta}_1 x_{12}$. Therefore,
the X_1-effects axis is a line connecting the origin to the point $(\hat{\beta}_1 x_{11}, \hat{\beta}_1 x_{12})$, projected
indefinitely. Similarly, the contribution of X_2 to $\hat{y}_1$ is $\hat{\beta}_2 x_{21}$ and its contribution to
$\hat{y}_2$ is $\hat{\beta}_2 x_{22}$, so the X_2-effects axis is a line connecting the origin to the point $(\hat{\beta}_2 x_{21}, \hat{\beta}_2 x_{22})$, again projected indefinitely.

Because there are two data points and two parameters, there are two Y axes and two
X-effects axes. Therefore in this case the X-effects axes define not just a subspace

Table 10.3   Three observations of an explanatory variable X and a response variable Y

i	X	Y
1	$x_1 = 10$	$y_1 = 15$
2	$x_2 = 20$	$y_2 = 21$
3	$x_3 = 15$	$y_3 = 9$

of the sample space, but the whole sample space – the two-dimensional plane of the paper. For the same reason, there is no distinction between the observed and fitted values of Y: just as the observed points lie exactly on the line of best fit in Figure 10.8, so the point representing the observed values of Y, lying in the plane defined by the Y axes, is the same as the point representing the fitted values of Y, lying in the plane defined by the X-effects axes. However, the distinction between observed and fitted values becomes apparent when the same model is fitted to a data set comprising three observations (Table 10.3). The relationship between these points and the fitted line is shown in Figure. 10.10. Each observation now has a residual component, represented by its distance from the line of best fit.

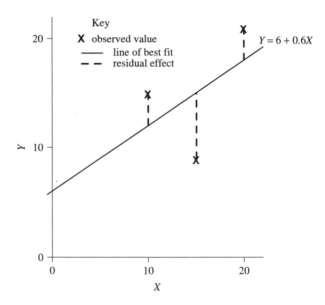

Figure 10.10   The line of best fit for the relationship between the explanatory variable X and the response variable Y when three pairs of observations have been made.

Once again, we re-express the relationship between the variables in the form of the general linear model (Table 10.4). The relationships between the sample space Y_1, Y_2, Y_3, the dimensions represented by X_1 and X_2 and the data points are now as shown in Figure. 10.11. The axes representing the X effects again define a plane, which is now a two-dimensional subspace of the three-dimensional sample space defined by the axes Y_1, Y_2 and Y_3. The point defined by the fitted values of Y, $\hat{\mathbf{y}} = (\hat{y}_1, \hat{y}_2, \hat{y}_3)'$,

Table 10.4   The three observations of an explanatory variable and a response variable presented in Table 10.3, expressed in the notation of the general linear model

i	X_1	X_2	Y
1	$x_{11} = 1$	$x_{21} = 10$	$y_1 = 15$
2	$x_{12} = 1$	$x_{22} = 20$	$y_2 = 21$
3	$x_{13} = 1$	$x_{23} = 15$	$y_3 = 9$

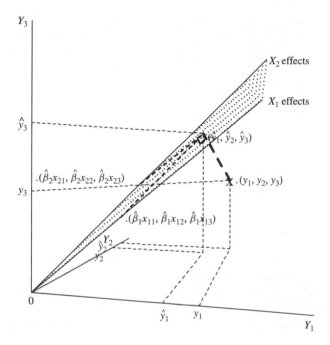

Figure 10.11   The relationship between the explanatory variables X_1 and X_2 and the response variable Y, expressed by representing the effects of the explanatory variables as dimensions in the sample space defined by three observations of Y.

lies in this plane, but the point defined by the observed values, $\mathbf{y} = (y_1, y_2, y_3)'$, does not.

10.8   Application of the REML criterion to the general linear model

The distance between the points representing the fitted and the observed values in Figure. 10.10 is related to the residual values. It is the square root of their sum of squares:

$$\text{distance}(\hat{\mathbf{y}}\mathbf{y}) = \sqrt{\sum_{i=1}^{n} \hat{\varepsilon}_i^2}, \qquad (10.29)$$

and the maximum likelihood estimate of σ is given by

$$\hat{\sigma} = \frac{\text{distance}(\hat{\mathbf{y}}\mathbf{y})}{\sqrt{n}} = \sqrt{\frac{\sum_{i=1}^{n} \hat{\varepsilon}_i^2}{n}}. \qquad (10.30)$$

This is the same formula as was obtained in the simpler case of the variation of a sample of observations about the mean value (Equation 10.17). In the present case, the residual values are as shown in Table 10.5. Hence

$$\text{distance}(\hat{\mathbf{y}}\mathbf{y}) = \sqrt{3^2 + 3^2 + (-6)^2} = \sqrt{54} = 7.348,$$

the maximum likelihood estimate of σ is

$$\hat{\sigma} = \sqrt{\frac{54}{3}} = 4.243$$

and the maximum likelihood estimate of σ^2 is

$$\hat{\sigma}^2 = \frac{54}{3} = 18,$$

However, as in the simpler case (Section 10.6, Figure 10.7), the point defined by the observed values is constrained to lie in an $(n - p)$-dimensional subspace of the sample space, orthogonal to (perpendicular to) the subspace defined by the X-effects axes. Hence the maximum likelihood estimate of σ can again be adjusted to give a REML estimate, replacing the divisor n by $(n - p)$ to give the familiar formulae

$$\hat{\sigma} = \frac{(\text{distance}(\hat{\mathbf{y}}\mathbf{y}))}{\sqrt{n - p}} = \sqrt{\frac{\sum_{i=1}^{n} \hat{\varepsilon}_i^2}{n - p}} \qquad (10.31)$$

Table 10.5 Fitted values and residuals from the line of best fit to the data presented in Table 10.3

i	y_i	$\hat{y}_i$	$\hat{\varepsilon}_i$
1	15	12	3
2	21	18	3
3	9	15	-6

and

$$\hat{\sigma}^2 = \frac{\sum\limits_{i=1}^{n} \hat{\varepsilon}_i^2}{n - p} \tag{10.32}$$

In the present case, the point y lies in a $(3 - 2)$-dimensional space, i.e. a line, perpendicular to the plane defined by the X_1-effects and X_2-effects axes. The REML estimate of σ^2 in this case is

$$\hat{\sigma}^2 = \frac{54}{3 - 2} = 54.$$

This unbiased estimate of the residual variance is substantially larger than the maximum likelihood estimate. When the number of parameters in the model (p) is close to the number of observations (n), the residual degrees of freedom are few, and the adjustment from maximum likelihood to REML is an important change (Patterson and Thompson, 1971).

10.9 Extension to models with more than one random-effect term

Finally, this approach can be extended to models that have more than one random-effect term – mixed models. In the case of an ordinary general linear model with only one random-effect term, the fixed and random effects can be estimated by the exact, analytic solution of a set of simultaneous equations: the model-fitting process is simply the dropping of a perpendicular from the observed values of the response variable to a subspace of the sample space, as we have seen in Sections 10.5 and 10.7. However, in a model with more than one random-effect term it is not generally possible to estimate the fixed and random effects analytically: REML algorithms proceed by obtaining initial estimates of the variance components for the random-effect terms, then using these as the basis for estimation of the fixed effects. These fixed-effect estimates are used to improve the estimates of variance components, and so on until convergence is achieved (Payne, 2003, Section 5.3.7, pp 554–556).

We have seen that the iterative process of fitting a mixed model using a REML algorithm is not guaranteed to succeed (Chapter 9, Section 9.3), and it may also require large amounts of computer time and data space. A straightforward approach to the application of the REML criterion requires the inversion of a matrix in which

number of rows = number of columns = number of fixed effects

+ number of random effects.

This is a serious computational challenge in the case of a model with a large number of effects. Consequently, in the more advanced applications of mixed-modelling methods, much effort is devoted to the application of techniques and specifications that will reduce this computational burden. These include the choice of realistic initial estimates for the variance components (Chapter 9, Section 9.3), the choice of the REML algorithm used, and other specifications that sometimes simplify the application of the algorithm chosen.

Two algorithms are available for applying the REML criterion, namely Fisher scoring and the average information (AI) algorithm. Their relative merits are described by Payne (2003, Section 5.3.1, pp 528–533 and Section 5.3.7, pp 554–556). The choice between the algorithms, and of other specifications within the Fisher-scoring algorithm, is largely concerned with reduction of the matrix-inversion burden. The AI algorithm, which uses *sparse matrix* methods for some of the model terms, is usually faster and requires less data space, but it does not permit the estimation of standard errors for differences between random effects. If Fisher scoring is used, one of the model terms may be specified as an *absorbing factor*. Matrix-inversion operations for the terms that involve the absorbing factor and those that do not are then performed separately, substantially reducing the size of the matrices that must be inverted. Again, there is a price to be paid in lost information, this time concerning the covariance estimates of parameters from the two sets.

This brief account will be sufficient to indicate that the serious application of mixed-modelling methods requires experience, experimentation and careful thought. Mixed modelling is not a process that is easily automated, and this limits its application to routine statistical analyses. However, any researcher engaged in the statistical analysis of data should consider whether mixed models would increase the realism of their work. The alternative is to remain in a world in which only residual variation is recognised as random – in which the multiple levels at which researchers *sample* reality are denied.

10.10   Summary

The maximum likelihood estimate of a parameter is the value that maximises the probability of the data.

The use of the maximum likelihood criterion for the estimation of the mean and variance of a normally distributed variable is illustrated.

The maximum likelihood estimate of the variance differs from the usual, unbiased estimate. However, it is shown that the criterion for obtaining the usual estimate can be viewed as residual or restricted maximum likelihood (REML).

A subspace of sample space that represents the fixed effects is first defined. The remaining subspace represents the residual effects, and the REML estimate is the maximum likelihood estimate within this subspace.

The REML criterion can be extended to the general linear regression model (which encompasses all ordinary regression models).

The REML criterion can be further extended to models with more than one random-effect term (mixed models). However, in this case it is not generally possible to estimate the fixed and random effects analytically.

REML algorithms obtain initial estimates of the variance components for the random-effect terms, then use these as the basis for estimation of the fixed effects. These fixed-effect estimates are used to improve the estimates of variance components, and so on until convergence is achieved.

The fitting of a mixed model using a REML algorithm is not guaranteed to succeed, and may require large amounts of computer time and data space.

The techniques that can be used to reduce this computational burden include realistic initial estimates of the variance components, and methods to reduce the size of the matrices that must be inverted.

Two REML algorithms are available, namely Fisher scoring and average information (AI). The AI algorithm is usually faster and requires less data space, but does not permit the estimation of standard errors for differences between random effects.

10.11 Exercises

1. Three observations were made of a random variable Y, namely

$$y_1 = 51, y_2 = 35, y_3 = 31.$$

(a) Assume that

$$Y \sim N(40, \sigma^2),$$

i.e. the mean is known but the variance must be estimated. Obtain the maximum likelihood estimate of σ^2 from this sample. Is this estimate unbiased?

(b) Now assume that

$$Y \sim N(\mu, \sigma^2),$$

i.e. both parameters must be estimated. Obtain the maximum likelihood estimates of μ and σ^2. Are these estimates unbiased? Obtain the residual maximum likelihood (REML) estimate of σ^2. Is this estimate unbiased?

(c) Make a sketch of the data space defined by this sample, corresponding to that presented in Figure 10.3 for a sample of two observations. Show the Y_1, Y_2 and Y_3 axes, the μ-effects axis and the observed values.

(d) Describe briefly how this geometrical representation can be used to obtain estimates of μ and σ^2, designated $\hat{\mu}$ and $\hat{\sigma}^2$. What is the distance from the point $\mathbf{y} = (y_1, y_2, y_3)'$ to the point $\hat{\boldsymbol{\mu}} = (\hat{\mu}, \hat{\mu}, \hat{\mu})'$?

When this graphical approach is used to represent a sample of two observations, and when any particular values, $\hat{\mu}$ and $\hat{\sigma}^2$, are postulated for μ and σ^2, the contours of the resulting probability distribution are circles, as shown in Figure 10.5.

(e) What is the shape of the corresponding contours for this sample of three observations?

(f) Sketch the contour at a distance $\sqrt{3}\hat{\sigma}$ from the point $(\hat{\mu}, \hat{\mu}, \hat{\mu})'$:

(i) when the postulated values $\hat{\mu}$ and $\hat{\sigma}^2$ are the maximum likelihood estimates,

(ii) when the postulated value $\hat{\mu}$ is above the maximum likelihood estimate and the postulated value $\hat{\sigma}^2$ is below the maximum likelihood estimate.

Now suppose that $\hat{\mu}$ is restricted to the maximum likelihood estimate.

(g) In what subspace of the data space must the point **y** then lie?

(h) Within this subspace, what will be the shape of the contours of the probability distribution?

2. Seven observations on two explanatory variates, X_1 and X_2, and a response variate Y, are presented in Table 10.6.

Table 10.6 Observations of two explanatory variates and a response variate

X_1	X_2	Y
42	7.3	128.0
58	9.2	145.0
93	3.9	101.6
70	4.1	90.9
35	8.4	135.6
61	3.7	85.0
29	8.0	108.7

(a) Fit the model

$$Y = \beta_0 + \beta_1 X_1 + \beta_2 X_2 + E$$

to these data, and obtain estimates of β_0, β_1 and β_2.

(b) When this model is fitted to these data, how many dimensions does each of the following have:

(i) the data space?

(ii) the fixed-effect subspace?

(iii) the random-effect subspace?

(c) Obtain the estimated value of Y, and the estimate of the residual effect, for each observation.

For Observation 5, the estimated value of $Y = 128.17$.

(d) What is the contribution to this value of:

(i) the constant effect?

(ii) the effect of X_1?

(iii) the effect of X_2?

It is assumed that

$$E \sim N(0, \sigma^2).$$

(e) Obtain the maximum likelihood estimate of σ^2 and the REML estimate of σ^2.

(f) What is the minimum number of observations required to obtain estimates of β_0, β_1, β_2 and σ^2? If the number of observations available is one less than this minimum, what estimates can be obtained? What is then the relationship between the estimated and observed values of Y?

3. Consider the final model fitted to the osteoporosis data in Chapter 7, Section 7.2.
   (a) When this model is fitted to these data, how many dimensions does each of the following have:
      (i) the data space?
      (ii) the fixed-effect subspace?
      (iii) the random-effect subspace?
   (b) What is the relationship between the number of dimensions of the random-effects subspace and the degrees of freedom of the deviance from this model?

References

Preface

Draper, N.R. and Smith, H. (1998) *Applied Regression Analysis*. 3rd edition. New York: John Wiley & Sons, Inc. 706 pp.

Mead, R. (1988) *The Design of Experiments. Statistical Principles for Practical Application*. Cambridge: Cambridge University Press. 620 pp.

Chapter 1

Draper, N.R. and Smith, H. (1998) *Applied Regression Analysis*. 3rd edition. New York: John Wiley & Sons, Inc. 706 pp.

Payne, R., Murray, D., Harding, S., Baird, D., Soutar, D. and Lane, P. (2003) *GenStat*® *for Windows*^{TM} *(7th Edition) Introduction*. Oxford: VSN International. 336 pp.

Venables, W.N., Smith, D.M. and the R Development Core Team (2004) *An Introduction to R. Notes on R: a Programming Environment for Data Analysis and Graphics. Version 2.0.0*. Available via R's Graphical User Interface (GUI).

Chapter 2

Bennett, C.A. and Franklin, N.L. (1954) *Statistical Analysis in Chemistry and the Chemical Industry*. New York: John Wiley & Sons, Inc. 724 pp.

Bulmer, M.G. (1979) *Principles of Statistics*. 2nd edition. New York: Dover. 252 pp.

Cochran, W.G. and Cox, G.M. (1957) *Experimental Designs*. 2nd edition. New York: John Wiley & Sons, Inc. 611 pp.

Mead, R. (1988) *The Design of Experiments. Statistical Principles for Practical Application*. Cambridge: Cambridge University Press. 620 pp.

Gower, J.C. (1975) Generalized Procrustes analysis. *Psychometrika* **40**: 33–51.

Welham, S.J. and Thompson, R. (1997) Likelihood ratio tests for fixed model terms using residual maximum likelihood. *Journal of the Royal Statistical Society, Series B* **59**: 701–714.

Wilkinson, G.N. and Rogers, C.E. (1973) Symbolic description of factorial models for analysis of variance. *Applied Statistics* **22**: 392–399.

Yates, F. (1937) *The Design and Analysis of Factorial Experiments.* Technical Communication No. 35 of the Commonwealth Bureau of Soils. Farnham Royal, UK: Commonwealth Agricultural Bureaux.

Chapter 3

Allard, R.W. (1960) *Principles of Plant Breeding.* New York: John Wiley & Sons, Inc. 485 pp.

Becker, W.A. (1992) *Manual of Quantitative Genetics.* 5th edition. Pullman, Washington: Academic Enterprises. 191 pp.

Crainiceanu, C.M. and Ruppert, D. (2004) Likelihood ratio tests in linear mixed models with one variance component. *Journal of the Royal Statistical Society B* **66**: 165–185.

Davies, O.L. and Goldsmith, P.L. (1984) *Statistical Methods in Research and Production with Special Reference to the Chemical Industry.* 4th edition. London: Longman. 478 pp.

Mokhtari, S., Galwey, N.W., Cousens, R.D. and Thurling, N. (2002) The genetic basis of variation among wheat F_3 lines in tolerance to competition by ryegrass (*Lolium rigidum*). *Euphytica* **124**: 355–364.

Snedecor, G.W. and Cochran W.G., (1967) *Statistical Methods.* 6th edition. Ames, Iowa: Iowa State University Press. 593 pp.

Snedecor, G.W. and Cochran, W.G. (1989) *Statistical Methods.* 8th edition. Ames, Iowa: Iowa State University Press. 503 pp.

Stram, D.O. and Lee, J.W. (1994) Variance components testing in the longitudinal mixed effects setting. *Biometrics* **50**: 1171–1177.

Chapter 4

Bennett, C.A. and Franklin, N.L. (1954) *Statistical Analysis in Chemistry and the Chemical Industry.* New York: John Wiley & Sons, Inc. 724 pp.

Moore, D.S. and McCabe, G.P. (1998) *Introduction to the Practice of Statistics.* 3rd edition. New York: Freeman. 825 pp.

Satterthwaite, F.E. (1946) An approximate distribution of estimates of variance components. *Biometrics Bulletin* **2**: 110–114.

Steel, R.G.D. and Torrie, J.H. (1981) *Principles and Procedures of Statistics. A Biometrical Approach.* International Student Edition. London: McGraw-Hill. 633 pp.

Taylor, J. (1950) The comparison of pairs of treatments in split-plot experiments. *Biometrika* **37**: 443–444.

Welham, S., Cullis, B., Gogel, B., Gilmour, A. and Thompson, R. (2004) Prediction in linear mixed models. *Australia and New Zealand Journal of Statistics* **46**: 325–347.

Chapter 5

Robinson, G.K. (1991) That BLUP is a good thing: the estimation of random effects. *Statistical Science* **6**: 15–51.

Chapter 6

Draper, N.R. and Smith, H. (1998) *Applied Regression Analysis.* 3rd edition. New York: John Wiley & Sons, Inc. 706 pp.

Mead, R. (1988) *The Design of Experiments. Statistical Principles for Practical Application.* Cambridge: Cambridge University Press. 620 pp.

Ridgman, W.J. (1975) *Experimentation in Biology.* Glasgow: Blackie. 233 pp.

Robinson, G.K. (1991) That BLUP is a good thing: the estimation of random effects. *Statistical Science* **6**: 15–51.

Snedecor, G.W. and Cochran, W.G. (1989) *Statistical Methods.* 8th edition. Ames, Iowa: Iowa State University Press. 503 pp.

Wilkinson, G.N. and Rogers, C.E. (1973) Symbolic description of factorial models for analysis of variance. *Applied Statistics* **22**: 392–399.

Chapter 7

Draper, N.R. and Smith, H. (1998) *Applied Regression Analysis.* 3rd edition. New York: John Wiley & Sons, Inc. 706 pp.

McCullagh, P. and Nelder, J.A. (1989) *Generalized Linear Models.* 2nd edition. London: Chapman and Hall. 511 pp.

McDonald, M.P., Galwey, N.W., Ellneskog-Staam, P. and Colmer, T.D. (2001) Evaluation of *Lophopyrum elongatum* as a source of genetic diversity to increase the waterlogging tolerance of hexaploid wheat (*Triticum aestivum*). *New Phytologist.* **151**: 369–380.

Ralston, S.H., Galwey, N., MacKay, I., Albagha, O.M., Cardon, L., Compston, J.E., Cooper, C., Duncan, E., Keen, R., Langdahl, B., McLellan, A., O'Riordan, J., Pols, H.A., Reid, D.M., Uitterlinden, A.G., Wass, J. and Bennett, S.T. (2005) Loci for regulation of bone mineral density in men and women identified by genome wide linkage scan: the FAMOS study. *Human Molecular Genetics* **14**: 943–951.

Si, P. and Walton, G.H. (2004) Determinants of oil concentration and seed yield in canola and Indian mustard in the lower rainfall areas of Western Australia. *Australian Journal of Agricultural Research* **55**: 367–377.

Spafford-Jacob, H., Minkey, D., Gallagher, R. and Borger, C. (2005) Variation in post-dispersal weed seed predation in a cropping field. *Weed Science* **54**: 148–155.

Chapter 8

Cochran, W.G. and Cox, G.M. (1957) *Experimental Designs.* 2nd edition. New York: John Wiley & Sons, Inc. 611 pp.

Cox, D.R. (1958) *Planning of Experiments.* New York: John Wiley & Sons, Inc. 308 pp.

John, J.A. and Williams, E.R. (1995) *Cyclic and Computer Generated Designs.* 2nd edition. London: Chapman and Hall. 255 pp.

Mead, R. (1997) Design of plant breeding trials. In *Statistical Methods for Plant Variety Evaluation*, ed. R.A. Kempton and P.N. Fox. London: Chapman and Hall. pp 40–67.

Patterson, H.D. and Williams, E.R. (1976) A new class of resolvable incomplete block designs. *Biometrika* **63**: 83–92.

Patterson, H.D., Williams, E.R. and Hunter, E.A. (1978) Block designs for variety trials. *Journal of Agricultural Science, Cambridge* **90**: 395–400.

Payne, R.W. and Tobias, R.D. (1992) General balance, combination of information and the analysis of covariance. *Scandinavian Journal of Statistics* **19**: 3–23.

Chapter 9

Abecasis, G.R., Cardon, L.R. and Cookson, W.O. (2000a) A general test of association for quantitative traits in nuclear families. *American Journal of Human Genetics* **66**: 279–292.

Abecasis, G.R., Cookson, W.O.C. and Cardon, L.R. (2000b) Pedigree tests of transmission disequilibrium. *European Journal of Human Genetics* **8**: 545–551.

Abecasis, G.R., Cherny, S.S., Cookson, W.O. and Cardon, L.R. (2002) Merlin – rapid analysis of dense genetic maps using sparse gene flow trees. *Nature Genetics* **30**: 97–101.

Brown, H. and Prescott, R. (1999) *Applied Mixed Models in Medicine.* Chichester: John Wiley & Sons, Ltd. 408 pp.

Bulmer, M.G. (1979) *Principles of Statistics.* 2nd edition. New York: Dover. 252 pp.

Falconer, D.S. and Mackay, T.F.C. (1996) *Introduction to Quantitative Genetics.* 4th edition. Harlow: Longman. 464 pp.

Finney, D.J. (1971) *Probit Analysis.* 3rd edition. Cambridge: Cambridge University Press. 333 pp.

Gilmour, A.R., Cullis, B.R. and Verbyla, A.P. (1997) Accounting for natural and extraneous variation in the analysis of field experiments. *Journal of Agricultural, Biological and Environmental Statistics* **2**: 269–293.

Goldin, L.R., Bailey-Wilson, J.E., Borecki, I.B., Falk, C.T., Goldstein, A.M., Suarez, B.K. and MacCluer, J.W. (eds) (1997) Genetic Analysis Workshop 10: Detection of Genes for Complex Traits. *Genetic Epidemiology* **14**, Issue 6.

Lee, Y. and Nelder, J.A. (1996) Hierarchical generalized linear models (with discussion). *Journal of the Royal Statistical Society, Series B* **58**: 619–678.

Lee, Y. and Nelder, J.A. (2001) Hierarchical generalized linear models: a synthesis of generalised linear models, random-effect models and structured dispersions. *Biometrika* **88**: 987–1006.

McCullagh, P. and Nelder, J.A. (1989) *Generalized Linear Models.* 2nd edition. London: Chapman and Hall. 511 pp.

Pinhero, J.C. and Bates, D.M. (2000) *Mixed-Effects Models in S and S-PLUS.* New York: Springer. 528 pp.

Robinson, G.K. (1991) That BLUP is a good thing: the estimation of random effects. *Statistical Science* **6**: 15–51.

Schall, R. (1991) Estimation in generalized linear models with random effects. *Biometrika* **78**: 719–727.

Sham, P. (1998) *Statistics in Human Genetics.* London: Arnold. 290 pp.

Snedecor, G.W. and Cochran, W.G. (1989) *Statistical Methods.* 8th edition. Ames, Iowa: Iowa State University Press. 503 pp.

Verbyla, A.P., Cullis, B.R., Kenward, M.G. and Welham, S.J. (1999) The analysis of designed experiments and longitudinal data by using smoothing splines. *Applied Statistics* **48**: 269–311.

Wilkinson, G.N. and Rogers, C.E. (1973) Symbolic description of factorial models for analysis of variance. *Applied Statistics* **22**: 392–399.

Chapter 10

Bulmer, M.G. (1979) *Principles of Statistics.* 2nd edition. New York: Dover. 252 pp.

Edwards, A.W.F. (1972) *Likelihood. An account of the statistical concept of likelihood and its application to scientific inference.* Cambridge: Cambridge University Press. 235 pp.

Patterson, H.D. and Thompson, R. (1971) Recovery of inter-block information when block sizes are unequal. *Biometrika* **58**: 545–554.

Payne, R.W. (ed.) (2003) *The Guide to GenStat Release 7.1. Part 2: Statistics.* Oxford: VSN International. 912 pp.

Index